Development and Application of Biomarkers

SERIES EDITOR

Roger L. Lundblad
Lundblad Biotechnology
Chapel Hill, North Carolina, U.S.A.

PUBLISHED TITLES

Application of Solution Protein Chemistry to Biotechnology
Roger L. Lundblad

Approaches to the Conformational Analysis of Biopharmaceuticals
Roger L. Lundblad

Development and Application of Biomarkers
Roger L. Lundblad

Development and Application of Biomarkers

Roger L. Lundblad

CRC Press
Taylor & Francis Group
Boca Raton London New York

CRC Press is an imprint of the
Taylor & Francis Group, an **informa** business

CRC Press
Taylor & Francis Group
6000 Broken Sound Parkway NW, Suite 300
Boca Raton, FL 33487-2742

First issued in paperback 2017

ISBN-13: 978-1-4398-1979-1 (hbk)
ISBN-13: 978-1-138-11405-0 (pbk)

Library of Congress Cataloging-in-Publication Data

Lundblad, Roger L.
    Development and application of biomarkers / Roger L. Lundblad.
      p. ; cm. -- (Protein science)
    Includes bibliographical references and index.
    Summary: "A comprehensive assessment of biomarkers, this book covers the history and current status of the application of biomarkers in diagnostics and prognostics. It explores the technology used for the study of biomarkers, and the validation of biomarkers including a comparison of the various technologies used to identify and measure biomarkers. The editors emphasize the technology underlying biomarkers and the translation of basic science to clinical laboratory technology, including the commercial development of biomarkers. The book also covers proteomics and proteomic technologies and their applications in the identification of biomarkers"--Provided by publisher.
    ISBN 978-1-4398-1979-1 (hardback : alk. paper)
    1. Biochemical markers. I. Title. II. Series: Protein science series.
    [DNLM: 1. Biological Markers. 2. Proteomics--methods. QW 541]

R853.B54L86 2011
610.28--dc22
                                            2010032787

Visit the Taylor & Francis Web site at
http://www.taylorandfrancis.com

and the CRC Press Web site at
http://www.crcpress.com

Contents

Preface

The term "biomarker" has assumed an identity of its own in the last decade. While it has been in use for a number of years, it has been most extensively used in geology. The term was first introduced in biomedical research in 1980 and has since found wide acceptance. While a more cynical approach could have been taken, I have tried to rationalize the current enthusiasm for biomarkers with the use of well-established clinical laboratory analytes in clinical medicine. As an example, O. Collinson (Cardiac markers, *Br. J. Hosp. Med.* (Lond) 70, M84–M87, 2009) notes that while scarcely a week passes by without the report of a new cardiac marker (biomarker), only two older analytes, troponins and B-type natriuretic peptide, are in routine use. This book tries to catalog the various existing biomarkers in clinical medicine and tries to match the expectations for advances in screening technologies with the realities of statistical analysis. Finally, it is hoped that this book will encourage consideration of biomarkers more as a concept than tangible analytes, and that biomarker research will equally enhance the understanding of disease and the development of a diagnostic.

Acknowledgments

First I thank Barbara Norwitz and Jill Jurgensen for their patience and encouragement during the preparation of the manuscript. Professor Keith Baggerly at University of Texas M.D. Anderson Cancer Center provided some useful insight into the issue of screening and biomarkers. Professor Ralph Bradshaw at the University of California at San Francisco brought about a sense of reality to both biomarkers and UNC basketball in 2010. Dr. Chris Burgess provided insight into pre-analytical issues from his castle in the U.K. Finally, Trish Maloney and other research librarians at University of North Carolina at Chapel Hill provided invaluable support.

Author

Roger L. Lundblad is a native of San Francisco, California. He received his undergraduate education at Pacific Lutheran University and his PhD in biochemistry at the University of Washington. After postdoctoral work in the laboratories of Stanford Moore and William Stein at The Rockefeller University, he joined the faculty of the University of North Carolina at Chapel Hill. He joined the Hyland Division of Baxter Healthcare in 1990. Currently, Dr. Lundblad works as an independent consultant at Chapel Hill, North Carolina, and writes on biotechnological issues. He is an adjunct professor of pathology at the University of North Carolina and the editor-in-chief of the Internet *Journal of Genomics and Proteomics*.

1 Introduction to Biomarkers

Biomarker research is an important category in current biomedical research. The number of journal citations (from SciFinder©) has increased from approximately 1000 in 2000 to more than 7000 (32,000 from PubMed) in 2008; the total number of citations from PubMed using biomarker was greater than 450,000. While the use of the term "biomarker" is highest in biomedical research, there is substantial use of this term in geology*; indeed the origin of the term appears to be in geology and it is not clear as to the first use in biomedical research. The first use in a journal article title (based on a PubMed search) was in 1980.[1,2] The term biomarker is formed from biological and marker, and while this seems straightforward enough, it has evolved to describe a broad range of chemicals and physiological phenomena. A definition for the term biomarker was proposed by a working group convened under the auspices of the National Institutes of Health (United States) in 2001.[3] A biomarker is defined as "a characteristic that is objectively measured and evaluated as an indicator of normal biological processes, pathogenic processes, or pharmacological responses to a therapeutic intervention." Thus, in the broad sense, a biomarker may be diagnostic and/or prognostic, and may or may not be useful as a screening tool for population studies. For example, prostate-specific antigen (PSA) was earlier used as a biomarker for prostate cancer and while more recently there has been some concern about diagnostic value,[4–6] it does retain great value as a prognostic biomarker.[7–10] Likewise, CA-125 is considered not useful as a screening tool but is useful for guiding treatment options in diagnosed patients and is a strong prognostic indicator.[11–18] Referring back to the above definition for biomarker, it might be useful to define "characteristic." The *Oxford Dictionary of the English Language* defines "characteristic" in a number of different ways; the one that appears to be most applicable for biomarker is "A distinctive mark, trait, or feature; a distinguishing or essential peculiarity or quality." Thus, a biomarker may be a molecule such as a protein or a peptide,[19–21] a modified protein such as through a process of oxidation,[21–23] a nucleic acid[24–26] or modified nucleic acid,[27–29] lipids,[30–34] and carbohydrates.[35–37] Biomarkers might reflect larger biological characteristics such as those

---

* There are hundreds of citations for the use of the term biomarker in the geochemical literature where it is used to describe organic material ("molecular fossils") in crude oil, which can be used to assign source. See Jacquot, F., Doumenq, P., Guiliano, M. et al., Biodegradation of the (aliphate + aromatic) fraction of Oural crude oil. Biomarker identification using GC/MS SIM and GC/MS/MS, *Talanta* 43, 319–330, 1996. See also Bauersachs, T., Kremer, B., Shouten, S. et al., A biomarker and delta *N*-15 study of thermally Silurian cyanobacterial mats, *Organic Geochem.* 40, 149–157, 2009; Ozcelik, O., Altunsoy, M., Acar, F., and Erik, N.Y., Organic-geochemical characteristics of the Milocene Lycian basin, western Taurides, Turkey, *Int. Geol. Rev.* 51, 77–93, 2009; Samuel, O.J., Cornford, C., Jones, M. et al., *Org. Geochem.* 40, 461–483, 2009.

detected at the cellular level by imaging such as positron emission tomography;[38–41] diffusion magnetic resonance;[42–44] metabolomics, which uses gas chromatography coupled with mass spectrometry for the identification of biomarkers;[45–49] and various other imaging technologies for the study of biomarkers at the cellular level.[50–52] There are some studies that describe blood pressure[53–56] and other fluid pressures[57] as biomarkers as well as arterial stiffness measured with pulse wave velocity.[58] One group has used inflammation as a biomarker for the prognosis of ischemic stroke.[59] There is interest in system biology, which identifies a biomarker as a pattern rather than a single characteristic.[60–64] It would appear that a biomarker is a characteristic that can be used either as a diagnostic or a prognostic while the ultimate goal is a characteristic, which could be used as a screening tool[65–68] for pathologies that tend to be somewhat "silent" prior to overt clinical display. The current work will focus on approaches to the identification, validation, and implementation of biopolymers, which can be used as biomarkers both for population screening and as surrogate endpoints for therapies.

As a start, I will try to trace the path to biomarkers. I have alluded to the evolution of the term from geological sources and will pursue that no farther. So, how have we moved from classical clinical and laboratory markers such as HbAlc, glucose, lipids, blood pressure, body mass index for type 2 diabetes to more sophisticated biomarkers?[69] My bias is that the development of analytical technologies such as mass spectrometry and ultra-high performance liquid chromatography together with computational ability to process large amounts of data have driven the search for biomarkers; thus, proteomics (see Chapter 4) has been the enabling technology underlying most of the searches for biomarkers. In that sense, the search for biomarkers has not been hypothesis-driven although there is progress in a hypothesis-driven approach.[70–73] It is also noted that biomarker research is also driven by the development of therapeutic approaches, which require diagnosis for application such as seen in breast cancer[74–77]; biomarkers that are developed in parallel with therapeutic products are referred to as companion diagnostics,[77–82] which seem to be related to theranostics.[83] As will be noted at several points in this book, the issue of nomenclature that has gotten somewhat out of control appears to be related more toward marketing than scientific content.

Biomarkers may be classified as to disease application such as cancer biomarkers, clinical application such as a screening biomarker, and system/organ biomarkers such as renal biomarkers. DeCaprio has discussed the classification of biomarkers in some detail,[84] and the reader is directed to this source for a larger discussion of this subject. DeCaprio suggests that the term biomarker has been diluted by overuse and I am of the same opinion. Referring to a laboratory analyte as a biomarker does not enhance value of the analyte and discovery of a biomarker by a sophisticated laboratory technique such as mass spectrometry does not endow the analyte with magical properties. The use of the term biomarker as a response to an environment stress or in response to a therapeutic intervention seems quite reasonable. The term biomarker then is more of a concept than a laboratory analyte, where the discovery of the biomarker is closely related to the biology underlying the production of the biomarker. I cannot emphasize this point too strongly as an understanding of the biology will increase the chances of developing an assay, which is robust and reproducible.

Dr. Keith Baggerly at M.D. Anderson notes that "Only if the data are reproducible can you talk about whether they are right."[85] The issue of reproducibility in the complex area of biomarker discovery and development has resulted in the emerging discipline of forensic bioinformatics[86] to evaluate high-throughput studies such as those involving microarray technology.

Various investigators have referred to the search for biomarker as "searching for a needle in a haystack."[87–93] Peck has recently suggested burning the haystack to find the needle within the context of eliminating unsuccessful drug candidates earlier in the development process.[94] More germane to the current discussion is a paper by James Bearden that appeared in 1972,[95] which I recommend for serious consideration within the context of current biomarker research; I also direct the reader to a recent Web site for a discussion of the expression "needle in a haystack."[96] The analogy here is how to find something that is significantly different from similar but not identical materials; in the case of biomarkers, how does one identify something abnormal when such material is only slightly abnormal when compared to normal. The understanding of "normal" presents its own problems. Murphy[97] lists seven definitions/meanings for the word "normal" including having probability density function, most representative of its class, most commonly encountered in its class, and commonly aspired to.

Much of the work on the development of biomarkers centers on the use of blood as a source (see Chapter 4), and there is much data on "normal" composition since blood is used so frequently for traditional diagnostic purposes [98–101] Palkuti[99] presents a good summary of the establishment of normal. A normal range (reference interval) must be established for each analyte (biomarker); this is value expected for 95% of the population.[101] The term "reference interval" has replaced the term "normal"; this is also referred to as reference range or reference values. At one time, a reference interval could be obtained from a pool of 30 individuals. The current approach to the determination of a reference interval uses a population of at least 120 individuals[102] although arguments are made for smaller sizes in unusual circumstances.[103] Ceriotti and colleagues[104] have recently discussed current concepts in the development of reference intervals. Reviews of importance for the consideration of reference intervals include that of Friedberg and colleagues,[105] Ricós and coworkers,[106] and the excellent chapter by Solberg in Tietz.[107]

The establishment of a reference standard and a reference interval is usually not considered until the process of transition from biomarker discovery to validated assay (see Chapter 9); however, the concept of reference intervals/normal should be considered in the discovery of biomarkers. It is worth noting that 198 (200) subjects are required to obtain a reference range with 99% confidence level.[108–110] The issue is further complicated by the suggestion that each significant subgroup (gender, age, etc.) contains 200 subjects.[111] This is likely necessary for the development of a screening assay but it is likely that less onerous requirements can be developed for developmental use. Research on biomarkers can be limited by smaller numbers of subjects and the cost of the experimental analyses. Some studies are relatively large[112] while others are embarrassingly small.[113]

The concept of normal is confounded by the related issues of biological variation and the "index of individuality." Biological variation includes issues such as gender, age, and ethnicity, which can be addressed by the identification of the subject and diurnal variation,[114] which can be addressed by sample collection time. Somewhat more difficult to assess are the effect of drugs such as that of oral contraceptives on C-reactive protein levels in women.[115] More complicated are effects of season[116–118] and occupation[116,117,119] (excluding exposure to environmental agents). There are several substantial works on biological variation.[120,121]

The index of individuality[122,123] is an approach to assessing the effect of intra-individual variation on the measurement of clinical analytes and is closely related to biological variability. This has been the subject of considerable discussion[124,125] over the years and continues to be of importance but does depend on the analyte.[126–129] The index of individuality is of importance more for the application of biomarker research (development of assays as in Chapter 8) than the actual discovery process as it is argued that a "one-hit" wonder should be seriously considered.[130] Jensen and colleagues[131] observed that there was extensive biological variability in bone biomarkers and a single measurement was of little value. There was a great difference in individual biomarkers; urinary hydroxyproline has a high index of individuality (1.23) such that 255 samples would be required to determine the true mean while osteocalcin in serum had a low index of individuality of 0.23 with six samples required for the determination of true mean. In general, individual (single subject) variation is suggested to be minor compared to intra-individual variation.[132,133] There are several studies that have attempted to "isolate" the variance in sample preparation from "downstream" variance in the analytical instrumentation and data analysis. Anderle and coworkers[134] have presented an analysis of proteomic expression profiling by HPLC-MS. The model sample was human serum. Sample preparation involved several steps: removal of albumin and IgG, reduction/alkylation, and tryptic digestion in a process known as shotgun proteomics.[135] Choe and Lee[136] observed a high CV in their studies on the reproducibility of two-dimensional gel electrophoresis and reported a method for assessing variability. Knudsen and colleagues evaluated pre-analytical variability and biological variability on IL-6 concentration in healthy subjects and patients with rheumatoid arthritis.[137] It was concluded that a change of greater than 60% is likely to indicate changes in disease activity.

The previous material is not meant to discourage investigators but is intended to emphasize that despite the large amount of published work, the search for biomarkers is not a trivial endeavor and considerable effort should be taken to carefully plan the work. The availability of user-friendly instrumentation has been both a blessing and a curse. The availability of instrumentation for surface plasmon resonance has enabled the rapid acquisition of data on molecular interactions and the study of biomarkers.[138,139] In the hand of skilled investigators,[140] good, reproducible data can be obtained. However, a critical evaluation of the literature[141] would suggest that there is a lack of careful investigators. The record of transition from discovered biomarker to clinical assay (approved FDA assay) is suggested to be one per year since 1998.[20,142] While some of the fault can be ascribed to the cost of obtaining data sufficient for approval,[143] there are likely other issues such as poor understanding and appreciation of pre-analytical issues, lack of understanding of the biology of the system in question, difficulty in migration from discovery platforms to commercial analytical platforms, and failure to understand that a new assay must add true value to the overall therapeutic process.

REFERENCES

1. Paone, J.F., Waalkes, T.P., Baker, R.R. et al., Serum UDP-galactosyl transferase as a potential biomarker for breast carcinoma, *J. Surg. Oncol.* 15, 59–66, 1980.
2. Webb, K.S. and Lin, G.H., Urinary fibronectin: Potential as a biomarker in prostatic cancer, *Invest. Urol.* 17, 401–404, 1980.
3. Downing, D.O., For the Biomarkers Definitions Working Group, NIH, Bethesda, MD, USA, Biomarkers and surrogate endpoints: Preferred definitions and conceptual framework, *Clin. Pharmacol. Ther.* 69, 89–95, 2001.
4. Pienta, K.J., Critical appraisal of prostate-specific antigen in prostate cancer screening: 20 years later, *Urology* 72(5 Suppl), S11–S20, 2009.
5. Schröder, F.H., Review of diagnostic markers for prostate cancer. Recent results, *Cancer Res.* 181, 173–182, 2009.
6. Kirby, R.S., Fitzpatrick, J.M., and Irani, J., Prostate cancer diagnosis in the new millennium: Strengths and weaknesses of prostate-specific antigen and the discovery and clinical evaluation of prostate cancer gene 3 (PCA3), *BJU Int.* 103, 441–445, 2009.
7. Mizuno, R., Nakashima, J., Mukai, M. et al., Tumour length of the largest focus predicts prostate-specific antigen-based recurrence after radical prostatectomy in clinically localized prostate cancer, *BJU Int.* 104, 1215–1218, 2009.
8. Choo, R., Donjoux, C., Gardner, S. et al., Efficacy of salvage radiotherapy plus 2-year androgen suppression for postradical prostatectomy patients with PSA relapse, *Int. J. Radiat. Oncol. Biol. Phys.* 75, 983–989, 2009.
9. Fleshner, N.E. and Lawrentschuk, N., Risk of developing prostate cancer in the future: Overview of prognostic biomarkers, *Urology* 73, S21–S27, 2009.
10. Park, Y.H., Hwang, I.S., Jeong, C.W. et al., Prostate specific antigen half-time and prostate specific antigen doubling time as predictors of response to androgen deprivation therapy for metastatic prostate cancer, *J. Urol.* 181, 2520–2524, 2009.
11. Gadducci, A., Tana, R., Cosio, S., and Genazzini, A.R., The serum assay of tumour markers in the prognostic evaluation, treatment monitoring and follow-up of patients with cervical cancer: A review of the literature, *Crit. Rev. Oncol. Hematol.* 66, 10–20, 2008.
12. Nossov, V., Amneus, M., Su, F. et al., The early detection of ovarian cancer from traditional methods to proteomics. Can we really do better than serum CA-125?, *Am. J. Obstet. Gynecol.* 199, 215–223, 2008.
13. Høgdall, E., Cancer antigen 125 and prognosis, *Curr. Opin. Obstet. Gynecol.* 20, 4–8, 2008.
14. Juriscova, A., Jurisica, I., and Kislinger, T., Advances in ovarian cancer proteomics: The quest for biomarkers and improved therapeutic interventions, *Expert Rev. Proteomics* 5, 551–560, 2008.
15. Argento, M., Hoffman, P., and Gauchez, A.S., Ovarian cancer detection and treatment: Current situation and future prospects, *Anticancer Res.* 28, 3135–3138, 2008.
16. Markman, M., The myth of measurable disease in ovarian cancer: Revisited, *Cancer Invest.* 27, 11–12, 2009.
17. Tung, C.S., Wong, K.K., and Mok, S.C., Biomarker discovery in ovarian cancer, *Women's Health* (London, England) 4, 27–40, 2008.
18. Medeiros, L.R., Rosa, D.D., da Rosa, M.I., and Bozzetti, M.C., Accuracy of CA 125 in the diagnosis of ovarian tumors: A quantitative systematic review, *Eur. J. Obstet. Gynecol. Reprod. Biol.* 142, 99–105, 2009.
19. Apple, F.S., Murakami, M.M., Pearce, L.A., and Herzog, C.A., Multi-biomarker risk stratification of N-terminal pro-B-type natriuretic peptide, high sensitivity C-reactive protein, and cardiac troponin T and I in end-stage renal disease for all-cause death, *Clin. Chem.* 50, 2279–2285, 2004.

20. Rifai, N., Gillette, M.A., and Carr, S.A., Protein biomarker discovery and validation: The long and uncertain path to clinical utility, *Nat. Biotechnol.* 24, 971–983, 2006.
21. Poli, G., Biasi, F., and Leonarduzzi, G., 4-Hydroxynonenal-protein adducts: A reliable marker of lipid oxidation in liver diseases, *Mol. Aspects Med.* 29, 69–71, 2008.
22. Zitnanová, I., Sumegová, K., Simko, M. et al., Protein carbonyls as a biomarker of foetal-neonatal hypoxic stress, *Clin. Biochem.* 40, 567–570, 2007.
23. Verhoye, E. and Langlois, M.R., For Asklepios Investigators, Circulating oxidized low-density lipoprotein: A biomarker of atherosclerosis and cardiovascular risk?, *Clin. Chem. Lab. Med.* 47, 128–137, 2009.
24. Pachner, A.R., Bertolotto, A., and Deisenhammer, F., Measurement of MxA mRNA or protein a biomarker of IFNβ bioactivity: Detection of antibody-mediated decreased bio-activity (ADB), *Neurology* 61(9 Suppl 5), S24–S26, 2003.
25. Ahmed, F.E., Role of miRNA in carcinogenesis and biomarker selection: A method-ological view, *Expert Rev. Mol. Diagn.* 7, 569–603, 2007.
26. Gormally, E., Caboux, E., Vineis, P. et al., Circulating free DNA in plasma or serum as biomarker of carcinogenesis: Practical aspects and biological significance, *Mutat. Res.* 635, 105–117, 2007.
27. Valavanidis, A., Vlachogianni, T., and Fiotakis, C., 8-Hydroxy-2'-deoxyguanosine (8-OHdG): A critical biomarker of oxidative stress and carcinogenesis, *J. Environ. Sci. Health C* 27, 120–139, 2009.
28. Rundle, A., Carcinogen-DNA adducts as biomarker for cancer risk, *Mutat. Res.* 600, 23–26, 2006.
29. Tost, J., DNA methylation: An introduction to the biology and the disease-associated changes of a promising biomarker, *Methods Mol. Biol.* 507, 3–20, 2009.
30. Helander, A. and Zheng, Y., Molecular species of the alcohol biomarker phosphati-dylethanol (PEth) in human blood measure by LC-MS, *Clin. Chem.* 55, 1395–1405, 2009.
31. Guyman, L.A., Adlercreutz, H., Koskela, A. et al., Urinary 3-(3,5-dihydroxyphenyl)-1-propanoic acid, an alkylresorcinol metabolite, is a potential biomarker of whole-grain intake in a U.S. population, *J. Nutr.* 138, 1957–1962, 2008.
32. Puri, B.K., Counsell, S.J., Ross, B.M. et al., Evidence from *in vivo* 31-phosphorus magnetic resonance spectroscopy phosphodiesters that exhaled ethane is a biomarker of cerebral n-3 polyunsaturated fatty acid peroxidation in humans, *BMC Psychiatry* 8 (Suppl 1), S2, 2008.
33. Hodson, L., Skeaff, C.M., and Fielding, B.A., Fatty acid composition of adipose tis-sue and blood in humans and its us a biomarker of dietary intake, *Prog. Lipid Res.* 47, 348–380, 2008.
34. Le Geudard, M., Schraauwers, B., Larrieu, I., and Bessoule, J.J., Development of a bio-marker for metal bioavailability: The lettuce fatty acid composition, *Environ. Toxicol. Chem.* 27, 1147–1151, 2008.
35. Kuiper, J.I., Verbeek, J.H., Frings-Dresen, M.H. et al., Keratan sulfate as a potential biomarker of loading of the intervertebral disc, *Spine* 23, 657–663, 1998.
36. Dehghan, M., Akhtar-Danesh, N., McMillan, C.R., and Thabane, L., Is plasma vitamin C an appropriate biomarker of vitamin C intake? A systematic review and meta-analysis, *Nutr. J.*, 6, 41, 2007.
37. Kissack, J.C., Bishop, J., and Roper, A.L., Ethylglucuronide as a biomarker for ethanol detection, *Pharmacotherapy* 28, 769–781, 2008.
38. Pupi, A., Mosconi, L., Nobili, F.M., and Sorbi, S., Toward the validation of functional neuroimaging as a potential biomarker for Alzheimer's disease: Implications for drug development, *Mol. Imaging Biol.* 7, 59–68, 2005.
39. Weber, W.A., Positron emission tomography as an imaging biomarker, *J. Clin. Oncol.* 24, 3283–3292, 2006.

40. Chen, D.L. and Schuster, D.P., Imaging pulmonary inflammation with positron emission tomography: A biomarker for drug development, *Mol. Pharm.* 3, 488–495, 2006.
41. Yu, E.Y. and Mankoff, D.A., Positron emission tomography imaging as a cancer biomarker, *Expert Rev. Mol. Diagn.* 7, 659–672, 2007.
42. Hamstra, D.A., Rehemtulla, A., and Ross, B.D., Diffusion magnetic resonance imaging: A biomarker for treatment response in oncology, *J. Clin. Oncol.* 25, 4104–4109, 2007.
43. Patterson, D.M., Padhani, A.R., and Collins, D.J., Technology insight: Water diffusion MRI—A potential new biomarker of response to cancer therapy, *Nat. Clin. Pract. Oncol.* 5, 220–233, 2008.
44. Padhani, A.R., Liu, G., Koh, D.M. et al., Diffusion-weighted magnetic resonance imaging as a cancer biomarker: Consensus and recommendations, *Neoplasia* 11, 102–125, 2009.
45. Kitteringham, N.R., Jenkins, R.E., Lane, C.S. et al., Multiple reaction monitoring for quantitative biomarker analysis in proteomics and metabolomics, *J. Chromatogr. B. Analyt. Technol. Biomed. Life Sci.* 877, 1229–1239, 2009.
46. Lewis, G.D., Asnani, A., and Gerszten, R.E., Application of metabolomics to cardiovascular biomarker and pathway discovery, *J. Am. Coll. Cardiol.* 52, 117–123, 2008.
47. Hu, C., van der Heijden, R., Wang, M. et al., Analytical strategies in lipidomics and applications in disease biomarker discovery, *J. Chromatogr. B. Analyt. Technol. Biomed. Life Sci.* 877, 2836–2846, 2009.
48. Riedmaier, I., Becker, C., Pfaffl, M.W., and Meyer, H.H., The use of omic technologies for biomarker development to trace functions of anabolic agents, *J. Chromatogr. A.* 1216, 6192–6199, 2009.
49. Li, X., Xu, Z., Lu, X. et al., Comprehensive two-dimensional gas chromatography/time-of-flight mass spectrometry for metabolomics: Biomarker discovery for diabetes mellitus, *Anal. Chim. Acta* 633, 257–262, 2009.
50. Fonović, M. and Bogyo, M., Activity based probes for proteases: Applications to biomarker discovery, molecular imaging and drug screening, *Curr. Pharm. Des.* 13, 253–261, 2007.
51. Laufer, E.M., Reuteliongsperger, C.P., Narula, J., and Hofstra, L., Annexin A5: An imaging biomarker of cardiovascular risk, *Basic Res. Cardiol.* 103, 95–104, 2008.
52. Eckelman, W.C., Reba, R.C., and Kelloff, G.J., Targeted imaging: An important biomarker for understanding disease progression in the era of personalized medicine, *Drug Discov. Today* 13, 748–759, 2008.
53. Lévy, B.I., Blood pressure as a potential biomarker of the efficacy angiogenesis inhibitor, *Ann. Oncol.* 20, 200–203, 2009.
54. Felker, G.M., Cuculich, P.S., and Gheorghiade, M., The Valsalva maneuver: A beside "biomarker" for heart failure, *Am. J. Med.* 119, 117–122, 2006.
55. Desai, M., Stockbridge, N., and Temple, R., Blood pressure as an example of a biomarker that functions as a surrogate, *AAPS J.* 8, E146–E153, 2006.
56. Giles, T.D., Blood pressure—The better biomarker: Delay in clinical application, *J. Clin. Hypertens.* (Greenwich) 9, 918–920, 2007.
57. Lunt, S.J., Fyles, A., Hill, R.P., and Milosevic, M., Interstitial fluid pressure in tumors: Therapeutic barrier and biomarker of angiogenesis, *Future Oncol.* 4, 793–802, 2008.
58. Wang, X., Keith, J.C. Jr., Struthers, A.D., and Feuerstein, G.Z., Assessment of arterial stiffness, a translational medicine biomarker system for evaluation of vascular risk, *Cardiovasc. Ther.* 26, 214–223, 2008.
59. McColl, B.W., Allan, S.M., and Rothwell, N.J., Systemic infection, inflammation and acute ischemic stroke, *Neuroscience* 158, 1049–1061, 2009.
60. Schrader, M. and Selle, H., The process chain for peptidomic biomarker discovery, *Dis. Markers* 22, 27–37, 2006.

61. de Roos, B. and McArdle, H.J., Proteomics as a tool for the modelling of biological processes and biomarker development in nutrition research, *Br. J. Nutr.* 99(Suppl 3), S66–S71, 2008.
62. Reckow, S., Gormanns, P., Holsboer, F., and Turck, C.W., Psychiatric disorders biomarker identification: From proteomics to systems biology, *Pharmacopsychiatry* 41(Suppl 1), S70–S77, 2008.
63. Nishijo, K., Hosoyama, T., Bjornson, C.R. et al., Biomarker system for studying muscle, stem cells, and cancer *in vivo*, *FASEB J.* 23, 2681–2690, 2009.
64. Lawlor, K., Nazarian, A., Lacomis, L. et al., Pathway-based biomarker search by high-throughput proteomics profiling of secretomes, *J. Proteome Res.* 8, 1489–1503, 2009.
65. Bencko, V., Use of human hair as a biomarker in the assessment of exposure to pollutants in occupational and environmental settings, *Toxicology* 101, 29–39, 1995.
66. Patel, A., Groopman, J.D., and Umar, A., DNA methylation as a cancer-specific biomarker: From molecules to populations, *Ann. N. Y. Acad. Sci.* 983, 286–297, 2003.
67. Bramer, S.L. and Kallungal, B.A., Clinical considerations in study designs that use cotinine as a biomarker, *Biomarkers* 8, 187–203, 2003.
68. Ilyin, S.E., Belkowski, S.M., and Plata-Salamán, C.R., Biomarker discovery and validation: Technologies and integrative approaches, *Trends Biotechnol.* 22, 411–416, 2004.
69. Pfützner, A., Weber, M.M., and Forst, T., A biomarker concept for assessment of insulin resistance, beta-cell function and chronic systemic inflammation in type 2 diabetes mellitus, *Clin. Lab.* 54, 485–490, 2008.
70. Payne, C.M., Holubec, H., Bernstein, C. et al., Crypt-restricted loss and decreased protein expression of cytochrome C oxidase subunit I as potential hypothesis-driven biomarkers of colon cancer risk, *Cancer Epidemiol. Prev.* 14, 2066–2075, 2005.
71. Schultz, K.R., Miklos, D.B., Fowler, D. et al., Towards biomarkers for chronic graft-versus-host disease: National Institutes of Health consensus development project on criteria for clinical trials in chronic graft-versus-host disease III. Biomarker Working Group Report, *Biol. Blood Marrow Transplant.* 12, 126–137, 2006.
72. Faratian, D. and Bartlett, J., Predictive markers in breast cancer—The future, *Histopathology* 52, 91–98, 2008.
73. Amonkar, S.D., Bertenshaw, G.P., Chen, T.H. et al., Development and preliminary evaluation of a multivariate index assay for ovarian cancer, *PLoS One* 4, e4599, 2009.
74. Ross, J.S., Fletcher, J.A., Linette, G.P. et al., The Her-2/neu gene and protein in breast cancer 2003: Biomarker and target of therapy, *Oncologist* 8, 307–325, 2003.
75. Pupa, S.M., Tagliabue, E., Ménard, S. et al., HER-2: A biomarker at the crossroads of breast cancer immunotherapy and molecular medicine, *J. Cell. Physiol.* 205, 10–18, 2005.
76. James, C.R., Quinn, J.E., Mullan, P.B. et al., BRCA1, a potential predictive biomarker in the treatment of breast cancer, *Oncologist* 12, 142–150, 2007.
77. Betucci, F. and Goncalves, A., Clinical proteomics and breast cancer: Strategies for diagnostic and therapeutic biomarker discovery, *Future Oncol.* 4, 271–287, 2008.
78. Blair, E.D., Assessing the value-added impact of diagnostic-type tests on drug development and marketing, *Mol. Diagn. Ther.* 12, 331–337, 2008.
79. Marrer, E. and Dieterie, F., Biomarkers in oncology drug development: Rescuers or troublemakers?, *Expert Opin. Drug Metabol. Toxicol.* 4, 1391–1402, 2008.
80. Jorgensen, J.T., Are we approaching the post-blockbuster era? Pharmacodiagnostics and rational drug development, *Expert Rev. Mol. Diagn.* 8, 689–695, 2008.
81. Knudsen, B.S., Zhao, P., Resau, J. et al., A novel multipurpose monoclonal antibody for evaluating human c-Met expression in preclinical and clinical settings, *Appl. Immunohistochem. Mol. Morphol.* 17, 57–67, 2009.

82. Hinma, L., Spear, B., Tsuchihashi, Z. et al., Drug-diagnostic codevelopment strategies: FDA and industry dialog at the 45th FDA/DIA/PhRMA/PWG/BIO pharmacogenomics workshop, *Pharmacogenomics* 10, 127–136, 2009.

83. Landais, P., Meresse, V., and Ghislain, J.-C., Evaluation and validation of diagnostic tests for guiding therapeutic decisions, *Therapie* 64, 195–201, 2009.

84. DeCaprio, A.P., Introduction to toxicological biomarkers, in *Toxicological Biomarkers*, A.P. DeCaprio (ed.), Taylor & Francis, New York, 2006.

85. Savage, L., Forensic bioinformatician aims to solve mysteries of biomarker studies, *J. Natl. Cancer Inst.* 100, 983–987, 2008.

86. Baggerly, K.A. and Coombes, K.R., Deriving chemosensitivity from cell lines: Forensic bioinformatics and reproducible research in high-throughput biology, *Ann. Appl. Stat.* 3, 1309–1334, 2009.

87. Bates, G. and Lehrach, H., The Huntington disease gene—Still a needle in a haystack?, *Hum. Mol. Genet.* 2, 343–347, 1993.

88. Reid, A.E. and Liang, T.J., Tumor marker for metastasis: Search for an abnormal needle in a haystack, *Hepatology* 20, 1631–1634, 1994.

89. Titus, K., Cancer algorithm may help find 'needle in a haystack', *CAP Today* 11, 26–28, 1997.

90. Atkin, W. and Martin, J.P., Stool DNA-based colorectal cancer detection: Finding the needle in the haystack, *J. Natl. Cancer Inst.* 93, 798–789, 2001.

91. Kumar, S., Mohan, A., and Guleria, R., Biomarkers in a cancer screening, research and detection: Present and future: A review, *Biomarkers* 11, 385–405, 2006.

92. Bednarczuk, T., Gopinath, B., Ploski, R., and Wall, J.R., Susceptibility genes in Graves' ophthalmopathy: Searching for a needle in a haystack?, *Clin. Endocrinol.* 67, 3–19, 2007.

93. Keers, R., Farmer, A.E., and Aitchison, K.J., Extracting a needle from a haystack: Reanalysis of whole genome data reveals a readily transplantable finding, *Psychol. Med.* 39, 1231–1235, 2009.

94. Peck, R.W., Driving earlier clinical attrition: If you want to find the needle, burn down the haystack. Considerations for biomarker development, *Drug Discov. Today* 12, 289–297, 2007.

95. Bearden, J.C. Jr., The needle in the haystack (a fable), *Perspect. Biol. Med.* 20, 355–359, 1977.

96. Zyra; http://www.zyra.org.uk/needle-haystack.htm

97. Murphy, E.A., The normal and the perils of the sylleptic argument, *Perspect. Biol. Med.* 15, 566–582, 1972.

98. Yasui, Y., Pepe, M., Thompson, M.L. et al., A data-analytic strategy for protein biomarker discovery: Profiling of high-dimensional proteomics data for cancer detection, *Biostatistics* 4, 449–463, 2003.

99. Palkuti, H.S., Specimen control and quality control, in *Hemostasis and Thrombosis in the Clinical Laboratory*, D.M. Corriveau and G.A. Fritsma (eds.), Lippincott, Philadelphia, PA, pp. 67–91, 1988.

100. Kratz, A., Ferraro, M., Sluss, P.M., and Lewandrowski, K.B., Laboratory reference values, *New Engl. J. Med.* 351, 1548–1563, 2004.

101. National Cancer Institute; http://www.cancer.gov/dictionary/?CdrID=635450

102. Horn, P.S. and Pesce, A.J., Reference intervals: An update, *Clin. Chim. Acta* 334, 5–23, 2003.

103. Geffré, A., Braun, J.P., Trumel, C., and Concorder, D., Estimation of reference intervals from small samples: An example using canine plasma creatinine, *Vet. Clin. Pathol.* 37, 477–484, 2009.

104. Ceriotti, F., Hinzmann, R., and Panteghini, M., Reference intervals: The way forward, *Ann. Clin. Biochem.* 46, 8–17, 2008.

105. Friedberg, R.C., Souers, R., Wagar, E.A. et al., The origin of reference intervals, *Arch. Pathol. Lab. Med.* 131, 348–357, 2007.
106. Ricós, C., Doménech, M.V., and Perich, C., Analytical quality specifications for common reference intervals, *Clin. Chem. Lab. Med.* 42, 858–862, 2004.
107. Solberg, H.E., Establishment and use of reference values, in *Tietz Textbook of Clinical Chemistry and Molecular Diagnostics*, C.A. Burtis, E.R. Ashwood, and D.E. Bruns (eds.), Elsevier Saunders, St. Louis, MO, Chapter 16, 2006.
108. Lott, J.A., Mitchell, L.C., Moschberger, M.L., and Sutherland, D.E., Estimation of reference ranges: How many subjects are needed?, *Clin. Chem.* 38, 648–650, 1992.
109. Smith, N.J., What is normal?, *Am. J. Electroneurodiagnostic Technol.* 40, 196–214, 2000.
110. Vasan, R.S., Biomarkers of cardiovascular disease—Molecular basis and practical considerations, *Circulation* 113, 2335–2362, 2006.
111. Sasse, E.A., Objective evaluation of data in screening for disease, *Clin. Chim. Acta* 315, 17–30, 2002.
112. Wadsworth, J.T., Somers, K.D., Cazares, L.H. et al., Serum protein profiles to identify head and neck cancer, *Clin. Cancer Res.* 10, 1625–1632, 2004.
113. Adkins, J.N., Varnum, S.M., Auberry, K.J. et al., Toward a human blood serum proteome. Analysis by multidimensional separation coupled with mass spectrometry, *Mol. Cell. Proteomics* 1, 947–955, 2002.
114. Møller, H.J., Petersen, P.H., Rejnmark, L., and Moestrup, S.K., Biological variation of soluble CD163, *Scand. J. Clin. Lab. Invest.* 63, 15–22, 2003.
115. Riese, H., Vrijkotte, T.G.M., Meijer, P. et al., Diagnostic strategies of C-reactive protein, *BMC Cardiovasc. Disord.* 2, 9, 2002.
116. Rossner, P. Jr., Svecova, V., Milcova, A. et al., Seasonal variability of oxidative stress markers in city bus drivers. Part I. Oxidative damage to DNA, *Mutat. Res.* 642, 14–20, 2008.
117. Rossner, P. Jr., Svecova, V., Milcova, A. et al., Seasonal variability of oxidative stress markers in city bus drivers. Part II. Oxidative damage to lipids and proteins, *Mutat. Res.* 642, 21–27, 2008.
118. Rudež, G., Meijer, P., Spronk, H.M.H. et al., Biological variation in inflammatory and hemostatic markers, *J. Thromb. Haemost.* 7, 1247–1255, 2009.
119. Lippincott, M.F., Desai, A., Zalos, G. et al., Predictors of endothelial function in employees with sedentary occupations in a worksite exercise program, *Am. J. Cardiol.* 102, 820–824, 2008.
120. Davenport, C.B., *Statistical Methods, with Special Reference to Biological Variation*, Wiley, New York, 1904.
121. Fraser, C.G., *Biological Variation: From Principles to Practice*, AACP Press, Washington, DC, 2001.
122. Williams, R.J., *Biochemical Individuality*, University of Texas Press, Austin, TX, 1956.
123. Harris, E.K., Effects of intra- and interindividual variation on the appropriate use of normal ranges, *Clin. Chem.* 20, 1535–1542, 1974.
124. González, C., Cava, F., Aylión, A. et al., Biological variation of interleukin-1beta, interleukin-8, and tumor necrosis factor-α in serum of healthy individuals, *Clin. Chem. Lab. Med.* 39, 836–841, 2001.
125. Petersen, P.H., Fraser, C.G., Jürgensen, L. et al., Combination of analytical quality specifications based on biological within- and between-subject variation, *Ann. Clin. Biochem.* 39, 543–550, 2002.
126. Tuxen, M.K., Sölétormos, G., Rustin, G.J. et al., Biological variation and analytical imprecision of CA 125 in patients with ovarian cancer, *Scand. J. Clin. Lab. Invest.* 60, 713–721, 2000.

127. Erden, G., Barazi, A.O., Tezcan, G., and Yildirimkaya, M.M., Biological variation and reference change values of CA 19–9, CEA, AFP in serum of healthy individuals, *Scand. J. Clin. Lab. Invest.* 68, 212–218, 2008.

128. Nguyen, T.V., Nelson, A.E., Howe, C.J. et al., Within-subject variability and analytic imprecision of insulin like growth factor axis and collagen markers: Implications for clinical diagnosis and doping tests, *Clin. Chem.* 54, 1268–1276, 2008.

129. Wu, A.H., Lu, Q.A., Todd, J. et al., Short- and long-term biological variation in cardiac troponin I measured with a high sensitivity assay: Implications for clinical practice, *Clin. Chem.* 55, 52–58, 2009.

130. Veenstra, T.D., Conrads, T.P., and Issaq, H.J., What to do with "one-hit wonders"?, *Electrophoresis* 25, 1278–1279, 2004.

131. Jensen, J.E.B., Kollerup, G., Sørensen, H.A. et al., A single measurement of biochemical markers of bone turnover has limited utility in the individual person, *Scand. J. Clin. Lab. Invest.* 57, 351–360, 1997.

132. Bocnish, O., Ehmke, K.D., Heddergott, A. et al., C-reactive protein and cytokine plasma levels in hemodialysis patients, *J. Nephrol.* 15, 547–551, 2002.

133. Molls, R.R., Ahluwalia, N., Fick, T. et al., Inter- and intra-individual variation in test of cell-mediated immunity in young and old women, *Mech. Ageing Dev.* 124, 619–627, 2003.

134. Anderle, M., Ray, S., Lin, H., Becker, C., and Joho, K., Quantifying reproducibility for differential proteomics: Noise analysis for protein liquid chromatography-mass spectrometry of human serum, *Bioinformatics* 20, 3575–3582, 2004.

135. McCormack, A.L., Schieltz, D.M., Goode, B. et al., Direct analysis and identification of proteins in mixtures by LC/MS/MS and database searching at the low-femtomole level, *Anal. Chem.* 69, 767–776, 1997.

136. Choe, L.H. and Lee, K.H., Quantitative and qualitative measure of intralaboratory two-dimensional protein gel reproducibility and the effects of sample preparation, sample load, and image analysis, *Electrophoresis* 24, 3500–3507, 2003.

137. Knudsen, L.S., Christensen, I.J., Lottenburger, T. et al., Pre-analytical and biological variability in circulating interleukin 6 in healthy subjects and patients with rheumatoid arthritis, *Biomarkers* 13, 59–78, 2008.

138. Ahmed, F.E., Mining the oncoproteome and studying molecular interactions for biomarker development by 2DE, ChIP and SPR technologies, *Expert Rev. Proteomics* 5, 469–496, 2008.

139. Arima, Y., Teramura, Y., Takiguchi, H. et al., Surface plasmon resonance and surface plasmon field-enhanced fluorescence spectroscopy for sensitive detection of tumor markers, *Methods Mol. Biol.* 503, 3–20, 2009.

140. Rich, R.L., Papalia, G.A., Flynn, P.J. et al., A global benchmark study using affinity-based biosensors, *Anal. Biochem.* 386, 194–216, 2008.

141. Rich, R.L. and Myszka, D.G., Survey of the year 2007 commercial optical biosensor literature, *J. Mol. Recognit.* 21, 355–400, 2008.

142. Kiernan, U.A., Biomarker rediscovery in diagnostics, *Expert Opin. Med. Diagn.* 2, 1391–1400, 2008.

143. Anon, Lost in validation, *Nat. Biotechnol.* 24, 869, 2006.

2 Application of Biomarkers in Diagnostics, Prognostics, Theranostics, and Personalized Medicine

INTRODUCTION

While the study of biomarkers has intrinsic value in systems biology,[1–3] most studies are related to therapeutics.[4–6] The following discussion will separate the use of biomarkers into diagnostic use, prognostic use, and/or theranostic use. There will also be mention of the use of biomarkers in screening[7–18] as this would be the most significant development from current biomarker research. It is recognized that the approach to using biomarkers for screening needs considerable thought before implementation.[19] Screening can be considered the application of a diagnostic test to a general population.

NOMENCLATURE/DEFINITIONS

Nomenclature is a challenge in science.[20–32,328] The following items are presented to define the terms used in this book; there is a potential problem of overlapping the use of terms. The reader is also referred to articles addressing this issue as such specifically relate to biomarker research.[33–35]

Biomarker: A characteristic of a biological system that can be measured in an objective manner and used as a metric of the behavior of such a system (see also Chapter 1).

Multiple biomarkers: A combination of biomarkers designed to provide more strength for the analytical procedure.[36–41] Multiple biomarkers[42] may be applied in parallel or in a serial manner as discussed below.

Clinical endpoint: An endpoint is a point at or near the end of a process or a period.[43] A clinical endpoint, which is a characteristic that reflects the function of a patient, can be determined. A clinical endpoint may be subjective in how a patient feels or objective such as survival. Clinical endpoints can be complex with disorders such as cystic fibrosis[44] or multiple sclerosis.[45] Clinical endpoint biomarkers for gene therapy of cystic fibrosis[46] include direct assay of cystic fibrosis transmembrane conductance regulator (CFTR) gene transfer by the measurement of CFTR mRNA, CFTR protein, and CFTR chloride channel activity; indirect assays of CFTR include bacterial colonization and lung function.

Surrogate endpoint: A surrogate endpoint substitutes for a clinical endpoint, which is generally less demanding and costlier than a clinical endpoint. A surrogate endpoint must predict clinical benefit and is linked to the clinical endpoint as a measure and an individual agreement between the two endpoints.[47-51] An auxiliary endpoint is another measurement (covariate) that can be used to strengthen the clinical endpoint.[49] Biomarkers are developed, in part, as surrogate endpoints.[51-56] Surrogate endpoints, as such, are used most often in clinical trials and are prognostic. It should be noted that the determination of surrogate biomarker can be somewhat complex.[57] Vascular endothelial growth factor (VEGF) is a proangiogenic growth factor secreted by tumor cells and is a therapeutic target.[58] Ebos and coworkers[57] showed that there was an inverse relationship between plasma levels of soluble VEGF receptor and tumor growth. Plasma levels of soluble VEGF receptor is, therefore, a suggested surrogate biomarker for tumor growth.

Diagnostic: A distinctive symptom or characteristic that identifies a specific condition or discriminates between several possibilities. For example, a prolonged blood clotting time is diagnostic of a bleeding disorder; however, more specific tests must be run to discriminate between several possible pathophysiological states. In this situation, the diagnostic assay (prolonged blood clotting time) is a screening assay and a subsequent prothrombin time determination would indicate, for example, warfarin intoxication. Compared to prognostic or screening, there is limited specific development of biomarkers for diagnostic purposes.[59-63] A biomarker such as procalcitonin may be used to differentiate between clinically similar conditions such as shown below for sepsis and systemic inflammatory response syndrome, but would not be a primary diagnostic tool.

Screening: Consultation of the Oxford English Dictionary provides a number of definitions for "screening"[43]; the one that is most useful within the present context is "Systematic examination of a large number of subjects, esp. for the detection of unwanted attributes or objects." Screening is therefore defined as the use of a specific analyte or group of analytes to examine a large population for the purpose of identifying a specific pathophysiology. The development of biomarkers for screening will likely be most useful for the identification of cancer prior to overt clinic symptoms.[64-66] The success of a screening assay is dependent on high sensitivity, high specificity and the availability of a beneficial therapeutic intervention.[67] Screening for ovarian cancer can receive considerable attention.[68-70] Screening patients with ischemia, renal and cardiovascular disease for Fabry disease provides an example of using a high-risk population.[71-73] Screening will be discussed in more detail in Chapter 3.

Sensitivity: Sensitivity can be defined as the proportion of positive test results to the total number of positive individuals; the true positive rate.[74]

Specificity: Specificity for a diagnostic test can be defined as the proportion of negative test results to the total number of negative individuals; the true negative rate.[74]

Prognostic/predictive: A clinical assay or finding what is used to predict the future course/outcome of a disease.[75-78] It is useful to separate prognostic from predictive.[79,80]

A prognostic biomarker would be defined as a biomarker that is a metric for the natural outcome of the disease or the outcome obtained with a standard treatment.[79] A predictive biomarker is a measurement associated with response to a therapeutic intervention.[80] In the context of the current discussion, a predictive biomarker would be a theranostic biomarker useful in personalized medicine (positive predictive probability, which is derived from sensitivity and specificity and disease prevalence[74]). A well-established prognostics (and diagnostic) biomarker for diabetes is glycated hemoglobin (HbA1c),[81] which proves useful for treatment of diabetes type 2 with novel therapeutics such as saxagliptin, which is dipeptidyl peptidase-4 inhibitor.[82] Glycated hemoglobin is derived from the nonenzymatic reaction of glucose with the amino groups of hemoglobin. This is an example of a protein biomarker derived from in vivo chemical modification; the oxidation of proteins provides other potential biomarkers.

Theranostic: A clinical assay or finding what is used to assess the effectiveness of a therapeutic approach with an emphasis on guiding future action.[83–86]

Theragnostic: Theragnostic is a descriptor developed to describe an imaging approach that maps the distribution of a tumor or tissue in three dimensions providing information about clinical response of a tumor or tissue to a therapeutic treatment.[87–89] Nanoparticles are being developed for theragnostic use as both imaging and therapeutic vehicles.[90,91] There does appear to be "crossover" in the use of theragnostics in the description of what might be considered a theranostic.[92,93]

Personalized medicine: Personalized medicine is therapy based on the individual where the genetic constitution is used to guide therapeutic approach[94–98]; in other words, the therapeutic approach is tailored to the individual patient. There are earlier works that address this issue. In vitro cell culture of patient cells can predict the success of chemotherapy.[99–104] The success of personalized medicine will depend on the development and effective use of biomarkers in evidence-based practice.[105–108]

Point-of-care: Describing a diagnostic test that is taken where the patient is located and data is obtained at site.[109–114] Point-of-care testing (POCT) is most useful when qualitative information and/or a defined cut-off point for analysis is valuable.[115] There is continuing interest in the development of POCT methods,[111,113,114] and recent work on the development of biosensors for the use of biomarkers in point-of-care has recently been reviewed.[113] There has been particular interest in cardiovascular applications.[116–123]

Clinical proteomics: Clinical proteomics can be described as the use of proteomic technologies to address issues in the diagnosis and treatment of disease.[124–128]

Translational medicine: An emerging activity that seeks to formalize the process of application of basic/social science findings into measurable clinical application.[129–132] The use of validated biomarkers is critical for the success of translational medicine.[133–136]

BIOMARKERS AND DIAGNOSTICS

Notwithstanding the above segmentation, the primary purpose for the identification of biomarkers is the development of metrics to be used in the treatment of disease. After validation, a biomarker will be measured in a central clinical laboratory or by

POCT technology. It is recognized that clinical laboratories are likely to undergo significant change in the next few years. Clinical laboratory test results are one of several aids that a physician uses in diagnosing and treating a disease,[137,138] and should have a positive effect on treatment[139] either by providing a timely diagnosis permitting early definitive treatment or a prognostic effect in therapy management.[140] A model for the use of proteomic technologies in clinical assay development can be provided by considering the use of nucleic acid technology in infectious disease[141,142] where assay technologies and molecular diagnostics,[143–146] not commonly used in the clinical laboratory, are used. Most clinical laboratory tests are based on the assay of an analyte in a biological fluid,[147] most often, blood. The use of serological markers for diagnosis and prognosis in oncology is not new[148] and it would be most useful for new investigators to consult earlier reference texts in this area[149,150] before "reinventing the wheel." Indeed, the use of Bence Jones protein as a biomarker for multiple myeloma dates back to 1848.[148]

It is extremely useful to have peer-reviewed publications that support the development of the biomarker assay.[151,152] Zolg and Langen have presented an overview[153] of biomarker research from an industrial perspective. These investigators emphasize the need for the biomarker to be useful in the sense of fulfilling an obvious medical need as mentioned above[139] as well as clearly defining what biomarkers may exist for this specific situation. Horton and colleagues reviewed the issues in the translation of new biomarkers into useful clinical assays.[154] Further comment was provided by Lippi and coworkers.[155] These types of decisions are driven by a combination of scientific merit and economic value. Evidence-based medicine as a concept underlying the overall practice of medicine will be of increasing importance in the adoption of new technologies.[156–163] The reader is also directed to several articles[164,165] that categorize biomarkers as "fit-for-purpose," exploration, demonstration, characterization, and/or surrogacy.

"Validation" is a term that is used extensively in the proteomics literature in the sense of validating a biomarker as opposed to validating an assay. The use of the term validation of a biomarker is unrelated to the concept of the validation of the assay for the biomarker, which is concerned with the accuracy, reproducibility, and robustness of the assay as well as its applicability for the determination of the particular analyte.[166,167] Thus, there is first the identification of a suitable analyte, the validation of the analyte/biomarker in a sufficiently large population,[168–171] and finally the validation of the assay in a clinical setting. The differentiation between pathological and normal is not a trivial issue and can require a large population.[172–175] Another consideration is the importance of the index of individuality/biological variation.[176–184] An issue related to biological variation is the validity of a single-point measurement[185–187]; it is recognized that a single-point measurement is common with point-of-care analyses.

BIOMARKERS AND SCREENING

The discovery of new biomarkers that could be used for early detection of disease such as cancer would be a major advance. If the cost of analysis was sufficiently low, then it would be possible to accept lower specificity since a positive result could be subjected to more rigorous analysis; low sensitivity presents a different type of problem, which is recognized and is under study.[188–192] There have been an increasing number of studies suggesting a cost benefit for screening although the analysis is complicated.[193–206] Any screening for a biomarker must have a strong statistical basis.[207] Gornall and coworkers[208] do make the point that there is little benefit in screening for early diagnosis unless there is a therapeutic option available for influencing outcome. Predictive probability is dependent largely on prevalence, and somewhat less on the sensitivity and specificity of the assay system.[209,210] Some studies on the identification of biomarkers for screening are provided in Table 2.1. In consideration of this information, it is clear that the identification of a unique biomarker is challenging, and the most productive approach is a combination of risk-factor assessment, clinical data, and sensitivity and specificity provided by the biomarker(s) for application to large populations. Positive predictive probability can be increased by considering parallel testing versus serial testing with several biomarkers.[238,239] Wright and Stringer[238] found that either a serial approach or a parallel approach is superior to a single screening assay in HIV-1 serodiagnosis; a parallel strategy yielded fewer incorrect results than a serial approach but was more expensive. Baron and coworkers[239] showed that parallel testing with fixed sEGFR (a soluble isoform for epidermal growth factor receptor) and CA125 cut-off limits maximized sensitivity for the detection of epithelial ovarian cancer while serial testing with age and sEGRF-dependent CA125 cut-off limits optimize specificity and discrimination of endothelial ovarian cancer from benign ovarian conditions. The reader is directed to an excellent discussion by Clarke-Pearson[240] of screening for ovarian cancer and the use of biomarkers.

BIOMARKERS AND PROGNOSIS

The current emphasis on early diagnosis/screening should not in any way diminish the need for the development of useful prognostic tests.[140] Prognostic analysis is a part of patient management in oncology.[241,242,77] As noted, the separation of prognostic from diagnostic can be artificial and a theranostic assay can also be considered to be prognostic. Examples of biomarkers used as prognostic indicators are shown in Table 2.2.

BIOMARKERS AND THERANOSTICS AND PERSONALIZED MEDICINE

Genetic analysis such as single nucleotide polymorphism (SNP),[268–272] or DNA microarray analysis,[273–277] is used extensively for personalized medicine. The use of biomarkers in personalized medicine is increasing.[278–281] It is recognized that SNPs and DNA expression analysis can be considered to be biomarkers,[282–285] so the separation is somewhat artificial. The reader is directed toward a concise consideration of biomarkers in guiding therapeutic options in cancer treatment,[286] which discusses the relationship of technologies and personalized medicine. Examples of

TABLE 2.1
Biomarkers Developed for Screening[a]

Study	References
A review of use of PSA in screening of prostate cancer	64
A review on the use of biomarker in breast cancer focusing on the use of blood-based markers and breast-based markers with emphasis on the identification of biomarkers for early detection of breast cancer in asymptomatic patients	18
A review of screening methods for colorectal cancer. Fecal occult blood testing (FOBT) remains the most useful screening method. Since colonic epithelial cells are replaced every 3–4 days, stool samples are of interest for the identification of DNA markers, which might be of value. Proteins such as calprotectin are stable in stool samples but are nonspecific indicators of inflammation	17
Proteomic analysis of exhaled break condensate as a source of biomarkers for lung cancer screening	211
A review of the current status of ovarian cancer screening based on the premise that a successful screening test must demonstrate a positive effect on ovarian cancer mortality while achieving the high specificity required to minimize down-side from false-positive results	212
A review of the use of biomarkers for the diagnosis of oral squamous cell carcinoma using negative/positive cell selection strategies for the enrichment of oral tumor cells and use of microarray technology to identify tumor biomarkers within the concept of lab-on-a-chip technology	213
DNA methylation as biomarker for screening for colorectal cancer using stool or serum samples	214
Fecal calprotectin as screening biomarker for colorectal cancer was evaluated in prospective study. The results suggest that fecal calprotectin was a poor biomarker for colorectal cancer; it is suggested that a tumor-based biomarker will be more useful than a blood-based biomarker	215
Development of a microfluidic chromatography system for biomarker screening. The microfluidic analysis of protein digest (shotgun/tryptic hydrolysis/prefractionation with strong cation exchange) permitted the identification of cancer-specific proteins	216
Use of multiplexed immunobead-based biomarker profiling for screening for head and neck cancer. The use of a multiplexed platform recognizing 25 biomarkers was found to be the most discriminating. Biomarkers included cytokines, growth factors, receptors, and tumor markers	217
Papillomavirus as a screening biomarker for cervical cancer screening	218
Autoantibody to alpha-2HS glycoprotein for breast cancer screening	219
Biomarkers (CRP, cell adhesion molecules, methylarginine) combined with clinical markers(flow-mediated vasodilation, carotid intima media thickness, arterial stiffness) to provide surrogate vascular markers to identify youth at risk for premature cardiovascular disease	220
Altered microRNA expression (RT-PCR) in sputum for diagnosis of non-small-cell lung cancer	221
B-type natriuretic peptide (BNP) is in routine use as a biomarker for cardiac failure.[b] This study demonstrates its usefulness for screening in a pediatric population	222
Elevation of serum metallothionein as a biomarker for cancer	223

TABLE 2.1 (continued)
Biomarkers Developed for Screening[a]

Study	References
Fucosylation as biomarker for heptocellular carcinoma (HCC). Fucosylated kininogen and fucosylated α-1-antitrypsin were significantly higher (P<0.0001) in HCC patients. The combination of these two biomarkers with Golgi protein 73 provided a sensitivity of 95% and a specificity of 70%.	224
Plasma levels of hyaluronidase elevated in subjects with coronary artery disease	225
Elevated chitotriosidase activity in subjects with coronary artery disease. Elevated chitotriosidase is presumed to be derived from activated macrophages during the development of atherosclerosis	226
Increased myeloperoxidase is associated with the development of heart failure in older (≥65 years) individuals. The association of independent of traditional cardiac risk factors	227
Use of gene expression analysis on rectal swab samples as biomarker for screening for colon cancer	228
Use of serum procalcitonin levels to screen for nosocomial infection in ICU patients	229
Evaluation of multiple tumor antigen-antibody interactions as biomarkers to be included in a "tumor-associated antigen array" assay system for oncology screening	230
A review of biomarkers used for screening for stroke	231
Screening for increased serum retinol concentration identifies patients with less aggressive prostate cancer	232
Use of multimodal screening (MMS) and ultrasound screening (USS) for ovarian tumors. MMS combined CA 125 and ultrasound. Sensitivity was similar in the two groups for all primary ovarian and tubal cancer; specificity was higher in the MMS group resulting in lower rates of repeat testing	233
Use of ion mobility spectroscopy to measure urine levels of acetone as a biomarker for disorders in lipid metabolism	234
Screening studies showed that there is an inverse relationship between body mass index and PSA concentration, which is explained by increased plasma volume	235
A review of existing and new biomarkers in urine for bladder cancer. BLCA-4, a nuclear matrix protein and member of ETS transcription family is demonstrated to have good specificity and sensitivity in previous studies.[c] Cystoscopy is still considered the "gold standard" for bladder cancer diagnosis[d]	236, 237
Use of urinary levels of globotriaosylceramide as a biomarker for Fabry disease	71

[a] These are selected from a total 176,016 citations obtained from PUBMED. Choice for inclusion is arbitrary and is by no means representative of the initial collection of citations. References 1–7 are review articles; other review articles are contained with the body of the table.

[b] Despite extensive work and publication on new cardiac biomarkers, only two are in common use; the cardiac troponins for diagnosis of acute coronary syndromes and BNP for differential diagnosis of cardiac failure (Collinson, P.O, Cardiac markers, *Br. J. Hosp. Med.* (Lond.) 70, M84–M87, 2009).

[c] Van Le, T.S., Miller, R., Barder, T. et al., Highly specific urine-based biomarker of bladder cancer, *Urology* 66, 1256–1260, 2005.

[d] See Cohan, R.H., Caoili, E.M., Cowan, N.C. et al., Urography: Exploring a new paradigm for imaging of bladder cancer, *AJR Am. J. Roentgenol.* 192, 1501–1508, 2009; Panebianco, V., Sciarra, A., Di Martino, M. et al., Bladder carcinoma: MDCT cystography and virtual cystoscopy, *Abdom. Imaging*, 2009. doi: 10.1007/s00261-009-9530-y.

TABLE 2.2
Some Biomarkers Useful for Prognostics

Biomarker	References
Use of acute-phase proteins and procalcitonin as biomarkers for prognosis of acute pancreatitis. Increases in the levels of procalcitonin could differentiate between severe and mild pancreatitis while CRP and serum amyloid A were elevated in both patient groups. A prognostic value was also seen for total calcium concentration and lactate dehydrogenase	243
Biopsy androgen receptor levels predict response in prostate cancer after recurrence	244
Gene expression profiles in cell lines derived from oral squamous cell carcinoma were compared with gene expression profiles from normal oral keratinocytes using Affymetrix GeneChip® technology. Expression of carcinoembryonic antigen-related cell adhesion molecule was correlated with poor prognosis	245
The ratio of gene expression for three genes (TM4SF1/PKM2, TMRSF1/ARHGDIA, and COBLL1/ARHGDIA) was determined by RT-PCR and has strong predictive value for survival in patients with malignant pleural mesothelioma	246
Procalcitonin identified as a biomarker for septic complications in trauma patients admitted to an ICU	247
A review on the use of PSA as a prognostic factor for the development of prostate cancer	248
Suggestion that UHRF1 (ubiquitin-like with PHD and ring-finger domain I) is a biomarker for bladder cancer, which can be detected by an immunohistochemical assay in urinary sediment	249
Serum lactate is a biomarker for mortality in severe sepsis	250
Review suggesting that there are no good independent blood biomarkers for ischemic stroke; while there are some promising studies, there are no biomarkers that are added to the existing validated clinical model	251
Tissue microarray for the identification of prognostic biomarkers in pancreatic cancer. MUC2 is identified as a useful complement to histological classification in predicting prognosis in cases of moderate to well-differentiated ductal adenocarcinoma	252
von Willebrand factor (vWF) and ADAMTS13 are useful in the prognosis of sepsis (thrombotic microangiopathy)	253
Prothrombin time ≥80% was a positive independent prognostic factor for patients with hepatocellular carcinoma (HCC) who had surgical resection of the tumor; elevated levels of des-γ-carboxyprothrombin was a negative prognostic factor for HCC patients who had radiofrequency ablation	254
Methylation of PCDH10 gene for protocadherin 10 was measured with bisulfite restriction analysis and bisulfite sequencing; methylation is an independent biomarker for poor prognosis in patients with gastric cancer	255
Epigenetic biomarkers for colorectal cancer are reviewed. CpG island methylation phenotype (CIMP) is suggested as an important biomarker for evaluation in prognosis	256
The expression of calreticulin as a biomarker for prognosis in gastric cancer. Increased expression (immunohistocytochemistry and cDNA microarray) results in angiogenesis and poor prognosis	257
Development of an immunohistochemical biomarker protein profile for rectal cancer prognosis. The presence of both receptor for hyaluronic acid–mediated motility (RHAMM) and CD8+ tumor-infiltrating lymphocytes (TIL) was predictive of poor prognosis	258

TABLE 2.2 (continued)
Some Biomarkers Useful for Prognostics

Biomarker	References
Biomarkers were evaluated in neuroblastoma patients. Elevated IL-6 in peripheral blood correlated with high-risk neuroblastoma and poor prognosis; decreased soluble IL-6 receptor correlated with the presence of metastatic disease	259
The combination of the serum levels of procalcitonin, total calcium, and lactate dehydrogenase provides a biomarker for the prognosis of acute pancreatitis	243
A panel of biomarkers from the kallikrein family combined with other biomarkers is prognostic for ovarian cancer	260
Plasma levels of vWF (arising from endothelial cell damage) is an independent biomarker for major adverse cardiovascular events in acute coronary syndrome	261
CRP and procalcitonin are independent prognostic biomarkers for pneumonia	262–266
Leukoaraiosis volume appears to be a composite prognostic marker for acutely ischemic tissue	267

the applications of biomarkers in personalized medicine are shown in Table 2.3. As noted above, most of the examples of the use of biomarkers are derived from molecular diagnostics.[273,282,298,301–303] The disciplines of pharmacogenetics*[304,305] and pharmacogenomics*[306,307] are an integral part of personalized medicine, which rely on biomarkers measured by molecular diagnostics.

BIOMARKERS AND SPECIFIC PATHOLOGIES

The following section will discuss the use of biomarkers in the screening, diagnosis, and prognosis of several specific pathologies. The reader is recommended to an excellent work edited by Trull,[308] which provides an extensive coverage of biomarkers by individual pathology. From consideration of this text and others,[309–313] it is easily observed that the majority of work has been done with cancer. Within oncology, most of the work has been done with ovarian cancer and to a lesser extent with lung cancer, prostate cancer, and bladder cancer. These are diseases where the development of a diagnostic, which would provide early identification of a tumor, could be an advance in therapy.[314–317] There is an extensive discussion of cancer biomarkers in Chapter 3. The selection of disease biomarkers is not so much an indication of importance of the disease in terms of morbidity or mortality as much an indication of the number of studies in that specific area.

A recurring theme in this book is the argument that biomarkers, as such, are not new and the term biomarker can, and is, used to describe biological analytes

---

* The following definitions are obtained from the *Oxford Dictionary of the English Language*, Oxford University Press, Oxford, U.K., 2009.

Pharmacogenetics: The branch of medical science concerned with the genetic factors involved in the responses of individuals to drug.

Pharmacogenomics: The branch of medical science concerned with the identification of genes involved in the individual responses to drugs.

TABLE 2.3

Biomarkers Developed for Personalized Medicine

Study	References
Development of biological markers for the diagnosis and treatment of neck pain. Muscle proteins in blood are being used to identify whiplash injury	287
Combination of biomarkers, nanotechnology, and bioinformatics for personalized cancer treatment	288
Use of human peripheral blood cells as surrogate biopsy sample in developing biomarkers for use in personalized medicine. This would permit expression analysis without requiring tissue biopsies	289
The combination of traditional biomarkers (prognostic and predictive) with emerging biomarkers will permit a better quantification of residual risk and the potential value of additional treatment for breast cancer. Traditional biomarkers include estrogen receptor and progesterone receptor, which guide endocrine therapy and HER2 (EGF receptor). Emerging biomarkers are represented by multigene expression profiling (DNA microarray analysis)	278
Discussion of the lack of progress in personalized medicine since the sequencing of the human genome. While gene expression is useful, there are also epigenetic factors that need to be considered	290
Use of metabolomics[a] for the determination of molecular phenotype and as theranostic platform for the elucidation of biomarker patterns is important for personalized medicine. The concept of molecular phenotyping on individual levels, in response to pharmacological intervention, provide the opportunity of clinical trials with $n=1$	291
ACE (angiotensin-converting enzyme) phenotyping may be more useful than ACE genotyping as an approach to personalized medicine for ACE inhibitors	292
Use of transcriptional analysis (mRNA analysis via microarray technology) as an approach to the development of a molecular profile (biomarker) for personalized medicine	273
Use of magnetic resonance imaging for identifying biomarkers for late-life depression as an approach to personalized medicine	293, 294
Use of single nucleotide polymorphisms for genotyping in developing a personalized medicine approach to acute macular degeneration	295, 296
The use of targeted imaging uses a radiolabeled probe that binds to specific structural component critical to a single disease or to a more general physiological response, such as inflammation in personalized medicine	297
Current use of chromosomal profile to guide chemotherapy in malignant gliomas is an example of personalized medicine. Future work on gene expression using microarray technology will identify other genetic markers useful in differentiating responder from non-responder subjects	298
The use of classification by ensembles from random partitions (CERP) and variable importance ranking to identify high-dimensional biomarkers[b] for personalized medicine	299
Future of personalized medicine in psychiatry will depend on pharmacogenomic testing, which, together with environmental variables that influence pharmacokinetic and pharmacodynamic response to therapy. Existing pharmacogenomic tests for CNS disorders include the CYP 450 Test (Roche), which measure CYP2D6 polymorphisms (CYP2D6 phenotype; CYP2D6 is an enzyme that metabolizes several antipsychotic and antidepressant drugs), clozapine-induced agranulocytosis, and a test for metabolic syndrome	276
Use of personalized medicine to guide the use of interferon therapy in multiple sclerosis	300

TABLE 2.3 (continued)
Biomarkers Developed for Personalized Medicine

[a] Metabolic profiling (metabolomics) using nuclear magnetic resonance spectroscopy, mass spectrometry, and statistical analysis provides new combination of metabolic biomarkers (Schlotterbeck, G., Ros, A., Dieterle, F., and Senn, H., Metabolic profiling technologies for biomarker discovery in biomedicine and drug development, *Pharmacogenomics* 7, 1055–1075, 2006).

[b] High-dimensional biology or high-dimensional data can refer to a large number of data points obtained from microarray analysis of the transcriptome or data from multiple technology sources. See Romero, R., Espinoza, J., Gotsch, F. et al., The use of high-dimensional biology (genomics, transcriptomics, proteomics, and metabolomics) to understand the preterm parturition syndrome, *BJOG* 113(Suppl 3), 118–135, 2006; Wang, Y., Miller, D.J., and Clarke, R., Approaches to working in high-dimensional data spaces: Gene expression microarrays, *Br. J. Cancer* 98, 1023–1028, 2008; Vivekanandan, P. and Singh, O.V., High-dimensional biology to comprehend heptocellular carcinoma, *Expert Rev. Proteomics* 5, 45–60, 2008; Wang, Y., Miller, D.J., and Clarke, R., Approaches to working in high-dimensional data spaces: Gene expression microarrays, *Br. J. Cancer* 98, 1023–1028, 2008.

discovered long before the formal use of the term biomarker or biological marker in the literature. Thus, a search for biomarkers and inflammation will yield a large number of citations for C-reactive protein (CRP); while CRP was not described as a biomarker at the time, such lack of mention does not prevent the various search engines from identifying CRP in such articles as a biomarker. In a similar manner, there is an enormous literature on rheumatoid arthritis while it could be argued that other pathologies are more significant.

There has been a proliferation of studies on biomarkers to include sepsis,[165,318] cardiovascular disease,[319,320] peritoneal dialysis,[321] vasculitis,[322,323] renal disease,[16,324,325] liver disease,[326–330] and neurology.[331–335] These efforts have resulted in some new biomarkers that may be of value and have validated the older analytes such as tumor necrosis factor (TNF), IL-6, and acute-phase reactants such as CRP. Finally, it is acknowledged that a new look at the existing information can provide a new biomarker such as the multiplication of neutrophil and monocyte counts to provide a prognostic biomarker for cervical cancer.[336]

It is clear that it is unlikely that a single biomarker will suffice for screening, diagnostic, prognostic, or predictive use, and interest has markedly increased in the use of multiple biomarkers with either combination of new markers with old markers or identification and use of new multiple marker strategies.[337–342]

RHEUMATOID ARTHRITIS

Rheumatoid arthritis is an inflammatory disorder, which is thought to proceed via the activation of macrophages by an unknown mechanism(s), which may involve viruses or bacteria. The activation of macrophages results in the release of TNF.[343–345] The consequence is a chronic inflammatory response with the destruction of bone and other tissue. There is a large genetic susceptibility component that influences the occurrence and severity of disease.[346–350] The heterogeneity of the course of disease creates an opportunity for the development of theranostic biomarkers.[349–352]

The differentiation of rheumatoid arthritis from osteoarthritis is critical for treatment. While there are treatment modalities for rheumatoid arthritis, the same cannot be said for osteoarthritis,[353] thus, treatment modalities for rheumatoid arthritis are not effective for osteoarthritis. There are a variety of therapeutic strategies for the treatment of rheumatoid arthritis.[354–360] Most of these therapeutic options are expensive and it is essential to have sufficient information about the disease for a cost-benefit analysis.[361] As noted above, rheumatoid arthritis has been an attractive target for biomarker research (almost 6000 PUBMED citations); some selected studies are shown in Table 2.4.

One study[380] used proteomics (electrophoresis/mass spectrometry) to identify a biomarker for rheumatoid arthritis in blood serum. The analysis of serum from

TABLE 2.4
Examples of Existing Diagnostics Developed for Rheumatoid Arthritis and Osteoarthritis

Study	Reference
Measurement of serum type IIa procollagen N-terminal propeptide in joint diseases	362
Autoantibodies to chromatin in juvenile rheumatoid arthritis	363
Anticyclic citrullinated peptide antibodies in juvenile idiopathic arthritis	364
General review of prognostic laboratory markers for rheumatoid arthritis	365
Serum fas in rheumatoid arthritis	366
CD-26(dipeptidyl peptidase IV) decreases in rheumatoid arthritis compared to osteoarthritis and is inversely correlated with CRP	367
Collagen 2-1 peptide and nitrated collagen 2-1 peptide in rheumatoid arthritis	368
Review of the use of protein microarrays in autoimmune disease	369
Autoantibodies to triosephosphate isomerase in osteoarthritis	370
Review of the diagnostic and prognostic use of autoantibodies in early rheumatoid arthritis. Rheumatoid factor is the most significant autoantibody	371
Differential gene expression in blood used to diagnosis mild osteosarthritis	372
Plasma levels of procalcitonin is a useful biomarker for osteomyelitis and differentiates from septic arthritis	373
Serum procalcitonin as a biomarker to differentiate between septic and non-septic arthritis (elevated in septic arthritis)	374
Diagnostic approaches to monarthritis (arthritis in one joint) such as inflammatory synovial fluid with negative Gram's stain	375
Blood leukocyte gene expression (microarray technology) to diagnose systemic onset juvenile idiopathic arthritis and response to IL-1 blockage	376
Use of multiple biomarkers for the diagnosis of rheumatoid arthritis. Evaluation of 131 biomarkers, it was concluded that anti-cyclic citrulline peptide or anti-cyclic citrulline peptide in combination with IL-6 had the highest power for established rheumatoid arthritis	377
Combination of multiplex test for rheumatoid factor in combination with anti-cyclic citrulline peptide increased sensitivity	378
Use of a radiolabeled ligand to the folate receptor on activated macrophages as method for the detection of disease activity in rheumatoid arthritis	379

rheumatoid arthritis but not osteoarthritis showed marked differences in fibrinogen degradation products and calprotectin (calgranulin A/calgranulin B, S100A9/
S100A8, MRP8/MRP14) suggesting that these proteins might be diagnostic biomarkers for rheumatoid arthritis. In subsequent work,[381] these investigators developed an immunoassay for calprotectin, which could be used to differentiate between
rheumatoid arthritis and osteoarthritis, permitting the identification of patients (rheumatoid arthritis) who would benefit from anti-TNFα therapy. This type of assay can
save a large amount of money in the health care system since the yearly cost of anti-
TNFα therapy is not trivial. Improvements in imaging technologies also permit the
facile differentiation between rheumatoid arthritis and osteoarthritis.[382,383] Another
group of investigators have used ProteinChip® technology developed by Ciphergen
(Fremont, CA)[384,385] to identify a protein found in synovial fluid from patients with
rheumatoid arthritis and not in osteoarthritis.[386] These studies were performed with a
hydrophobic (C_{16}) and a strong anion exchange (SAX) matrix. These studies identified a protein consistent with the properties of MRP8 (S100A8) as described above.
It is recognized that Berntzen and coworkers[387] using ELISA technology earlier
reported elevated levels of leukocyte protein L1 in plasma and synovial fluids from
patients with rheumatoid arthritis, while normal levels of protein L1 are found in
patients with osteoarthritis. More recently, MRP8/MRP14 has been suggested as
marker for relapse in inactive juvenile idiopathic arthritis[388] based on its previous
identification in rheumatoid arthritis.[387,389,390] It is recognized that the terms S100A8
and S100A9 are new terminology for MRP8 and MRP14 and calgranulin A and calgranulin B[390–392]; thus, calprotectin is the complex of S100A8 (MRP8) and S100A9
(MRP14).[393] Although the linage is a little difficult to those not of the cognoscenti, it
would appear that the parent term is L1 (L1 heavy chain and L1 light chain), which
is released from leukocytes following immunological injury and is thought to be
involved with leukocyte trafficking.[394] Elevated levels of these proteins is seen in
the plasma/serum from individuals with a variety of inflammatory diseases.[395–397]
The nomenclature of the S100 proteins is confusing and the reader is directed to a
recent clarification of this problem.[398] Calprotectin has continued to be of interest as
a biomarker for rheumatoid arthritis,[399,400] but is not considered to a primary marker
for either diagnosis or prognosis (Table 2.5). There is a considerable discussion of
calprotectin as a biomarker for inflammation in cancer below.

Several other biomarkers have proven to be more useful for rheumatoid arthritis. The
first is CRP, an acute phase protein,[412,413] which has been known to be in association
with inflammation for many years.[414–416] Acute phase proteins are elevated in response
to a variety of physiological stimuli[417] including psychological factors.[418,419] Acute
phase proteins are proposed for monitoring acute phase proteins as biomarkers of
inflammation (SIRS, systemic inflammatory response syndrome) during preclinical
studies[420] in an effort to prevent disasters[421] during early clinical trials.[422–424]

CRP is a primary biomarker of rheumatoid arthritis used in evaluating therapeutic approach.[350,351,425–427] CRP has received considerable interest as a cardiac
biomarker.[428–430] To an extent, CRP is a secondary marker since its expression by
hepatocytes is stimulated by IL-6[431,432] derived, presumably, from activated macrophages and, in the case of obesity, from white fat tissue.[433] Some examples of the use
of CRP as a biomarker are presented in Table 2.6.

TABLE 2.5

The Promiscuity of Calprotectin as a Biomarker; Various Indications for Calprotectin[a] as a Biomarker

Pathology	References
Differentiation of rheumatoid arthritis and osteoarthritis	380, 381, 386, 387
Reactive arthritis	390
Juvenile idiopathic arthritis	388
Ovarian carcinoma	401
Colorectal carcinoma	402
Intra-amniotic infection	397
Head and neck tumors	403
Nasal lavage fluid from chemical (dimethylbenzylamine, epoxy) exposure	404
Inflammatory bowel disease	395, 405–409
Type 1 diabetes	410
Periodontal disease	411

[a] Calprotectin is a leukocyte protein originally referred to as leukocyte protein 1 (LP1), see Brandtzaeg, P., Gabrielsen, T.O., Dale, I. et al., The leukocyte protein L1 (calprotectin): A putative nonspecific defense factor at epithelial surfaces, *Adv. Exp. Med. Biol.* 371A, 201–206, 1995; Berntzen, H.B., Munthe, E., and Fagerhol, M.K., A longitudinal study of the leukocyte protein L1 as an indicator of disease activity in patients with rheumatoid arthritis, *J. Rheumatol.* 16, 1416–1420, 1989. For a comprehensive discussion of the current nomenclature for S100 proteins see Marenholz, I., Lovering, R.C., and Heizmann, C.W., Au update of the S100 nomenclature, *Biochim. Biophys. Acta* 1763, 1282–1283, 2006.

TABLE 2.6

Examples of the Use of CRP as a Biomarker for Various Indications[a]

Indication	References
Cardiovascular disease[b]	434–436
Continuous exercise training decreased CRP levels provided non-pharmacological management of erectile dysfunction	437
hsCRP (high-sensitivity CRP) is a member of a biomarker panel (β-2-microglobulin, cystatin C, and glucose) combined with other risk factors (age, diabetes status, smoking status) predictive of peripheral artery disease	438
CRP as a biomarker for inflammation in atherosclerosis	439
CRP as a member of a panel of biomarkers (placental growth factor, pro-B-type natriuretic peptide, cardiac troponin I, metalloproteinase-9) were predictive of increased risk of adverse events in acute coronary syndrome	440

TABLE 2.6 (continued)
Examples of the Use of CRP as a Biomarker for Various Indications[a]

Indication	References
Significant reduction of CRP (as IL-6, soluble IL-2 receptor, soluble E-selectin and soluble intracellular adhesion molecule-1) with abatacept in patients with rheumatoid arthritis	441
CRP is a predictive biomarker for stroke	442–444
CRP (and fibrinogen) is a prognostic biomarker for metastatic renal cell carcinoma treated with subcutaneous IL-2 immunotherapy	445
CRP is highly associated with osteoarthritis of the knee but its high association with obesity limits utility as an exclusive biomarker for osteoarthritis of the knee	446
The ability of erythrocytes to aggregate correlated with high-sensitivity CRP (or fibrinogen or erythrocyte sedimentation rate) and is suggested to provide an inexpensive method for assessing low levels of inflammation, which would result in atherosclerosis	447
Use of lab-on-a-chip for measuring a panel of CRP, myoglobin, and myeloperoxidase is complementary to EEG for acute myocardial infarction	116
Elevated CRP in cardiac syndrome X and coronary artery disease	448
Elevated CRP in colorectal adenoma is not modified by smoking or obesity; suggested role of inflammation in colorectal adenoma	449
CRP is increased in obesity; CRP levels doubled with each increase in weight class. Also, direct association between obesity, diabetes, and hypertension	450
CRP is a prognostic indicator (mortality) in hemodialysis patients	451
CRP is a predictive indicator in patients with metastatic renal cell carcinoma treated with IL-2-based immunotherapy	445
CRP as biomarker for bloodstream infections; IL-8 is also suggested as biomarker for infection. Neither CRP nor IL-8 is considered as strong as procalcitonin	452

[a] More recent work has emphasized the importance of high-sensitivity CRP assays for measuring low concentrations of C-reactive protein. High-sensitivity CRP (1–3 mg/L see Ockene, I.S., Matthews, C.E., Rifai, N. et al., Variability and classification accuracy of serial C-reactivity protein measurements in healthy adults, *Clin. Chem.* 47, 444–450, 2001).

[b] The use of CRP in cardiovascular disease extends back to the 1950s before the term biomarker came into use with respect to results from clinical laboratories. Early studies include Ziegra, S.R. and Kuttner, A.G., Reappearance of abnormal laboratory findings in rheumatic patients following withdrawal of ACTH or cortisone; with special reference to the C-reactive protein, *Am. J. Med. Sci.* 222, 516–522, 1951; Kroop, I.G. and Shackman, N.H., Level of C-reactive protein as a measure of acute myocardial infarction, *Proc. Soc. Exp. Biol. Med.* 86, 95–97, 1954; Ladue, J.S., Laboratory aids in diagnosis of myocardial infarction; changes in muscle enzymes, erythrocyte sedimentation rate, and C-reactive protein in myocardial infarction, *J. Am. Med. Assoc.* 165, 1776–1781, 1957; Phear, D. and Stirland, R., The value of estimating fibrinogen and C-reactive protein levels in myocardial ischaemia, *Lancet* 273, 270–271, 1957.

Rheumatoid factor (RF) describes autoantibodies[453,454] that react with immuno-globulin in patients with rheumatoid arthritis.[455–457] The value of RF as a biomarker in the diagnosis of rheumatoid arthritis was recognized fifty years ago[458–460] and remains of major value.[461–463] As with many biomarkers, the presence of RF is not uniform in expression as a biomarker in rheumatoid arthritis.[464] Erythrocyte sedimentation rate (ESR) is another long-recognized biomarker for rheumatoid arthritis,[417] which is of great value as a theranostic biomarker.[465–468] The change in ESR is a reflection of the change of plasma protein composition[469] in the acute-phase response.[470]

Antibodies against citrullinated proteins[471–475] or, more recently, antibodies against cyclic citrullinated peptide (CCP)[476,477] are proving to be extremely useful biomarkers for rheumatoid arthritis.[360,425,478–480] While antibodies to CCP are reasonably specific for rheumatoid arthritis, they are present in other immunological disorders as well.[481–483] Citrulline[484,485] is the result of the posttranslational modification[486] of arginine residues in proteins[487] (Figure 2.1). Current practice[470] uses antibodies against a CCP.[477] Antibodies to CCP are extremely useful in diagnosis when used with other markers[488] but may not be useful in personalized medicine.[489]

Arthritis is associated with the degradation of connective tissue and bone, and degradation products[353] are used as biomarkers.[490–494] Current biomarker research for rheumatoid arthritis and osteoarthritis is presented in Table 2.7. It is clear that there is no single marker at this time for arthritis but a combination of clinical evaluation and laboratory finding is still the appropriate approach.[488]

INFLAMMATION BIOMARKERS IN DIFFERENT PATHOLOGIES

Inflammation is a confounding aspect of many pathologies and must be considered in biomarker research. Inflammatory biomarkers (Table 2.8) are of importance in the study of arthritic disease, as discussed above, and there is considerable interest in inflammatory biomarkers in cancer, as discussed below. The biomarkers in Table 2.8 have been developed with classical technology and, the assays, in general, are based on ELISA technology. This material is presented to show the wide variety of disease states where the identification and measurement of inflammation biomarkers is considered relevant. Table 2.9 describes biomarkers of inflammation that have been identified with proteomic technology and may be of value in personalized medicine. The following section presents information on the characteristics of inflammation biomarkers in various syndromes.

Some selected biomarkers for inflammation in atherosclerosis are shown in Table 2.10; these biomarkers are also prognostic of the development of coronary artery disease (CAD). Atherosclerosis is also considered to be a risk factor for myocardial infarction so that the biomarkers in Table 2.10 could also be considered as screening biomarkers for cardiac disease.

Diabetes mellitus type 1 results from the loss of pancreatic islet beta cells by an immune-mediated process,[549] and has a strong genetic component. Autoantibodies are suggested as biomarkers for development of diabetes mellitus type 1[550,551] but may not measure T-cell function in beta cell destruction.[552] Inflammation is an issue

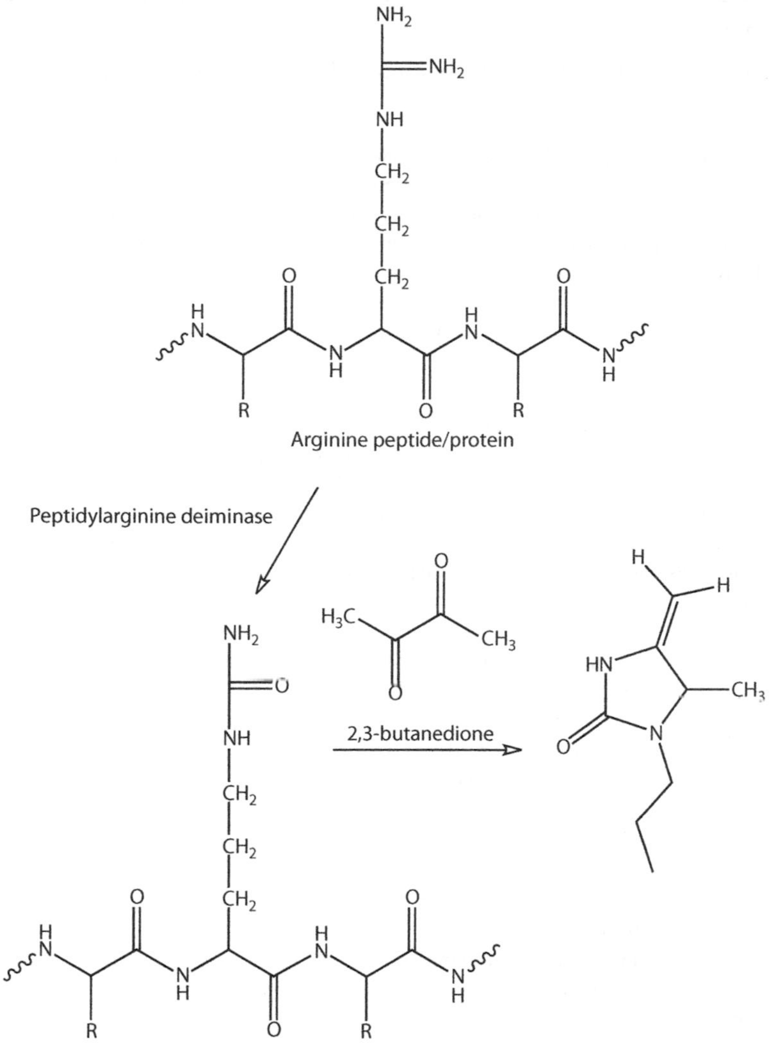

FIGURE 2.1 The posttranslational modification of an peptide-bound arginine to citrulline and the reaction with 2,3-butanedione. Arginine is modified to form citrulline by peptidylarginine deiminase (Vssenaar, E.R., Zendman, A.J., van Venrooij, W.J. et al., PAD, a growing family of citrullinating enzymes: Genes, features and involvement in disease, *Bioessays* 25, 1106–1118, 2003; Suzuki, A., Yamada, R., and Yamamoto, K., Citrullination by peptidylarginine deiminase in rheumatoid arthritis, *Ann. N. Y. Acad. Sci.* 1108, 323–339, 2007).

with diabetes mellitus type 1 as a result of advanced glycation end products (AGEs) but is of a much greater problem with diabetes mellitus type 2.[553] Diabetes mellitus type 2 has a more complex pathophysiology than diabetes mellitus type 1 with the involvement of skeletal muscle and adipose tissue in addition to pancreatic islet beta cells,[554] and is responsible for 90% of clinical diabetes.[555] Biomarkers for

TABLE 2.7
Recent Biomarker Research for Rheumatoid Arthritis and Osteoarthritis[a]

Study	Reference
Use of SELDI-TOF mass spectrometry to identify new serum biomarkers for rheumatoid arthritis. Samples were obtained from rheumatoid arthritis patients; the inflammation control group consisted subjects in psoriatic arthritis, asthma, and Crohn's disease. The control (no inflammation) group was composed of normal subjects and subjects with osteoarthritis. A pattern of protein peaks specific for rheumatoid arthritis was identified in this study	495
Differential proteome analysis was used to study synovial fluid and blood samples from patients with rheumatoid arthritis, osteoarthritis, and reactive arthritis. Protein samples obtained from gel electrophoresis were characterized by mass spectrometry. Fibrinogen degradation products were present in synovial fluid samples from all patient populations; calgranulin B (MRP14) was found in synovial fluid samples from patients with rheumatoid arthritis but not osteoarthritis	380
Use of two-dimensional electrophoresis to identify proteins differentially expressed in peripheral blood mononuclear cells of patient with rheumatoid arthritis compared to normal individuals. The differentially expressed proteins were characterized by mass spectrometry. It is suggested that the expression pattern could be used for diagnosis	496
Two-dimensional electrophoresis followed by linear ion trap-Fourier transform ion cyclotron resonance mass spectrometry (LTQ-FT/MS) is used to characterize the cartilage proteome. Differentially expressed proteins were obtained from cartilage obtained from human osteoarthritic knee	497
Two-dimensional liquid chromatography tandem mass spectrometry used to identify differentially expressed proteins in rheumatoid arthritic patients being treated with anti-TNFα antibody	498
Albumin-depleted plasma from patients with rheumatoid arthritis was examined by mass spectrometry following tryptic digestion and HPLC separation. Proteins previously associated with RA such as calgranulins and CRP were identified as well as new protein biomarkers such as actin, tubulin, vimentin, and thymosin β4.	499
Proteins from synovial membranes obtained from normal individuals and subjects with rheumatoid arthritis, osteoarthritis, and ankylosing spondylitis were separated by two-dimension gel electrophoresis and differentially expressed proteins identified by mass spectrometry. Plasma samples also were evaluated	500
Use of multiplex bead cytokine analysis (Luminex) and peptide microarray for candidate autoantibodies, and conventional ELISA assays for development of a predictive biomarker profile for anti-TNFα therapy (etanercept)	501

[a] This table contains data from proteomic analysis. It is acknowledged that there is genetic basis for susceptibility and there is work on gene expression analysis (Strietholt, S., Mauer, B., Peters, M.A. et al., Epigenetic modifications in rheumatoid arthritis, *Arthritis Res. Ther.* 10, 219, 2008; Toonen, E.J., Barrera, P., Radstake, T.R. et al., Gene expression profiling in rheumatoid arthritis: Current concepts and future directions, *Ann. Rheum. Dis.* 67, 1663–1669, 2008) and pharmacogenomics (Danila, M.I., Hughes, L.B., and Bridges, S.L., Pharmacogenomics of etanercept in rheumatoid arthritis, *Pharmacogenomics* 9, 1011–1015, 2008; Ranganathan, P., An update on pharmacogenomics in rheumatoid arthritis, *Curr. Opin. Mol. Ther.* 10, 562–567, 2008). The ability to predict susceptibility of an

TABLE 2.7 (continued)
Recent Biomarker Research for Rheumatoid Arthritis and Osteoarthritis[a]

individual to rheumatic arthritis would be a great benefit (Woolf, A.D., Hall, N.D., Goulding, N.J. et al., Predictors of the long-term outcome of early synovitis: A 5-year follow-up study, *Br. J. Rheumatol.* 30, 251–254, 1991; Verhoeven, A.C., Boers, M., te Koppele, J.M. et al., Bone turnover, joint damage and bone mineral density in early rheumatoid arthritis treated with combination therapy including high-dose prednisolone, *Rheumatology* (Oxford) 40, 1231–1237, 2001; Tran, C.N., Lundy, S.K., and Fox, D.A., Synovial biology and T cells in rheumatoid arthritis, *Pathophysiology* 12, 183–189, 2005; Tanaka, E., Taniguchi, A., Urano, W. et al., Pharmacogenetics of disease-modifying anti-rheumatic drugs, *Best Pract. Res. Clin. Rheumatol.* 18, 233–247, 2004). The identification of quantitative trait loci for rheumatoid arthritis would be a useful approach to identifying diagnostic genetic biomarkers (Johannesson, M., Hultqvist, M., and Holmdahl, R., Genetics of autoimmune diseases: A multistep process, *Curr. Top. Microbiol. Immunol.* 305, 259–376, 2006; Rosenlöf, L.W., Gene expression profiling as a tool for positional cloning of gene-shortcut or the longest way around, *Curr. Genomics* 9, 494–499, 2008).

TABLE 2.8
Biomarkers for the Inflammatory Response in Various Clinical Situations

Biomarker	References
Increase in leukocytes, in IL-6 and CRP are reported in subjects with periodontal disease while there is a decrease in total hemoglobin. Most of the systemic markers for periodontitis are also markers for cardiovascular disease	502
A review on biomarkers for cardiac inflammation. Biomarkers include CRP, IL-6,[a] IL-10, lipoprotein-associated phospholipase A_2 (Lp-PLA$_2$), selectins (e.g., soluble ICAM-1), serum amyloid A (SAA), and CD40 ligand	503
A review on inflammatory biomarkers useful in screening for stroke. The most important biomarkers include CRP (including high-sensitivity CRP,[b] IL-6, selectins, and leukocyte count	504
Inflammatory biomarkers (IL-6, TNFα, IL-1β) increase with aging while testosterone levels decrease. It is suggested that the increases in inflammatory biomarkers is casually linked to testosterone levels in aging men. Subsequent work showed that testosterone levels is negatively associated with metabolic syndrome[c] and CRP and selectins are elevated in metabolic syndrome[d]	505
Fecal calprotectin is a biomarker for intestinal inflammation in inflammatory bowel disease (IBD). Fecal calprotectin correlated with histological inflammation (colonoscopy/biopsies) and discriminated between IBD and irritable bowel syndrome and was shown to have prognostic value	506
Statins reduced levels of inflammatory biomarkers (e.g., CRP), which were suggested to be associated with reduction of cardiovascular risk	507–510
YKL-40 (CH13L1 chitinase 3-like 1; cartilage glycoprotein 39) is biomarker for inflammation in cardiovascular disease and other disorders	511–513
Eosinophil count in induced sputum and exhaled nitric oxide as markers of airway inflammation	514

(continued)

TABLE 2.8 (continued)

Biomarkers for the Inflammatory Response in Various Clinical Situations

Biomarker	References
Review of biomarkers for mucosal inflammation in Crohn's disease and correlation with clinical observations	515
Myeloperoxidase and lipoprotein-associated phospholipase A_2 are biomarkers for coronary artery disease	516
Consumption of nuts is shown to reduce cardiovascular mortality as well as decreasing the concentration of inflammation biomarkers such as CRP, IL-6, and endothelial biomarkers	517
A review of nucleic acid biomarkers of inflammation focusing on products resulting from nitration, oxidation, peroxidation, halogenation, nitrosation, and alkylation. Lead biomarkers include 8-oxo-deoxyguanosine and products from oxidative deamination by nitrosation	518
Inflammatory biomarkers observed in obesity include CRP, TNFα, IL-6, and leptin, which are reduced by weight loss	519
Use of biomarkers for inflammation in point-of-case (POC) testing. Leukocyte/ neutrophil count, CRP, procalcitonin (PCT) are useful for POC testing in children and adults while IL-6 and IL-8 are useful for the detection of sepsis in neonates. It is noted that POCT permits the testing of biomarkers in ascites fluid, pleural fluid, and abscess materials	111
CRP and albumin as biomarkers of systemic inflammation in cancer; these biomarkers are prognostic indicators in cancer	520
Inverse relationship between biomarkers for inflammation such as CRP, fibrinogen, erythrocyte sedimentation rate, and level of education	521

[a] IL-6 is a cytokine produced by macrophages (Cavaillon, J.M., Cytokines and macrophages, *Biomed. Pharmacother.* 48, 445–453, 1994) and endothelial cells (Krishnaswamy, G., Kelley, J., Yerra, L. et al., Human endothelium as a source of multifunctional cytokines: Molecular regulation and possible role in human disease, *J. Interferon Cytokine Res.* 19, 91–104, 1999) and other cells (Chudek, J. and Wiecek, A., Adipose tissue, inflammation and endothelial dysfunction, *Pharmacol. Rep.* 58 Suppl, 81–88, 2006) which is responsible for increased hepatic synthesis of acute phase proteins (Sehgal, P.B., Interleukin-6: A regulator of plasma protein gene expression in hepatic and non-hepatic tissues, *Mol. Biol. Med.* 7, 117–130, 1990; Carter, A.M., Inflammation, thrombosis and acute coronary syndromes, *Diab. Vasc. Dis. Res.* 2, 113–121, 2005).

[b] High-sensitivity CRP refers to CRP measured at low concentration (≤ 0.5 mg/L); changes in the concentration of CRP at this level is considered to be diagnostic of increased risk for cardiovascular disease (see Tarkkinen, P., Palenius, T., and Lövgren, T., Ultrarapid, ultrasensitive one-step kinetic immunoassay for C-reactive protein (CRP) in whole blood samples: Measurement of the entire CRP concentration range with a single sample dilution, *Clin. Chem.* 48, 269–277, 2002; Musunuru, K., Kral, B.G., Blumenthal, R.S. et al., The use of high-sensitivity assays for C-reactive protein in clinical practice, *Nat. Clin. Pract. Cardiovasc. Med.* 5, 621–635, 2008).

[c] Maggio, M., Lauretani, F., Ceda, G.P. et al., Association between hormones and metabolic syndrome in older Italian men, *J. Am. Geriatr. Soc.* 54, 1832–1838, 2006.

[d] Ingelsson, E., Hultha, J., and Lind, L., Inflammatory markers in relation to insulin resistance and the metabolic syndrome, *Eur. J. Clin. Invest.* 38, 502–509, 2008.

TABLE 2.9
Recent Biomarkers for Inflammation Identified with Proteomic Technology

Study	Reference
Use of two-dimensional electrophoresis/mass spectrometry to evaluate inflammation biomarkers in albumin/IgG-depleted plasma samples from patients with severe acute respiratory syndrome (SARS). α_1-acid glycoprotein, haptoglobin, α_1-antichymotrypsin, and fetuin were elevated; transferring, apolipoprotein A-I, and transthyretin were decreased	522
More than 20,000 unique peptides from four different peptide populations (cysteinyl peptides, non-cysteinyl peptides, *N*-glycopeptides and non-glycopeptides) were identified by LC-MS/MS. The analysis of these peptides allowed the identification of more than 3,000 proteins. A panel of proteins comprised proteins involved in inflammation and immune responses	523
SELDI[a]-TOF/MS used to assess changes in bronchoalveolar lavage protein after challenge with local bronchial lung endotoxin	524
Use of proteomic technology to identify disease biomarkers in plasma from pediatric patients undergoing cardiopulmonary bypass surgery. Two-dimensional electrophoresis (fluorescent detector) and analysis by mass spectrometry. More than 70 protein species were altered (more than 50%) by cardiopulmonary bypass, which could be used as biomarkers	525
Biomarkers of endothelial cell dysfunction prognostic for cardiovascular disease patients with chronic kidney disease evaluated. Suggested evaluation of vasorelaxation, oxidative stress and defenses, inflammatory biomarkers and coagulation biomarkers	526
Predictive biomarkers for dialysis in end-stage renal disease include protein carbonyl, AGEs, tyrosine modified by reactive oxygen species/reactive nitrogen species (3-chlorotyrosine, 3-nitrotyrosine, dityrosine)	527
Degradation products of acute-phase proteins (apolipoprotein A-I, α_1-antitrypsin) are biomarkers for gastric inflammation (irrespective of infection by *Helicobacter pylori*). Degradation products were identified by two-dimensional gel electrophoresis followed by mass spectrometry and/or Western blot analysis	528
Specific urine biomarkers permitted distinguishing pancreatic cancer from chronic inflammation; analysis used solid phase extraction with lipophilic resin followed by MALDI MS analysis	529
The Random Forests method was used to filter the most important variables from 1442 genetic variables and 108 proteomic variables to develop model for adverse events secondary to smallpox vaccination; a hypothesis involving prolonged stimulation of inflammatory pathways has been developed	530
Use of the HDL proteome inflammation biomarkers; shotgun proteomic analysis[b] was used to identify inflammation biomarkers in HDL diagnostic of coronary artery disease. Inflammatory biomarkers in HDL include chlorotyrosine and nitrotyrosine; apolipoprotein E is also found in HDL from coronary artery disease subjects	531
Biomarkers for perinatal and neonatal sepsis include defensin-2, defensin-1, S100A12, S100A8, S100A9, and insulin-like growth factor. The presence of amniotic fluid biomarkers of inflammation is associated with increased inflammatory status of the fetus at birth	532

[a] Hydrophobic matrix.

[b] Shotgun proteomic analysis is a technique where a mixture of proteins is subjected to proteolytic digestion and the resulting digest is analyzed by LC/MS/MS (see McDonald, W.H. and Yates III J.R., Shotgun proteomics and biomarker discovery, *Dis. Markers* 18, 99–105, 2005; Swanson, S.K. and Washburn, M.P., The continuing evolution of shotgun proteomics, *Drug Discov. Today* 10, 719–725, 2005).

TABLE 2.10
Biomarkers for Inflammation in Atherosclerosis

Biomarker	Reference
Microparticles determined by flow cytometry[a] as biomarkers for inflammation in atherosclerosis	533
Higher levels of fibrinogen and hsCRP (inflammatory biomarkers) in woman with early carotid atherosclerosis (ultrasonography)	534
Use of inflammatory biomarkers (e.g., hs-CRP, cytokines) to identify atherosclerotic plaques subject to rupture	535
CRP is a strong indicator of atherosclerotic disease (carotid intima-media thickness and plaque by ultrasound) in patients with recent onset rheumatoid arthritis	536
Inflammation biomarkers are used to monitor atherosclerosis	537
Inflammatory biomarkers such as CRP, fibrinogen, erythrocyte sedimentation rate, and IL-6 are increased in type 2 diabetic patients with atherosclerotic vascular disease; homocysteine is also elevated	538
Principal factor analysis is used to investigate clustering of variables associated with elevated CRP levels (CRP is used as a biomarker for low-grade inflammation); low grade inflammation is considered a predisposing factor for atherosclerosis	539
Studies suggest that inflammatory biomarkers (i.e., CRP, fibrinogen, IL-6) are more useful as prognostic biomarkers in individuals with established coronary artery disease rather than as predictors of future coronary heart disease	540
New imaging techniques such as PET with 18-FDG may be more useful in predicting atherosclerotic plaque progression than the use of blood biomarkers	541
Fibrinogen, ceruloplasmin, α_1-antitrpsin (inflammation-sensitive plasma proteins) served a prognostic biomarkers for the long-term incidence of hospitalization for heart failure in middle-aged men	542
Nucleic acid biomarkers for inflammation	518
Common cytokines involved in rheumatoid arthritis and atherosclerosis	543
Fibrinogen and CRP are prognostic for subclinical and clinical arthritis	544
YLK-40[b] is a biomarker elevated in type 1 diabetes and associated with atherosclerosis.	545
Biomarkers of inflammation and hemostasis (fibrinogen, Factor VIIIc, vWF) are associated with greater risk of kidney function decrease in chronic kidney disease.	546
Nonfried fish consumption inversely related to CRP and IL-6	547
Soluble CXCL16[c] is a biomarker for inflammation and atherosclerosis with importance for acute coronary syndrome	548

[a] Flow cytometry is the accepted method for the determination of microparticles as biomarkers (see Arteaga, R.B., Chrinos, J.A., Soriano, A.O. et al., Endothelial microparticles and platelet and leukocyte activation in patients with the metabolic syndrome, *Am. J. Cardiol.* 98, 70–74, 2006; Shah, M.D., Bergeron, A.L., Dong, J.F., and López, J.A., Flow cytometric measurement of microparticles: Pitfalls and protocol modifications, *Platelets* 19, 365–372, 2008; Lal, S., Brown, A., Nguyen, L. et al., Using antibody arrays to detect microparticles from acute coronary syndrome patients based on cluster of differentiation (CD) antigen expression, *Mol. Cell. Proteomics* 8, 799–804, 2009).

[b] YLK-40 is also known as cartilage glycoprotein 39 (see Roslind, A. and Johansen, J.S., YKL-40: A novel marker shared by chronic inflammation and oncogenic transformation, *Methods Mol. Biol.* 511, 159–184, 2009.

TABLE 2.10 (continued)
Biomarkers for Inflammation in Atherosclerosis

[c] Soluble CXCL16 is a cytokine derived from the proteolysis of CXCL16, a scavenger receptor expressed on macrophages (see Sheikine, Y. and Hansson, G.K., Chemokines and atherosclerosis, *Ann. Med.* 36, 98–118, 2004; Ludwig, A. and Weber, C., Transmembrane chemokines: Versatile 'special agents' in vascular inflammation, *Thromb. Haemost.* 97, 694–703, 2007; Li, X., Conklin, L. and Alex, P., New serological biomarkers of inflammatory bowel disease, *World J. Gastroenterol.* 14, 5115–5124, 2008).

inflammation in diabetes are shown in Table 2.11. It is clear that diabetes mellitus type 1 is a genetic disease while diabetes mellitus type 2 is a lifestyle disease, with obesity as a major contributing factor.[563] Obesity is mostly an acquired disease, which is approaching epidemic proportions in the United States and other developed countries[564–566] and is responsible for approximately 10% of health care

TABLE 2.11
Biomarkers for Inflammation in Diabetes

Biomarker	Reference
Serum resistin may be a link between inflammation, acute phase response, and insulin resistance in ICU patients; resistin is not linked to type 2 diabetes or obesity	556
Copeptin is a 39-amino acid glycopeptides derived from the C-terminal domain of arginine vasopressin precursor. Copeptin is used in the diagnosis of diabetes insipidus	557
Potential biomarkers are found in saliva from subjects with type 2 diabetes	558
CRP and fibrinogen are biomarkers for obesity class with increasing concentration with increasing obesity class with the highest levels in class 3; CRP and fibrinogen are increased in diabetes	450
Superoxide release from phobol ester-stimulated peripheral blood polymorphonuclear leukocytes is used as biomarker for oxidative stress and inflammation in type 2 diabetes	559
Neopterin is a biomarker for inflammation in diabetes; TNFα is also elevated in diabetic patients	560
Serum levels of soluble tumor necrosis factor-like weak inducer of apoptosis (sTWEAK) are reduced in type 2 diabetes and end-stage renal disease; the effect is additive and sTWEAK is suggested as a potential biomarker for atherosclerosis	561
YKL-40[a] is a biomarker for inflammation that is elevated in type 2 diabetes and is related to insulin resistance. YKL-40 is not related to hsCRP is an independent biomarker for inflammation in diabetes	562

[a] YKL-40 is HHI3l1 chitinase 3-like; cartilage glycoprotein-39; see the following for more discussion of the role of YKL-40 as a biomarker in diabetes and insulin resistance: Rathcke, C.N., Vestergaard, H., YKL-40, a new inflammatory marker with relation to insulin resistance and with a role in endothelial dysfunction and atherosclerosis, *Inflamm. Res.* 55, 221–227, 2006; Rathcke, C.N., Holmkvist, J., Jørgensen, T. et al., Variation in CHI3LI in relation to type 2 diabetes and related quantitative traits, *PLoS One* 4, e5469, 2009.

TABLE 2.12
Biomarkers in Obesity and Metabolic Syndrome[a]

Biomarker	Reference
Biomarkers for inflammation change with obesity weight class.[b] CRP doubles for each increase in class and fibrinogen increases as well	450
Adipokines[c] as biomarkers for obesity. Adiponectin, retinol binding protein 4, and resistin are three adipokines that are important biomarkers for inflammation	573
NFκB (nuclear factor kappa B) is a biomarker for metabolic syndrome and obesity-related inflammation; weight loss showed reduction of NFκB expression	574
Gender differences in biomarkers in metabolic syndrome. Women had higher levels of hs-CRP and IL-1Ra (interleukin-1 receptor antagonist) than men; there were no gender differences in a population without metabolic syndrome; adiponectin was lower in subjects with metabolic syndrome	575
Unique interaction of biomarkers for adiposity, inflammation, and cardiovascular risk in seriously obese subjects	576
Adiponectin serves as a biomarker for metabolic syndrome and is independent of adiposity, insulin resistance, and inflammation	577
Review of inflammation biomarkers (TNFα, monocyte chemoattractant protein-1 and IL-6) and hemostasis biomarkers (plasminogen activator inhibitor-1) for diagnosis of metabolic syndrome	578
Endothelial lipase is a biomarker for metabolic syndrome that is positively correlated with hs-CRP, soluble tumor necrosis factor receptor II, and IL-6, and negatively correlated with adiponectin	579
Ethnic differences in the expression of inflammation biomarkers in pediatric subjects with metabolic syndrome	580
Plasminogen activator inhibitor-1 (PAI-1) is identified as an important biomarker for risk of metabolic syndrome; CRP, IL-6, and fibrinogen are biomarkers that are also associated with metabolic syndrome	581
CRP as a biomarker for cardiovascular risk in individuals with metabolic syndrome	582
Inflammation biomarkers (Il-6, TNFα, CRP) and dyslipidemia biomarkers (HDL cholesterol, triglyceride, non-esterified fatty acid) were not significant in predicting progression to type 2 diabetes in subjects with metabolic syndrome while obesity (BMI as biomarker), insulin resistance and β-cell function (insulin, proinsulin) were prognostic biomarkers for progression to type 2 diabetes in subjects with metabolic syndrome	583
Soluble CXCL16[d] is a biomarkers for inflammation, atherosclerosis, and acute coronary syndrome	548
sCD163[e] (soluble CD163) is a biomarker for mature macrophages associated with increasing fat mass and may be prognostic for renal disease	584
Review of inflammatory biomarkers in the metabolic syndrome. The predictive biomarker effects for TNFα (i.e., decreases with Simvastin), IL-6 (decreases with Mediterranean diet)—other biomarkers are discussed	585

TABLE 2.12 (continued)
Biomarkers in Obesity and Metabolic Syndrome[a]

[a] Metabolic syndrome is a cluster of risk factors including abdominal obesity, elevated blood pressure, insulin resistance, and atherogenic dyslipidemia, which predispose to type 2 diabetes, atherosclerosis, chronic heart disease, and chronic kidney disease (Timóteo, A., Santos, R., Lima, S. et al., Does the new International Diabetes Federation definition of metabolic syndrome improve prediction of coronary artery disease and carotid intima-media thickening?, *Rev. Port. Cardiol.* 28, 173–181, 2009; Huang, P.L., A comprehensive definition for metabolic syndrome, *Dis. Model. Mech.* 2, 231–237, 2009; Natali, A., Pucci, G., Boldrini, B., and Schillaci, G., Metabolic syndrome: At the crossroads of cardiorenal risk, *J. Nephrol.* 22, 29–38, 2009; Koutsovasilis, A., Protopsaltis, J., Triposkiadis, F. et al., A comparative performance of three metabolic syndrome definitions in the prediction of acute coronary syndrome, *Intern. Med.* 48, 179–187, 2009).

[b] Normal, BMI (body mass index) > 25.0; overweight = 25.0–29.9; obesity class 1 = 30.0–34.9; obesity class 2 = 35.0–39.9; and obesity class 3 ≥ 40.0.

[c] Adipokines are biologically active materials secreted by adipose tissue such as leptins, adiponectin (see Chaldakov, G.N., Stankulov, I.S., Hristova, M., and Ghenev, P.I., Apidobiology of disease: Adipokines and adipokine-targeted pharmacology, *Curr. Pharm. Des.* 9, 1023–1031, 2003; Wiecek, A., Adamczak, M., and Chudek, J., Adiponectin—An adipokine with unique metabolic properties, *Nephrol. Dial. Transplant.* 22, 981–988, 2007).

[d] CXCL16 (CXC chemokines ligand 16) is identical with the scavenger receptor SR-PSOX (Shimaoka, T., Nakayama, T., Kume, N. et al., Cutting edge: SR-PSOX/CXC chemokines ligand 16 mediates bacterial phagocytosis by APCs through its chemokines domain, *J. Immunol.* 171, 1647–1651, 2003). CXCL16 can be cleave to yield a soluble form, which acts as a chemoattractant (Abel, S., Hundhausen, C., Mentlein, R. et al., The transmembrane CXC-chemokine ligand 16 is induced by IFN-γ and TNF-α and shed by the activity of the disintegrin-like metalloproteinase ADAM10, *J. Immunol.* 172, 6362–6372, 2004; Ludwig, A. and Weber, C., Transmembrane chemokines: Versatile 'special agents' in vascular inflammation, *Thromb. Haemost.* 97, 694–703, 2007). There is suggestion that soluble CXCL16 is a prognostic biomarker for acute coronary syndrome (Jansson, A.M., Aukrust, P., Ueland, T. et al., Soluble CXCL16 predicts long-term mortality in acute coronary syndrome, *Circulation* 119, 3181–3188, 2009).

[e] CD163 is a hemoglobin scavenger receptor, which is an exclusively expressed mature tissue macrophages (see Moestrup, S.K. and Møller, H.J., CD163: A regulated hemoglobin scavenger receptor with a role in the anti-inflammatory response *Ann. Med.* 36, 347–354, 2004; Farbriek, B.O., Dijkstra, C.D., and van den Berg, T.K., The macrophage scavenger receptor CD163, *Immunobiology* 210, 153–160, 2005).

expenditures in the United States.[567] Obesity is considered a component of metabolic sydrome.[568] Metabolic syndrome is a group or cluster of risk factors prognostic of cardiovascular disease.[569,570] Insulin resistance is a component of metabolic sydrome,[571] providing a link to diabetes mellitus type 2. The comorbidities of metabolic syndrome include hypertension, cardiomyopathy, and coronary heart disease.[572] Biomarkers for inflammation in obesity and metabolic syndrome are shown in Table 2.12.

Gastrointestinal diseases such as Crohn's disease, which can involve the entire digestive tract and ulcerative colitis, which is confined to the colon, are characterized

TABLE 2.13
Biomarkers for the Inflammation in Gastrointestinal Disease[a]

Biomarker	Reference
New biomarkers for inflammatory bowel disease (IBD) (reported since 2007) are reviewed. Anti-glycan antibodies are biomarkers that differentiate Crohn's disease, ulcerative colitis, and other non-IBD diseases. Other new serological biomarkers for IBD are reported; some are nonspecific such as CXCL16, resistin, apolipoprotein A-IV, and ubiquityinylation factor E4A	341
Urinary biomarkers (determined by NMR spectroscopy) distinguish between inflammatory bowel disease, Crohn's disease, and ulcerative colitis	591
Anti-glycan antibodies are biomarkers for differentiating between Crohn's disease and ulcerative colitis. IgA cell wall polysaccharide antibodies were directed against laminarin or chitin	592
Use of perinuclear antineutrophil cytoplasmic antibodies and anti-*Saccharomyces cerevisiae* antibodies as biomarkers in the diagnosis of inflammatory bowel disease	593
IL-18 is a biomarker for acute phase reactivity in Crohn's disease.	594
Gene expression analysis (real-time PCR) of CXCL2 (macrophage inflammatory protein 2), CXCL8 (IL-8), CXCL10 (INFγ inducible protein 10), and calgranulin in colonic biopsies provides biomarkers for disease activity (bowel inflammation, clinical activity index) in ulcerative colitis	595
The combined use of fecal calprotectin and fecal lactoferrin provide a biomarker for inflammatory bowel disease in a pediatric population	589
CRP and erythrocyte sedimentation rate were biomarkers for proximal lesions in ulcerative colitis while clinical symptoms were associated with the activity of distal colonic lesions	596
Human neutrophil peptides 1–3[b] are novel diagnostic and predictive biomarkers for ulcerative colitis	597
The gene expression of NFκB, A20 (a negative regulator of NFκB), polymeric IgG receptor, TNF, and IL-8 in mucosal biopsies provides predictive biomarkers for Crohn's disease. The use of the biomarkers divided the patient populations into three groups	598
Perinuclear antineutrophil cytoplasmic antibodies (pANCA) and anti-*Saccharomyces cerevisiae* antibodies (ASCA) are biomarkers for diagnosis and prognosis for ulcerative colitis and Crohn's disease	599
Soluble CXCL16[c] is a biomarker for Crohn's disease and ulcerative colitis and is associated with CRP in this group of patients with inflammatory bowel disease	600
Fecal calprotectin is a surrogate predictive biomarker for treatment outcome in patients with IBC	601
sCD14, TNFα, and sTNFRII, CRP, and fecal calprotectin were theranostic biomarkers for the use of atorvastatin in Crohn's disease. sCD14, TNFα, and sTNFRII were slightly decreased with treatment while the decrease in CRP was more significant	602
Serum hsCRP, IL-6, fecal calprotectin, and lactoferrin were biomarkers for Crohn's disease and correlated with the endoscopic measurement of disease activity	588

[a] The search for biomarkers and gastrointestinal disease yielded more that 24,000 citations; restricting the search to biomarkers and inflammatory bowel disease yield approximately 3,000 citations. It is intended that the material selected is representative of the field. It does provide an example of the use of sample selection to provide analytical specificity.

TABLE 2.13 (continued)
Biomarkers for the Inflammation in Gastrointestinal Disease[a]

b Human neutrophil peptides 1–3 are also known as α-defensins, which are released from the granules of activated neutrophils (see Henie, R.P., Wiesenfeld, H., Mortimer, L, and Greig, P.C., Amniotic fluid defensins: Potential markers of subclinical intrauterine infection, *Clin. Infect. Dis.* 27, 513–518, 1998; Tanska, S., Edberg, J.C., Chatham, W. et al., FcγRIIIb allele-sensitive release of α-defensins: Anti-neutrophil cytoplasmic antibody-induced release of chemotaxins, *J. Immunol.* 171, 6090–6096, 2003).

c CXCL16 contains a transmembrane domain and both membrane-bound and soluble forms are produced (Matloubian, M., David, A., Engel, S. et al., A transmembrane CXC chemokines is a ligand for HIV-coreceptor Bonzo, *Nat. Immunol.* 1, 298–304, 2000).

---

by inflammation.[586] Biomarkers, which are derived from leukocytes such as calprotectin and lactoferrin,[409,587–590] have been of particular interest as diagnostic/prognostic tools for inflammatory bowel disease. Biomarkers for inflammation in gastrointestinal disease are shown in Table 2.13. These various studies emphasize that the inflammation and expression of biomarkers of inflammation such as the acute-phase proteins is common to a number of disease states.

Systemic inflammatory response is seen in a variety of clinical situations including traumatic brain injury; sepsis is SIRS secondary to a bacterial infection while septic shock and multiorgan dysfunction are further complications of SIRS/sepsis. Sepsis can be defined as the host response to an infection.[603] There are approximately 500,000 cases of sepsis per year in the United States and is responsible for 3%–5% of ICU admissions with a mortality of 35%–50%. Early recognition of sepsis is critical for successful treatment[604]; and biomarkers may be selected from plasma/serum or cells.[605] Infection is a prerequisite for sepsis; thus the treatment for sepsis must include consideration of the underlying bacterial infection as well as the immunological problems in SIRS. Some of the biomarkers such as procalcitonin[606] provide the differentiation between sepsis and SIRS, which is critical for effective therapeutics.[607–609] Inflammation biomarkers for sepsis and SIRS are shown in Table 2.14.

I would be remiss to not mention age, both as a confounding factor (Table 2.15) and as a biomarker for prognostic analysis. Under normal circumstances, there are substantial changes in three major N-glycan structures.[641] AGEs increase as a function of age[642–644] and are suggested to be a biomarker for age as well as being prognostic for cardiac surgery.[634,637,645]

BIOMARKERS IN OPHTHALMOLOGY

The eye is "isolated" from the rest of the body, and work on circulating biomarkers of retinal disorders is quite limited. The extensive use of imaging technologies has presented opportunities for the development of biomarkers in theragnostics[92] in ophthalmology. Biomarkers for ophthalmology are presented in Table 2.16.

TABLE 2.14
Biomarkers for Sepsis and SIRS[a]

Biomarker	References
A review on biomarkers for inflammation in SIRS, sepsis, and septic shock. Biomarkers are described with a differentiation between SIRS and sepsis (CRP, IL-6, procalcitonin (PCT), and LPS-binding protein). PCT also differentiates between bacterial and viral infection	431
CRP used with SIRS parameters to diagnose infection in ICU patients. A decrease in CRP levels between admission and day 4 was the best prognostic biomarker	610
Regulatory T cells (Treg cells) were higher in patients with sepsis compared to SIRS patients; soluble CD25 was also higher in patients with sepsis	611
IL-6 was elevated in sepsis compared to acute pancreatitis patients	612
Lipopolysaccharide-binding protein (LBP) is a nonspecific acute phase-response protein, which is elevated in both SIRS and sepsis and does not differentiate between SIRS and sepsis	613
Generation of leukotriene C_4 from isolated/stimulated (calcium ionophore) leucocytes decreased in patients with sepsis and was prognostic. Leucotriene C_4 generation correlated with a sepsis severity score.[b] More recent work has shown that leukotriene C_4 synthase is downregulated in mononuclear phagocytes by TNFα[c]	614
Procalcitonin is elevated in patients with sepsis and is considered to differentiate between sepsis and SIRS. Evidence suggests good use in diagnosis but not prognosis. Recent work shows elevation in patients with abdominal trauma	615–617
Soluble CD163, a macrophage-specific serum marker, is demonstrated to be a prognostic biomarker for pneumococcal bacteremia. Analysis was performed using a previously developed ELISA assay[d]	618, 619
High mobility group-box 1 protein (HMGB1) is proinflammatory cytokine elevated in both gram-positive and gram-negative bacteremia. HMGB1 correlated with CRP, procalcitonin, and WBC (white blood cell) count	619, 620
8-Hydroxy-2′-deoxyguanosine (8-OHdG) is a biomarker for oxidative stress and is elevated in patients with ARDS (adult respiratory distress syndrome)[e]; 8-OHdG correlated with HMGB1. These biomarkers were used in the evaluation of the effectiveness of polymyxin B-immobilized fiber hemoperfusion	620
Biomarkers are being developed for anastomotic leakage; anastomotic leakage can result in peritonitis, which in turn can result in sepsis. It is suggested that biomarkers can be measured in fluid from peritoneal drainage	621
Selenoprotein P declines in septic shock are associated with a poor prognosis. Selenoprotein P is determined with immunoluminometric sandwich assay with two polyclonal antibodies	622
Protein C is decreased in sepsis and the level of protein C is a theranostic marker for the use of activated protein C in the treatment of sepsis	623, 624
Fluid balance is considered as a biomarker for sepsis (acute kidney injury) (SOAP study; sepsis occurrence in acutely ill patients)	625
Multidimensional native chromatography is used to identify new biomarkers for septic shock and Alport syndrome; 33 new candidates for severe inflammatory disease (sepsin) were identified including α1B-glycoprotein, cysteine-rich secretory protein, and a low-molecular-weight albumin variant	626

TABLE 2.14 (continued)
Biomarkers for Sepsis and SIRS[a]

Biomarker	References
Glyco-isoforms (glycoprotein isoforms based on differences in glycosylation patterns) of apolipoprotein C3 were used as a biomarkers for sepsis. Changes in isoform distribution were also observed in liver disease (hepatitis C and alcoholic cirrhosis)	627
YKL-40 (CH13L1 chitinase 3-like 1; cartilage glycoprotein 39) is identified as a possible biomarker for sepsis by the combination of HPLC and electrophoresis following the depletion of plasma of major proteins. The presence of YKL-40 was established by Western blotting	628
Neopterin is a prognostic indicator in sepsis/septic shock. Neopterin was measured in urine and found to be correlated with APACHE II (Acute Physiology and Chronic Health Evaluation II) scores	318

[a] SIRS is characterized by a combinant of hyperthermia/hypothermia, tachycardia, hyperventilation (tachypnea), and leucocytosis/leucopenia; sepsis is SIRS caused by infection (bacterial or viral). See American College of Chest Physicians/Society of Critical Care Medicine Consensus Conference, Definitions for sepsis and organ failure and guidelines for the use of innovative therapies in sepsis, *Crit. Care Med.* 20, 864–874, 1992. See also Dremsizov, T. T., Kellum, J.A., and Angus, D.C., Incidence and definition of sepsis and associated organ dysfunction, *Int. J. Artif. Organs* 27, 352–359, 2004; Robertson, C.M. and Coopersmith, C.M., The systemic inflammatory response syndrome, *Microbes Infect.* 8, 1382–1389, 2006; Marik, P.E. and Lipman, J., The definition of septic shock: Implications for treatment, *Crit. Care Resusc.* 9, 101–103, 2007.

[b] Elebute, E.A. and Stoner, H.B., The grading of sepsis, *Br. J. Surg.* 70, 29–31, 1983.

[c] Serio, K.J., Luo, C., Luo, L. et al., TNFα downregulates the leukotrienes C4 synthase gene in mononuclear phagocytes, *Am. J.Physiol. Lung Cell. Physiol. Mol. Physiol.* 292, L215–L222, 2007.

[d] Moller, H.J., Hald, K., and Moestrup, S.K., Characterization of an enzyme-linked immunosorbent assay for soluble CD163, *Scand. J. Clin. Lab. Invest.* 62, 293–299, 2002.

[e] ARDS is a complication of sepsis (Balk, R.A. and Bone, R.C., The septic syndrome. Definition and clinical implications, *Crit. Care Clin.* 5, 1–8, 1989) which can occur after treatment with pegylated interferon α-2b and ribavirin, *Heart Lung*, 37, 153–156, 2008.

TABLE 2.15
The Effect of Aging on Biomarker Expression

Biomarker	Reference
sRAGE[a] decreases with age	629
Expression of the tumor repressor P16(INK4a) is a biomarker of aging found in peripheral blood T-lymphocytes. Expression of P16(INK4a) is also correlated with tobacco use and physical inactivity	630
Arterial stiffness (pulse wave velocity[b]) as biomarker of aging	631
Transformed osseographic score (TOSS) as a biomarker	632
Lens opacity (cataract formation) as a biomarker of aging	633

(*continued*)

TABLE 2.15 (continued)
The Effect of Aging on Biomarker Expression

Biomarker	Reference
AGEs as prognostic marker for senescence (and clinical outcome in cardiac surgery)	634
Low serum HSP70 (heat shock protein 70) is a longevity biomarker	635
Dityrosine as a biomarker of aging	636
Estrogen receptor as an age biomarker in breast cancer	637
Luminol-induced chemiluminescence from phorbol ester-stimulated polymorphonuclear leukocytes (PMNs)	638
Carboxymethylethanolamine as a potential biomarker for aging.	639
VO$_2$max is a biomarker for aging	640

[a] sRAGE, soluble receptor for advanced glycation end products. sRAGE is thought to act as a decoy receptor for AGEs thus acting in a protective manner in inflammation (see Yan, S.F., D'Agati, V., Schmidt, A.M. and Ramasamy, R., Receptor for advanced glycation end products (RAGE): A formidable force in the pathogenesis of the cardiovascular complications of diabetes & aging, *Curr. Mol. Med.* 7, 699–710, 2007).

[b] See Davies, J.I. and Struthers, A.D., Pulse wave analysis and pulse wave velocity: A critical review of their strengths and weaknesses, *J. Hypertens.* 21, 463–472, 2003.

TABLE 2.16
Biomarkers for Age-Related Macular Degeneration and Other Retinal Disorders

Biomarker	References
Activated forms of the VEGF receptor are found in the vitreous fluid and serve as a predictive biomarker for anti-VEGF therapy	646
Single nucleotide polymorphisms of tenomodulin can serve a biomarker in women for age-related macular degeneration	647
The concentration of cystatin C is a biomarker for age-related macular degeneration (the presence of chronic kidney disease is also a biomarker)	648
VEGF and ICAM-1 are biomarkers in vitreous fluid for diabetic macular edema	649
The optical density ratio in subretinal fluid (determined with high resolution optical coherence tomography) is a biomarker for exudative macular disease	650
Use of Raman spectroscopy for the measurement of AGEs as biomarker for retinal aging and prognostic biomarker for age-related macular degeneration	651
High-sensitivity CRP (hsCRP) is a biomarker for age-related macular degeneration[a]	652, 653
Antibodies against retinal antigens are possible biomarkers for "wet" age-related macular degeneration	654
Endogenous soluble receptor for advance glycation end products as a biomarker for susceptibility to diabetic retinopathy	655
Biomarkers in subretinal fluid from patients with retinal detachment	656
IL-6 as biomarker in vitreous fluid for uveitis	657

TABLE 2.16 (continued)
Biomarkers for Age-Related Macular Degeneration
and Other Retinal Disorders

[a] Another study has suggested that there is no consistent pattern for the presence of biomarkers for inflammation in age-related maculopathy (Wu, K.H., Tan, A.G., Rochtchina, E. et al., Circulating inflammatory markers and hemostatic factors in age-related maculopathy: A population-based case-control study, *Invest. Ophthalmol. Vis. Sci.* 48, 1983–1988, 2007). There are other articles supporting the relationship of inflammation biomarkers and age-related macular degeneration and still others which suggest no independent relationship (see Seddon, J.M. George, S., Rosner, B., and Rifai, N., Progression of age-related macular degeneration: Prospective assessment of C-reactive protein, interleukin 6, and other cardiovascular markers, *Arch. Ophthalmol.* 123, 774–782, 2005; Klein, R., Klein, B.E., Knudtson, M.D. et al., Systemic markers of inflammation, endothelial dysfunction, and age-related maculopathy, *Am. J. Ophthalmol.* 140, 35–44, 2005.

BIOMARKERS IN NEUROLOGY

Neurology is a complicated discipline and accurate diagnosis is a challenge. Most diagnoses are made on the basis of clinical signs; the argument has been made that clinical evaluation is a biomarker in Alzheimer's disease.[658] The difficulty with biomarkers in neurology is illustrated by studies on serum S100B as a biomarker for ischemic stroke.[659] Some examples of biomarkers in neurology are given in Table 2.17.

CARDIAC BIOMARKERS

Cardiac biomarkers are like the most prominent biomarkers in use at present. Biomarkers for atherosclerosis are prognostic of coronary heart disease and have been discussed above. As an example, CRP is prognostic for serving cardiac events in unstable angina.[670] Current cardiac biomarkers are shown in Table 2.18 while Table 2.19 provides some examples of cardiac biomarkers under development using proteomic technology. Gene expression is also considered a biomarker, which can be studied by ex vivo techniques as discussed above. Table 2.20 describes some gene expression biomarkers for cardiac function. The reader is also directed to a work edited by David Marrow,[744] which discusses cardiac biomarkers in detail. A sense of reality is provided by a recent comment by O. Collinson,[745] which notes that despite considerable activity for the discovery of cardiac biomarkers, there are only two cardiac biomarkers in extensive use. Two troponins, cardiac troponin T and cardiac troponin I, are specific biomarkers for cardiac muscle damage, while natriuretic peptides (B-type natriuretic peptide (BNP) and N-terminal pro-B-type natriuretic peptide) measure cardiac failure (ventricular dysfunction).[210,698,744,746–749] Most cardiovascular diseases are preceded by other pathologies such as atherosclerosis/diabetes, which establishes an at-risk population, which would, in principle, increase the predictive value of

TABLE 2.17
Biomarkers for Neurological Disorders

Biomarker	Reference
Neuronal pentraxin receptor (NPR)[a] may be a specific biomarker for Alzheimer's disease in cerebrospinal fluid(CSF); NPR, α-dystroglycan, and NCAM-120 (neural cell adhesion molecule 1) may be biomarkers for Alzheimer's disease and Parkinson's disease	660
Use of saccadic[b] movements as a biomarker for Parkinson's disease	661
Truncated tau forms are a biomarker for supranuclear palsy	662
Use of proteomics to identify biomarkers for Alzheimer's disease and Parkinson disease	663
Serum urate is a prognostic biomarker for Parkinson disease	664
Midbrain iron content determined by magnetic resonance imaging[c] as a potential biomarker for Parkinson disease prognosis	665
Grooved pegboard test score is a biomarker for prognosis in Parkinson disease. Grooved pegboard test score is inversely correlated with dopamine transporter (DAT) positron emission tomography	666
Use of proteomic technology to identify potential biomarkers in cerebrospinal fluid for neurodegenerative disease	667
Oligomeric forms of α-synuclein protein in plasma as a biomarker for Parkinson disease	668
Urinary 8-hydroxydeoxyguanosine is a prognostic biomarker for Parkinson disease	669

[a] Neuronal pentraxin receptor (see Doods, D.C., Omels, I.A., Cushman, S.J. et al., Neuronal pentraxin receptor, a novel putative integral membrane pentraxin that interacts with neuronal pentraxin 1 and 2 and taipoxin-associated calcium-binding protein 49, *J. Biol. Chem.* 272, 21488–21494, 1997.

[b] Having to do with eye movement during neurological examinations.

[c] Schenck, J.F. and Zimmerman, E.A., High-field magnetic resonance imaging of brian iron: Birth of a biomarker?, *NMR Biomed.* 17, 433–445, 2004.

screening biomarkers.[210,749] Finally, the amount of work on cardiac diagnostics is overwhelming, and blood biomarkers have a role when compared to the physical examination and history combined with biomarkers obtained with imaging technologies.

Cardiac biomarkers are also used for the diagnostic, prognostics, and predictive management of acute coronary syndrome and heart failure. Acute coronary syndrome is a complex syndrome with a variety of phenotypic expressions resulting from multiple potential etiologies.[750] Braunwald[751] has examined the etiologies of unstable angina that, with myocardial infarction, are acute coronary syndromes. These include non-occlusive thrombus or preexisting plaque, dynamic obstruction, progressive mechanical constriction, inflammation/infection, and secondary unstable angina. The point here is that the biomarkers for inflammation, atherosclerosis, diabetes, and obesity can be considered screening biomarkers for acute coronary syndrome, while the cardiac biomarkers are diagnostic, prognostic, and predictive for acute coronary syndrome and myocardial infarction. Lee and Cannon discuss the role of biomarkers in the treatment of the patient with chest pain noting the importance of creatine kinase isozymes, troponins, and myoglobin.[752]

TABLE 2.18
Some Established Cardiac Biomarkers[a]

Biomarker	References
CK-MB (creatine kinase muscle BL, an isozyme of creatine kinase). CK-MB is an isozyme of creatine kinase preferentially found in cardiac muscle as compared to skeletal muscle. Elevation of CK-MB is considered a sign of cardiac muscle necrosis (myocardial infarction). CK-MB is more specific than some of the earlier enzymes assays for tissue necrosis (see below for LDH and AST). Measurement of CK-MB mass by immunoassays is preferred to measurement of activity[b]	319, 671–678
Myoglobin (a protein found in muscle that is responsible for intracellular oxygen transport; see Wittenberg, J.B., Myoglobin-facilitated oxygen diffusion: Role of myoglobin in oxygen entry into muscle, *Physiol. Rev.* 50, 559–636, 1970). While somewhat nonspecific, the long history of study in myocardial infarction provides a strong basis for the continuing use of myoglobin as a cardiac biomarker.[c] Myoglobin is considered to be an early biomarker for myocardial infarction and a candidate for point-of-care analysis (Freiss, U. and Stark, M., Cardiac markers: A clear cause for point-of-care testing, *Anal. Bioanal. Chem.* 393, 1453–1462, 2009	679–685
Cardiac troponins cTnT (cardiac troponin T, type 2) and cTnI (cardiac troponin I type 3). The troponin complex is composed of troponin c, which binds calcium ions, troponin I, which is the inhibitory subunit, and troponin T, the subunit, which binds tropomyosin.[d] Cardiac troponin I (cTnI) and cardiac troponin T (cTnT) are released with myocardial necrosis elevated serum levels persist for a considerable period of time. cTnT and cTnI have diagnostic and prognostic value allowing for risk stratification	686–693
B type—Natriuretic Peptide (BNP) and N-terminal proB-type natriuretic peptide (NT-proBNP)—precursor to BNP.[e] The presence of the natriuretic peptide is one of the indications of the neuroendocrine function of the heart. BNP peptides are found in both the atrium and the ventricle; elevation of BNP in the circulation is usually taken as an indication of left ventricular dysfunction/heart failure	694–702
Lactate Dehydrogenase (LDH)—The use of LDH isoenzyme patterns for the diagnosis of myocardial infarction dates to 1964, While LDH isoenzymes continue to be useful for the diagnosis of myocardial infarction, it is a somewhat nonspecific measurement and has significant current use in oncology.[f] LDH is no longer recommend for the diagnosis of myocardial infarction[g] but is still reported in clinical cardiology reports	703–709
Aspartate aminotransferase (AST)—No longer recommended for use in diagnosis of myocardial infarction[g] but there is some sporadic continued use.	710–713
SGOT (serum glutamic-oxaloacetic acid transferase)—No longer recommended for use in diagnosis of myocardial infarction.[g]	714

[a] Biomarkers are included for myocardial infarction (muscle necrosis) and for heart failure (cardiac failure). The reader is directed to Morrow, D.A. (ed.), *Cardiovascular Biomarkers Pathophysiology and Disease Management*, Humana Press, Totowa, NJ, 2006.

[b] See Murthy, V.V. and Karmen, A., Activity concentration and mass concentration (monoclonal antibody immunenzymometric method) compared for creatine kinase MB isoenzyme in serum, *Clin. Chem.* 32, 1956–1959, 1986; Youens, J.E., Calvin, J., and Price, C.P., Clinical and analytical validation of an enzymometric assay for creatine kinase-MB isoenzyme, *Ann. Clin. Biochem.* 23, 463–469, 1986; Weber, M.,

(*continued*)

TABLE 2.18 (continued)
Some Established Cardiac Biomarkers[a]

Rau, M. Madlener, K. et al., Diagnostic utility of new immunoassays for the cardiac markers cTnI, myoglobin and CK-MB mass, *Clin. Biochem.* 38, 1027–1030, 2005. State-of-the-art cardiac biomarker assays are mass assays and results are reported in units of mass (Christenson, R.H. and Azzazy, H.M.E., Biomarkers of myocardial necrosis, in *Cardiovascular Biomarkers*, D.A. Morrow (ed.), Humana Press, Totowa, NJ, 2006, Chapter 1, pp. 3–25.).

[c] See Roberts, R., Myoglobinemia as index to myocardial infarction, *Ann. Intern. Med.* 87, 788–789, 1977; Straface, A.L., Myers, J.H., Kirchick, H.J. et al., A rapid point-of-care cardiac marker testing strategy facilitates the rapid diagnosis and management of chest pain patients in the emergency department, *Am. J. Clin. Pathol.* 129, 788–795, 2008; Hayes, M.A., Petkus, M.M., Garcia, A.A. et al., Demonstration of sandwich and competitive modulated supraparticle fluoroimmunoassay applied to cardiac protein biomarker myoglobin, *Analyst* 134, 533–541, 2009.

[d] See Solaro, R.J. and Rarick, H.M., Troponin and tropomyosin: Proteins that switch on and tune in the activity of cardiac myofilaments, *Circ. Res.* 83, 471–480, 1998; Liou, Y.M. and Chang, J.C.H., Differential pH effect on calcium-induced conformational changes of cardiac troponin C complexed with cardiac and fast skeletal isoforms of troponin I and troponin T, *J. Biochem.* 136, 683–692, 2004; Kobayashi, T. and Solaro, R.J., Calcium, thin filaments, and the integrative biology of cardiac contractility, *Ann. Rev. Physiol.* 67, 39–67, 2005; Adamcova, M., Sterba, M., Simunek, T. et al., *Eur. J. Heart Fail.* 8, 333–342, 2006.

[e] See Kragelund, C. and Omland, T., Biology of the natriuretic peptides, in *Cardiovascular Biomarkers*, D.A. Morrow (ed.), Humana Press, Totowa, NJ, Chapter 21, pp. 347–372, 2006.

[f] Schwartz, M.K., Enzymes as prognostic markers and therapeutic indicators in patients with cancer, *Clin. Chim. Acta* 206, 77–82, 1992; Sadamori, N., Clinical and biological significance of serum tumor markers in adult T-cell leukemia, *Leuk. Lymphoma* 22, 415–419, 1996; Morra, E., The biological markers of non-Hodkin's lymphomas: Their role in diagnosis, prognostic assessment and therapeutic strategy, *Int. J. Biol. Markers* 14, 149–153, 1999; Duffy, M.J. and Crown, J., A personalized approach to cancer treatment: How biomarkers can help, *Clin. Chem.* 54, 1770–1779, 2008.

[g] Alpert, J.S., Antman, E., Apple, F. et al., Myocardial infarction redefined—A consensus document of The Joint European Society of Cardiology/American College of Cardiology Committee for the redefinition of myocardial infarction, *J. Am. Coll. Cardiol.* 36, 959–969, 2000.

BIOMARKERS IN RENAL DISEASE

There are diverse interests in biomarkers for renal disease, which are separate from oncology considerations or the use of urine as an analytical biofluid like blood or saliva, as discussed in Chapter 4. Biomarkers for renal disease include renal transplantation[753]; some selected biomarkers for renal disease are presented in Table 2.21. BNP peptides, which were discussed above as biomarkers for cardiovascular disease, have also been suggested as biomarkers for prognosis in chronic kidney disease.[779–782] The more recent thinking suggests that natriuretic peptides are primary biomarkers for cardiovascular function.[782] There is also considerable interest in the development of biomarkers for prognostic use in kidney transplantation.[753,783] Urine is the biofluid of choice for the study of biomarkers for toxicology/ exposure to environmental agents (see Table 2.23). Urine also serves a source of biomarkers for the measurement of glomerular filtration rate.[784–788]

TABLE 2.19
Emerging Biomarkers for Cardiac Function[a]

Biomarker	References
Heart-type fatty acid–binding protein (HFABP). Fatty acid–binding proteins (FABP) are low-molecular-weight proteins, which function as lipid chaperones and found in the cytoplasm of many cell types.[b] FABP are released from the cell during necrosis and can serve as a biomarker for tissue damage. In myocardial infarction, HFABP are released early in the same time frame as myoglobin	685, 715–718
Ischemia-modified albumin (IMA). Albumin is modified as a result of ischemia and no longer binds cobalt as effectively as native albumin. The assay is based on the determination of cobalt not bound to ischemia-modified albumin in the sample.[c] The presence of IMA is not unique to cardiovascular disease.[d]	719–723
Carbonic anhydrase. Human muscle carbonic anhydrase III is found in skeletal muscle but not in cardiac muscle. As such, carbonic anhydrase III can be used as a correction factor for biomarkers, such as myoglobin, which are released from both skeletal muscle and cardiac muscle. It has been suggested that the ratio of myoglobin to carbonic anhydrase III could be used to improve the specificity of myoglobin as a biomarker for myocardial infarction	724–727
A review focusing on use of proteomic technologies to identify new biomarkers for cardiovascular disease with discussion of the issues of sample collection, reduction of sample complexity, selection of technology platforms and transfer to immunoassay platform	728
Effect of fish oil on serum biomarkers of inflammation important in the development of coronary artery disease. Proteomic technology (2-D electrophoresis/mass spectrometry) was used to identify proteins downregulated with supplementation of fish oil. Proteins downregulated included apolipoprotein A1, HDL particle size, haptoglobin precursor, and haptoglobin	729
Use of stable isotope dilution mass spectrometry coupled with peptide immunoaffinity enrichment to evaluate biomarkers for cardiovascular disease in human plasma. IL-33 and troponin (cTnI) peptides were enriched from tryptic digests of plasma by immunoaffinity (antipeptide antibodies)	730
Evaluation of multiple biomarker strategies for heart failure (ST-segment depression, BNP, cTnI)	42
Extracellular matrix fibrotic biomarkers (collagen-derived peptides)	731

[a] These are biomarkers that have been identified in the last decade and are in the process of validation. There is skepticism as to the value of new biomarkers for cardiac function (Ilva, T., Lund, J., Porela, P. et al., Early markers of myocardial injury: cTnI is enough, *Clin. Chim. Acta* 400, 82–85, 2009; Collinson, P.O, Cardiac Markers, *Brit. J. Hosp. Med.* 70, M84–M87, 2009; Berridge, B.R., Pettit, S., Walker, D.B. et al., A translational approach to detecting drug-induced cardiac injury with cardiac troponins: Consensus and recommendations from the Cardiac Troponins Biomarker Working Group of the Health and Environmental Sciences Institute, *Am. Heart J.* 158, 21–29, 2009.

[b] See Noiri, E., Doi, K., Negishi, K. et al., Urinary fatty acid-binding protein 1: An early predictive biomarker of kidney injury, *Am. J. Physiol. Renal Physiol.* 296, F669–F679, 2009; Storch, J. and McDermott, L., Structural and functional analysis of fatty acid-binding proteins, *J. Lipid Res.* 50(Suppl), S126–S131, 2008; Furuhashi, M. and Hotamisligil, G.S., Fatty acid-binding proteins: Roles in metabolic

(continued)

TABLE 2.19 (continued)
Emerging Biomarkers for Cardiac Function[a]

diseases and potential as drug targets, *Nat. Rev. Drug Discov.* 7, 489–503, 2008; Ono, T., Studies of the FABP family: A restrospective, *Mol. Cell Biochem.* 277, 1–6, 2005; Zanotti, G., Muscle fatty acid-binding protein, *Biochim. Biophys. Acta* 1441, 94–105, 1999; Said, B. and Schulz, H., Fatty acid binding protein from rat heart. The fatty acid binding protein from rat heart and liver are different proteins, *J. Biol. Chem.* 259, 1155–1159, 1984; Ockner, R.K., Manning, J.A., and Kane, J.P., Fatty acid binding protein. Isolation from rat liver, characterization, and immunochemical quantification, *J. Biol. Chem.* 257, 7872–7878, 1982.

[c] The ability of human plasma albumin to bind cobalt is decreased in ischemia reflecting a modification of the amino-terminal region. This allowed the development of an assay based on the colorimetric determination of cobalt not binding to an albumin sample (see Bar-Or, D., Lau, E., and Winkler, J.V., A novel assay for cobalt-albumin binding and its potential as a marker for myocardial ischemia-a preliminary report, *J. Emerg. Med.* 19, 311–315, 2000; Bar-or, D., Curtis, G., Rao, N. et al., Characterization of the Co(2+) and Ni(2+) binding amino-acid resides of the N-terminus of the human albumin. An insight into the mechanism of a new assay for myocardial ischemia, *Eur. J. Biochem.* 268, 42–47, 2001).

[d] Picowar, A., Knapik-Kordecka, M., and Warwas, M., Ischemia-modified albumin level in type 2 diabetes mellitus—Preliminary report, *Dis. Markers* 24, 311–317, 2008; Turedi, S., Gunduz, A., Mentese, A. et al., The value of ischemia-modified albumin compared with d-dimer in the diagnosis of pulmonary embolism, *Respir. Res.* 9, 49, 2008; Abboud, H., Labreuche, J., Mesequer, E. et al., Ischemia-modified albumin in acute stroke, *Cerebrovasc. Dis.* 23, 216–220, 2007.

TABLE 2.20
Gene Expression Biomarkers for Cardiac Function

Study	Reference
Gene polymorphism analysis of the angiotensin converting enzyme (ACE)	732
cDNA microarrays used to study cytokine expression in peripheral blood mononuclear cells	733
cDNA microarrays used to study gene expression in human stenotic bypass grafts (retrieved in re-do surgeries). cMyc expression identified; blocking cMyc expression reduces neointima formation	734
Use of cDNA microarrays to provide expression profiles from peripheral blood (whole blood) as a screening test of allograft rejection (cardiac transplant patients). Studies of immunosuppression suggests the gene expression may be more useful than endomyocardial biopsy for screening for allograft rejection	735
Myocardial transcriptome as a biomarker for cardiomyopathy	736
Custom cDNA microarrays used study gene expression in left ventricular myocardial samples from patients with coronary artery disease (CAD_ and dilated cardiomyopathy transplants. Differential expression gene expression was observed in 153 genes (CAD) and 147 in the transplant group. The best classifiers for end-stage heart failure were MMP 3, fibulin 1, ABC subfamily B, Iroquois homeobox protein 5	737

TABLE 2.20 (continued)
Gene Expression Biomarkers for Cardiac Function

Study	Reference
cDNA microarray technology was used to study identify genes, which are upregulated in endomyocardial samples from heart transplant patients before and rejection. ELISA assays were used to determined if the products from upregulated genes were also increased in serum samples, CXCL9 was markedly upregulated during cardiac rejection but this change was not seen in serum concentration	738
Gene expression in peripheral blood mononuclear cells obtained from heart transplant patients in clinical quiescence was studied by cDNA microarray technology. Gene expression was correlated with serum BNP (B-type natriuretic peptide) levels. 54 unique genes were correlated with BNP concentration. Gelsolin, matrix metalloproteinase (MMP8, MMP9), thrombospondin, platelet factor 4, plasminogen activator 2 are examples of over-expressed genes; inflammation biomarkers (B-cell lymphokine, heat shock protein) under-expressed	739
cDNA microarray used to study gene expression in macrophages from patients with atherosclerosis and patients with coronary heart disease. Increased expression of IRS2 (insulin receptor substrate 2) in macrophages may be associated with increased risk of coronary heart disease.	740
cDNA microarray analysis of gene expression in endometrial biopsy samples identified 46 over-expressed genes associated with good prognosis in new-onset heart failure	741
Real-time PCR analysis (left ventricular myocardium) showed over-expression of follistatin-related genes (FSTL1 and FSTL3) in heart failure. FSTL3 correlated with disease severity while FLST1 correlated with CD31 suggesting a role in angiogenesis. cDNA microarray analysis FSLT1 association with extracellular matrix and calcium-binding proteins while FSLT3 was associated with cell signaling and transcription	742
cDNA microarray analysis showed over-expression of apolipoprotein D in left ventricular issue from heart failure patients; ELISA assays showed elevated levels of apolipoprotein D in plasma suggesting use as a circulating biomarker	743

TABLE 2.21
Biomarkers for Renal Disease

Study	References
Kidney Injury Molecule-1 (KIM-1) is a type 1 membrane protein with a complex ectodomain containing IgG motifs and extensively O-glycosylated motifs as well as numerous N-glycosylation sites. As with other membrane proteins, the ectodomain can be shed from the parent cell. KIM-1 has been demonstrated to be a biomarker for acute kidney injury (acute renal failure)	754–758
A discussion of the use of multiple biomarkers for acute kidney injury. The candidate biomarkers are neutrophil gelatinase-associated lipocalin (NGAL) and cystatin C for use in plasma and the use of NGAL, IL-18, and KIM-1 in urine	759
Proteinuria; mostly the presence of albumin in urine reflecting vascular permeability issues. Albumin in urine as a biomarker of association between chronic kidney disease (CKD) and cardiovascular disease	325, 760–762

(continued)

TABLE 2.21 (continued)
Biomarkers for Renal Disease

Study	References
Urinary calcium as biomarker for renal dysfunction from exposure to cadmium	763
Plasma levels of Cu/Zn superoxide dismutase as a biomarker for oxidative stress in end-stage renal disease (ESRD)	764
AGEs as determined by skin autofluorescence is a biomarker for chronic renal transplant dysfunction	765
Interleukin-18 (IL-18) as a biomarker in urine for acute kidney injury	766
A study on the use of mass spectrometry analysis of urine and blood plasma to discover biomarkers for kidney disease	767
Use of urinary levels of liver fatty acid–binding protein as a biomarker for acute kidney injury after cardiac surgery	768
The concentration of Smad1[a] in urine is a biomarker for diabetic nephropathy	769, 770
Connective tissue growth factor is a biomarker in urine for chronic allograft nephropathy	771
Angiotensinogen in urine is a biomarker for the intrarenal renin-angiotensin system; urinary angiotensinogen/creatine was correlated with systolic blood pressure and diastolic blood pressure	772
Plasma hepcidin is a biomarker for iron status and erythropoietin resistance in chronic kidney disease	773
Neutrophil gelatinase-associated lipocalin (NGAL)[b] is a biomarker in urine for acute kidney injury	774, 775
Cystatin c[c] is a biomarker in urine for kidney disease	776, 777
Discussion of the urinary proteome and biomarker discovery in pediatric renal disease	778
Identification of biomarkers for tolerance and rejection for kidney transplantation	753

[a] Smad proteins are transducer proteins for the TGF superfamily (see Massaqué, J., TGF-beta signal transduction, *Annu. Rev. Biochem.* 67, 753–791, 1998; Miyazono, K., TGF-beta signaling by Smad proteins, *Cytokine Growth Factor Rev.* 11, 15–22, 2000).

[b] Neutrophil gelatinase-associated lipocalin is also known as lcn2 or siderocalin and is found in the storage granules of neutrophils (Borregäard, N. and Cowland, J.B., Neutrophil gelatinase-associated lipocalin: A siderophore-binding eukaryotic protein, *Biometals* 19, 211–215, 2006). Lipocalins have been associated with a variety of disease processes (Xu, S. and Venge, P., Lipocalins as biochemical markers of disease, *Biochim. Biophys. Acta* 1482, 296–307, 2000).

[c] Cystatin c is a low-molecular-weight (ca. 15 kDa) inhibitor of cysteine proteases that is used as a biomarker in urine for the measurement of glomerular filtration rate (Brzin, J., Popoic, T., Turk, V. et al., Human cystatin, a new protein inhibitor of cysteine proteinases, *Biochem. Biophys. Res. Commun.* 118, 103–109, 1984; Grubb, A., Simonsen, O., Sturfelt, G. et al., Serum concentration of cystatin C, factor D and beta 2-microglobulin as a measure of glomerular filtration rate, *Acta Med. Scand.* 218, 499–503, 1985; Grubb, A.O., Cystatin C—properties and use as diagnostic marker, *Adv. Clin. Chem.* 35, 63–99, 2000; Séronie-Vivien, S., Delanaye, P., Piéroin, L. et al., Cystatin C: Current position and future prospects, *Clin. Chem. Lab. Med.* 46, 1664–1686, 2008).

BIOMARKERS FOR BONE

It is generally considered that the best biomarkers for bone are derived from imaging technologies but that there is a need for biochemical markers for earlier diagnosis.[789–797] It is noted that there is scant mention of biomarkers as such in the bone literature as of 2002,[798] but this is changing, as shown by examples in Table 2.22. It is noted that there is considerable literature on the use of bone biomarkers for arthritis.

BIOMARKERS FOR EXPOSURE TO ENVIRONMENTAL TOXINS

Methods for the evaluation of exposure to environmental toxins are of considerable importance in occupational health. The reader is directed to an excellent collection of articles on the use of biomarkers in toxicology edited by Anthony DeCaprio.[810]

TABLE 2.22
Biomarkers for Bone

Study	References
Lead as biomarker for environmental exposure in bone as determined by K x-ray fluorescence[a]	799
Keratan sulfate in serum as biomarker for the mechanical loading of the spine	800
The cross-linked nonisomerized form of C-telopeptide of collagen type I (alpha CTX) as biomarker in urine for breast cancer metastases to bone	801
Serum cartilage oligomeric matrix protein as a biomarker for osteoarthritis	802
Review of biomarkers for diagnostic and predictive use in osteoarthritis	794, 803
Use of serum bone markers for the identification of bone metastasis in prostate cancer	804
Fragments of interalpha-trypsin-inhibitor heavy-chain H4 precursor as serum biomarker presumably derived from increased osteoclast activity for high bone turnover and bone mineral density (BMD)	805
Fibroblast growth factor 23 (FGF23) and 1,25-dihydroxyvitamin D3 as biomarkers for oncogenic osteomalacia	806
Deoxypyridinoline in urine as biomarker for lung metastases in lung cancer	807
Macrophage inflammatory protein-1α is a biomarker in saliva for bone loss in children with periodontal disease	808
Use of C-terminal telopeptides of type I collagen as biomarkers for bone resorption and measurement of effectiveness of a therapeutic monoclonal antibody directed against receptor activator of NF-κB ligand	809

[a] See Todd, A.C. and Chettle, D.R., In vivo X-ray fluorescence of lead in bone: Review and current issues, *Environ. Health Perspect.* 102, 172–177, 1994; Todd, A.C., L-shell x-ray fluorescence measurements of lead in bone: Theoretical considerations, *Phys. Med. Biol.* 47, 491–505, 2002.

TABLE 2.23

Biomarkers for Exposure to Environmental Agents[a]

Study	Reference
Confounding factors for the study of biomarker response in exposure to vinyl chloride or petroleum emissions. Age, sex, and lifestyle are examples of such confounding factors	811
Urinary biomarkers for measurement of exposure to alkylating agents; examples include 3-alkyladenine derivatives	812
Metallothionein[b] isoform IA expression in peripheral blood lymphocytes can be used as a biomarker for cadmium exposure	813
Blood lead levels as biomarker for exposure of lead	814
Urine levels of naphthalene and phenanthrene as biomarkers for occupational exposure to polycyclic aromatic hydrocarbons	815
Urinary benzene is a biomarker for low-level exposure to benzene; urinary benzene concentration was lower in nonsmokers than in smokers	816
Mandelic acid concentration in urine is a biomarker for exposure to styrene	817
Urinary concentration of ethylenethiourea as a biomarker for exposure to fungicides such as ethylenebisdithiocarbamate. Analysis used liquid chromatography coupled with MS/ MS	818
Urine and plasma levels of aniline can be used as a biomarker for exposure to phenylisocyanate. Analysis of plasma provided higher sensitivity. The samples were analyzed by gas chromatography after hydrolysis	819
Blood levels of malondialdehyde are biomarker for exposure to mineral wool exposure	820
Phenylmercapturic acid, benzylmercapturic acid, and *o*-methylbenzylmercapturic acid in human urine as biomarkers for exposure to benzene, toluene, and xylene. Analysis used liquid chromatography with tandem mass spectrometry (LC-MS/MS)	821
Plasma β-glucuronidase as biomarker for monitoring exposure to anticholinesterase pesticides	822

[a] The articles selected are limited to human studies.
[b] Metallothionein is a family of proteins that binds metal ions (see Karin, M. and Richards, R.I., The human metallothionein gene family: Structure and expression, *Environ. Health Perspect.* 54, 111–115, 1984.

Some biomarkers of interest in occupational health are shown in Table 2.23. Most of these biomarkers are low-molecular-weight organic compounds, which are, for the most part, derived from the environmental agent and are found in urine. There have been a limited number of studies using proteomic technologies for toxicological purposes and a selection of such studies is presented in Table 2.24. There is a discipline referred to as toxicoproteomics, which concerns both environmental toxicology and drug toxicity.[830,832–836]

TABLE 2.24

Proteomic Biomarkers for Environment Exposure and Toxicology[a]

Study	Reference
Proteomic analysis (MALDI-TOF) of plasma from individuals exposed to benzene showed up-regulation of T cell receptor β-chain, FK506-binding protein, MMP-13. Comet analysis was used to determine DNA damage in lymphocytes	823
Proteomic analysis of cytosolic fraction from kidneys obtained from mice (Swiss Webster) exposed to aerosolized JP-8 jet fuel. 2-D gel electrophoresis (scanned/digitized) showed significant changes associated with exposure	824
Changes in rat bronchoalveolar lavage fluid after exposure to an oil mist[a] were examined by HPLC/electrospray ionization/MS/MS. Changes were observed in 29 proteins. Notable was a marked decrease in surfactant-associated protein A while there were increases in TGFα and S100 proteins (e.g., calgranulin, calreticulin)	825
Review on the applications of proteomics to the study of plant response to metals such as copper, zinc, lead, and mercury	826
Use of proteomic analysis (2D electrophoresis of cell lysates) to study the changes of cultured cells (MCF-7 cells) to okadaic acid or gambierol or a combination of the two agents	827
Decrease in urinary β-defensin 1 as a biomarker for exposure to arsenic. This study contains an excellent discussion of the issue of sample stability	828
Proteomic analysis (2D electrophoresis/MS) is used to study the effect of arsenic on proteins in rice roots. S-Adenosylmethionine synthetase, cysteine synthase, glutathione transferases and tyrosine-specific protein phosphatases were upregulated in response to arsenic	829
Review on the use of proteomics in toxicology research (toxicoproteomics)	830
Development of a monoclonal antibody for the measurement of an acrylamide-hemoglobin adduct as a biomarker for dietary acrylamide exposure	831

[a] The animals were exposed to oil mist (cutting oil mist from a thread rolling process) in a factory setting for 21 days. A control group was also in the factory but distant (inventory area) from the actual manufacturing area.

BIOMARKERS AND THE DEVELOPMENT OF BIOPHARMACEUTICALS

Biomarkers as an entity are critical for the drug development process.[55] It is not possible to develop a pharmaceutical product without a method to measure a surrogate outcome. As such, it is optimal for the development of biomarker and therapeutic to occur in parallel.[837] Proteomics and biomarkers are also part of the preclinical toxicology program in pharmaceutical development.[838–841]

The success of a clinical trial depends on having a method to measure the outcome*. The best method would be a direct measure of the syndrome treated (clinical endpoint). The great majority of trials depend on the measurement of a biomarker, which serves as a surrogate.[55,842–844] Clinical trials that use an actual clinical endpoint instead of a surrogate endpoint tend to be longer and more expensive.[845] The development of biomarkers for use as surrogate endpoints can be a significant challenge in syndromes such as Alzheimer disease,[846,847] where the clinical endpoint is complex.[848]

The validation of a surrogate biomarker is not a trivial process; surrogate biomarkers must have an accurate measure of true clinical outcome and must accurately reflect the effect of intervention in disease prognosis.[56,849] The validation of biomarkers for clinical trials has been discussed by several authors.[838,850,851]

It is the author's sense that he would be remiss if he did not mention companion diagnostics. The author would note that he is not clear as to the meaning of companion diagnostics. Searches with electronic databases collected an interesting assortment of items ranging from astrophysics to biotechnology. It would seem that a companion diagnostic is a method that is developed in concert with a therapeutic product[852,853] such as a method to determine which tumors contain the genetic alterations of the target appropriate for the application of a specific chemotherapeutic drug.[854,855] Companion diagnostics seem be predictive by category and most often use the methods of molecular diagnostics.[856,857] A recent article discusses the development of a variety of assays in the development of companion diagnostics for the use of statin.[858] It is suggested that theranostics (see above) is the use of a companion diagnostic to guide treatment.[859] It is obvious that there is an overlap of terminology with what appears to be attempts to show scientific progress by terminology rather than technology. I think that it is useful to consider a companion diagnostic as the use of a biomarker to identify individuals most likely to respond to a given therapeutic. Also, the development of a companion diagnostic is of great economic value.[852]

---

* Successful clinical trials require significant input from statisticians during the planning of the trials. The reader is directed to the following sources for more information:

Cook, T.D. and DeMets, D.L., *Introduction to Statistical Methods for Clinical Trials*, Chapman & Hall/CRC Press, Boca Raton, FL, 2008.

Chow, S.-C. and Liu, J.-P., *Design and Analysis of Bioavailability and Bioequivalence Studies*, CRC Press, Boca Raton, FL, 2009.

Peace, K.E. (ed.), *Design and Analysis of Clinical Trials with Time-to-Event Endpoints*, CRC Press, Boca Raton, FL, 2009.

Julious, S.A., *Sample Sizes for Clinical Trials*, CRC Press, Boca Raton, FL, 2009.

Meinert, C.L. and Tonascia, S., *Clinical Trials: Design, Conduct, and Analysis*, Oxford University Press, New York, 1986.

Davis, J.R., *Assuring Data Quality and Validity in Clinical Trials for Regulatory Decision Making*, National Academy Press, Washington, DC, 1999.

Everitt, B. and Pickles, A., *Statistical Aspects of the Design and Analysis of Clinical Trials*, Imperial College Press, London, U.K., 2004.

Prorok, P., Andride, R., Bresalier, S. et al., Design of the prostate, lung, colorectal, and ovarian (PLCO) cancer screening trial, *Control. Clin. Trials* 21, 273S–309S, 2000.

REFERENCES

1. van der Greef, J., Martin, S., Juhasz, P. et al., The art and practice of systems biology in medicine: Mapping patterns of relationships, *J. Proteome Res.* 6, 1540–1559, 2007.
2. Malmström, J., Lee, H., and Abersold, R., Advances in proteomic workflow for systems biology, *Curr. Opin. Biotechnol.* 18, 378–384, 2007.
3. Mullassery, D., Horton, C.A., Wood, C.D. et al., Single live-cell imaging for systems biology, *Essays Biochem.* 45, 121–133, 2008.
4. Reckow, S., Gormanns, P., Holboer, F., and Turck, C.W., Psychiatric disorders biomarker identification: From proteomics to systems biology, *Pharmacopsychiatry* 41(Suppl 1), S70–S77, 2008.
5. Laaksonen, R., Katajamaa, M., Päivä, H. et al., A systems biology strategy reveals biological pathways and plasma biomarker candidates for potentially toxic statin-induced changes in muscle, *PLoS ONE* 1, e97, 2006.
6. Perco, P., Wilflingseder, J., Bernthaler, A. et al., Biomarker candidates for cardiovascular disease and bone metabolism disorders in chronic kidney disease: A systems biology perspective, *J. Cell. Mol. Med.* 12, 1177–1187, 2008.
7. Rosenthal, A.N., Menon, U., and Jacobs, I.J., Screening of ovarian cancer, *Clin. Obstet. Gynecol.* 49, 433–447, 2006.
8. Nijhuis, E.R., Reesink-Peters, N., Wisman, G.B. et al., An overview of innovative techniques to improve cervical cancer screening, *Cell. Oncol.* 28, 233–246, 2006.
9. Munkarah, A., Chatterjee, M., and Tainsky, M.A., Update on ovarian cancer screening, *Curr. Opin. Obstet. Gynecol.* 19, 22–26, 2007.
10. Linkov, F., Yurkovetsky, Z., and Lokshin, A., Biomarker approaches to the development of cancer screening tests: Can cancer blood tests become a routine health check-up?, *Future Oncol.* 3, 295–298, 2007.
11. Wentzensen, N. and von Knebel Doeberitz, M., Biomarkers in cervical cancer screening, *Dis. Markers* 23, 315–330, 2007.
12. Esserman, L.J., Shieh, Y., Park, J.W., and Ozanne, E.M., A role for biomarkers in the screening and diagnosis of breast cancer in younger women, *Expert Rev. Mol. Diagn.* 7, 533–544, 2007.
13. Safaeian, M., Solomon, D., and Castle, P.E., Cervical cancer prevention—Cervical screening: Science in evolution, *Obstet. Gynecol. Clin. North Am.* 34, 739–760, 2007.
14. Midthun, D.E. and Jett, J.R., Update on screening for lung cancer, *Semin. Respir. Crit. Care Med.* 29, 233–240, 2008.
15. Svatek, R.S. and Lothan, Y., Is there a rationale for bladder cancer screening?, *Curr. Urol. Rep.* 9, 339–341, 2008.
16. Wu, I. and Parikh, C.R., Screening for kidney diseases: Older measures versus novel biomarkers, *Clin. J. Am. Soc. Nephrol.* 3, 1895–1901, 2008.
17. Goodbrand, S.A. and Steele, R.J., An overview of colorectal cancer screening, *Scott. Med. J.* 53, 31–37, 2008.
18. Brooks, M., Breast cancer screening and biomarkers, *Methods Mol. Biol.* 472, 307–321, 2009.
19. Etzioni, R., Statistical issues in the evaluation of screening and early detection modalities, *Urol. Oncol.* 26, 308–315, 2008.
20. Lo, C.W., Genes, gene knockouts, and mutations in the analysis of gap junctions, *Dev. Genet.* 24, 1–4, 1999.
21. Ashford, R.W., Current usage of nomenclature for parasitic diseases, with special reference to those involving arthropods, *Med. Vet. Entomol.* 15, 121–125, 2001.
22. Park, J., Park, H.J., Lee, H.J., and Ernst, E., What's in a name? A systematic review of the nomenclature of Chinese medical formulae, *Am. J. Chin. Med.* 30, 419–427, 2002.

23. Erdman Jr. J.W., Badger, T.M., Lampe, J.W. et al., Not all soy products are created equal: Caution needed in interpretation of research results, *J. Nutr.* 134, 1229S–1233S, 2004.
24. Bernard, H.J., The clinical importance of the nomenclature, evolution and taxonomy of human papillomaviruses, *J. Clin. Virol.* 32(Suppl 1), S1–S6, 2005.
25. Young, J.M., An overview of bacterial nomenclature with special reference to plant pathogens, *Syst. Appl. Microbiol.* 31, 405–424, 2008.
26. Moran, C.A., Suster, S., Coppola, D., and Wick, M.R., Neuroendocrine carcinomas of the lung: A critical analysis, *Am. J. Clin. Pathol.* 131, 206–221, 2009.
27. Shostak, S., (Re)defining stem cells, *Bioessays* 28, 301–308, 2006.
28. Ogino, S., Gulley, M.L., den Dunnen, J.T., and Wilson, R.B., Standard mutation nomenclature in molecular diagnostics—Practical and educational challenges, *J. Mol. Diagn.* 9, 1–6, 2007.
29. Qin, L., Gilbert, P.B., Corey, L. et al., A framework for assessing immunological correlates of protection in vaccine trials, *J. Infect. Dis.* 196, 1304–1312, 2007.
30. Laurin, M., The splendid isolation of biological nomenclature, *Zoolog. Scripta* 37, 223–233, 2008.
31. Cunningham, S.C., Klein, R.V., and Kavic, S.M., A nomenclature of nomenclature. The source of terminologic uncertainty and confusion and the value of communication, *Arch. Surg.* 144, 104–106, 2009.
32. Chung, K.F., Bolser, D., Davenport, P. et al., Semantics and types of cough, *Pulm. Pharmacol. Ther.* 22, 139–142, 2009.
33. Downing, G.J. and the Biomarkers Definitions Working Groups, Biomarkers and surrogate endpoints: Preferred definitions and conceptual framework, *Clin. Pharmacol. Ther.* 69, 89–95, 2001.
34. Winchester, B., How are biomarkers defined and validated?, *Clin. Ther.* 30, S88–S89, 2008.
35. Lassere, M.N., The biomarker-surrogacy evaluation schema: A review of the biomarker-surrogate literature and a proposal for a criterion-based, quantitative, multidimensional hierarchical levels of evidence schema for evaluating the status of biomarkers as surrogate endpoints, *Stat. Methods Med. Res.* 17, 303–340, 2008.
36. Kochanek, P.M., Berger, R.P., Bayir, H. et al., Biomarkers of primary and evolving damage in traumatic and ischemic brain injury: Diagnosis, prognosis, probing mechanisms, and therapeutic decision making, *Curr. Opin. Crit. Care* 14, 135–141, 2008.
37. Apple, F.S., Smith, S.W., Pearce, L.A., and Murakami, M.M., Assessment of the multiple-biomarker approach for diagnosis of myocardial infarction in patients presenting with symptoms suggestive of acute coronary syndrome, *Clin. Chem.* 55, 93–100, 2009.
38. Vaidya, V.S., Waikar, S.S., Ferguson, M.A. et al., Urinary biomarkers for sensitive and specific and specific detection of acute kidney injury in humans, *Clin. Transl. Sci.* 1, 200–208, 2008.
39. Narain, V.S., Gupta, N., Sethi, R. et al., Clinical correlation of multiple biomarkers for risk assessment in patients with acute coronary syndrome, *Indian Heart J.* 60, 536–542, 2008.
40. Lee, J.G., Lee, S., Kim, Y.J. et al., Multiple biomarkers and their relative contributions to identifying metabolic syndrome, *Clin. Chim. Acta* 408, 50–55, 2009.
41. Nozaki, T., Sugiyama, S., Koga, H. et al., Significance of a multiple biomarkers strategy including endothelial dysfunction to improve risk stratification for cardiovascular events in patients at high risk for coronary heart disease, *J. Am. Coll. Cardiol.* 54, 601–608, 2009.
42. Allen, L.A., Use of multiple biomarkers in heart failure, *Curr. Cardiol. Rep.* 12, 230–236, 2010.
43. *Oxford English Dictionary*, on-line dictionary, www.oed.com

44. Davis, S.D., Brody, A.S., Emond, M.J. et al., Endpoints for clinical trials in young children with cystic fibrosis, *Proc. Am. Thorac. Soc.* 4, 418–430, 2007.
45. Bosma, L., Kragt, J., Brieva, L. et al., The search for responsive clinical endpoints in primary progressive multiple sclerosis, *Mult. Scler.* 15, 715–720, 2009.
46. Griesenbach, U. and Boyd, A.C., Pre-clinical and clinical endpoint assays for cystic fibrosis gene therapy, *J. Cyst. Fibros.* 4, 89–100, 2005.
47. Prentice, R.L., Opportunities for enhancing efficiency and reducing cost in large scale disease prevention trials: A statistical perspective, *Stat. Med.* 9, 161–170, 1990.
48. Boissel, J.P., Collet, J.P., Moleur, P., and Haugh, M., Surrogate endpoints: A basis for a rational approach, *Eur. J. Clin. Pharmacol.* 43, 235–244, 1992.
49. Fleming, T.R., Prentice, R.L., Pepe, M.S., and Glidden, D., Surrogate and auxiliary endpoints in clinical trials, with potential applications in cancer and AIDS research, *Stat. Med.* 13, 955–968, 1994.
50. Buyse, M., Molenberghs, G., Burzykowski, T. et al., The validation of surrogate endpoints in meta-analyses of randomized experiments, *Biostatistics* 1, 49–67, 2000.
51. Gilbert, P.B. and Hudgens, M.G., Evaluating candidate principle surrogate endpoints, *Biometrics* 64, 1146–1154, 2008.
52. Lassere, M.N., Johnson, K.R., Boers, M. et al., Definitions and validation criteria for biomarkers and surrogate endpoints: Development and testing of a quantitative hierarchical levels of evidence schema, *J. Rheumatol.* 34, 607–615, 2007.
53. Cummings, J., Ward, T.H., Greystoke, A. et al., Biomarker method validation in anticancer drug development, *Br. J. Pharmacol.* 153, 646–656, 2007.
54. Hess, V., Glimelius, B., Grawa, P. et al., CA 19-9 tumour-marker response to chemotherapy in patients with advanced pancreatic cancer enrolled in a randomized controlled trial, *Lancet Oncol.* 9, 132–138, 2008.
55. McShane, L.M., Hunsberger, S., and Adjej, A.A., Effective incorporation of biomarkers into phase II trials, *Clin. Cancer Res.* 15, 1898–1905, 2009.
56. Cook, T.D. and Demets, D.L. (eds.), *Introduction to Statistical Methods for Clinical Trials*, Chapman & Hall/CRC Press, Boca Raton, FL, 2008.
57. Ebos, J.M., Lee, C.R., Bogdanovic, E. et al., Vascular endothelial growth factor-mediated decrease in plasma soluble vascular endothelial growth factor receptor-2 levels as a surrogate biomarker for tumor growth, *Cancer Res.* 68, 521–529, 2008.
58. Lien, S. and Lowman, H.B., Therapeutic anti-VEGF antibodies, *Handb. Exp. Pharmacol.* 181, 131–150, 2008.
59. Zamora-Ros, R., Urpí-Sardà, M., Lamuela-Raventós, R.M. et al., Diagnostic performance of urinary resveratrol metabolites as a biomarker of moderate wine consumption, *Clin. Chem.* 52, 1373–1380, 2006.
60. Khan, T.K. and Alkon, D.L., Early diagnostic accuracy and pathophysiologic relevance of an autopsy-confirmed Alzheimer's disease peripheral biomarker, *Neurobiol. Aging* 31, 889–900, 2010.
61. Paci, M., Maramotti, S., Bellesia, E. et al., Circulating plasma DNA as diagnostic biomarker in non-small cell lung cancer, *Lung Cancer* 64, 92–97, 2009.
62. Koh, O.L., Yip, T.T., Ho, M.F. et al., The distinctive gastric fluid proteome in gastric cancer reveals a multi-biomarker diagnostic profile, *BMC Med. Genomics* 1, 54, 2008.
63. Ellinger, J., Albers, S., Müller, S.C. et al., Circulating mitochondrial DNA in the serum of patients with testicular germ cell cancer as novel noninvasive diagnostic biomarker, *BJU Int.* 104(1), 48–52, 2009.
64. Pienta, K.J., Critical appraisal of prostate-specific antigen in prostate cancer screening: 20 years later, *Urology* 73(5 Suppl), S11–S20, 2009.
65. Havrilesky, L.J., Maxwell, G.L., and Myers, E.R., Cost-effectiveness analysis of annual screening strategies for endometrial cancer, *Am. J. Obstet. Gynecol.* 240, 640, e1–e8, 2009.

66. Partridge, E., Kreimer, A.R., Greenlee, R.T. et al., Results from four rounds of ovarian cancer screening in a randomized trial, *Obstet. Gynecol.* 113, 775–782, 2009.
67. Hu, P. and Zelen, M., Experimental design issues for early detection of disease: Novel designs, *Biostatistics* 3, 299–313, 2002.
68. Rosenthal, A. and Jacobs, I., Familial ovarian cancer screening, *Best Pract. Res. Clin. Obstet. Gynaecol.* 20, 321–338, 2006.
69. Menon, U., Skates, S.J., Lewis, S. et al., Prospective study using the risk of ovarian cancer algorithm to screen for ovarian cancer, *J. Clin. Oncol.* 23, 7919–7926, 2005.
70. Sharma, A. and Menon, U., The value of ovarian cancer screening, *Br. J. Hosp. Med.* 67, 314–317, 2006.
71. Auray-Blais, C., Millington, D.S., Young, S.P. et al., Proposed high-risk screening for Fabry disease in patients with renal and vascular disease, *J. Inherit. Metab. Dis.* 32, 303–308, 2009.
72. Pitt, A.R. and Spickett, C.M, Mass spectrometric analysis of HOCl- and free-radical-induced damage to lipids and proteins, *Biochem. Soc. Trans.* 36, 1077–1082, 2008.
73. Verhoye, E., Langlois, M.R., and Asklepios Investigators, Circulating oxidized low-density lipoprotein: A biomarker of atherosclerosis and cardiovascular risk?, *Clin. Chem. Lab. Med.* 47, 128–137, 2009.
74. Sharples, L.D., Statistical approaches to rational biomarker selection, in *Biomarkers in Disease. An Evidence-Based Approach*, A.K. Trull, L.M. Demers, D.W. Holt, A. Johnston, J.M. Tredger, and C.P. Price (eds.), Cambridge University Press, Cambridge, U.K., Chapter 3, pp. 24–31, 2002.
75. Davicioni, E., Wai, D.H., and Anderson, M.J., Diagnostic and prognostic sarcoma signatures, *Mol. Diagn. Ther.* 12, 359–374, 2008.
76. Briggs, C.D., Neal, C.P., Mann, C.D. et al., Prognostic molecular markers in cholangiocarcinoma: A systematic review, *Eur. J. Cancer* 45, 33–47, 2009.
77. Culine, S., Prognostic factors in unknown primary cancer, *Semin. Oncol.* 36, 60–64, 2009.
78. Riley, R.D., Sauerbrei, W., and Altman, D.G., Prognostic markers in cancer: The evolution of evidence from single studies to meta-analysis, and beyond, *Br. J. Cancer* 100, 1219–1229, 2009.
79. Clark, G.M., Zborowski, D.M., Culbertson, J.L. et al., Clinical utility of epidermal growth factor receptor expression for selecting patients with advanced non-small cell lung cancer for treatment with erlotinib, *J. Thorac. Oncol.* 1, 837–846, 2006.
80. Clark, G.M., Prognostic factors versus predictive factors: Examples from a clinical trial of erlotinib, *Mol. Oncol.* 1, 406–412, 2008.
81. ten Brinke, R., Dekker, N., de Groot, M., and Ikkersheim, D., Lowering HbA1c in type 2 diabetics results in reduced risk of coronary heart disease and all-cause mortality, *Prim. Care Diabetes* 2, 45–49, 2008.
82. Crepaldi, G., Carruba, M., Comaschi, M. et al., Dipeptidyl peptidase 4 (DPP-4) inhibitors and their role in type 2 diabetes management, *J. Endocrinol. Invest.* 30, 610–614, 2007.
83. Picard, F.J. and Bergeron, M.G., Rapid molecular theranostics in infectious diseases, *Drug Discov. Today* 7, 1092–1101, 2002.
84. Philip, R., Murthy, S., Kroakover, J. et al., Shared immunoproteome for ovarian cancer diagnostics and immunotherapy: Potential theranostic approach to cancer, *J. Proteome Res.* 6, 2509–2517, 2007.
85. Arbustini, E. and Gambarin, F.I., Theranostic strategy against plaque angiogenesis, *JACC Cardiovasc. Imaging* 1, 635–637, 2008.
86. Gorelik, B., Ziv, I., Shohat, R. et al., Efficacy of weekly docetaxel and bevacizumab in mesenchymal chondrosarcoma: A new theranostic method combining xenografted biopsies with a mathematical model, *Cancer Res.* 68, 9033–9040, 2008.

87. Bentzen, S.M., Theragnostic imaging for radiation oncology: Dose-painting by numbers, *Lancet Oncol.* 6, 112–117, 2005.

88. Bentzen, S.M., Dose painting and theragnostic imaging: Towards the prescription, planning and delivery of biologically targeted dose distributions in external beam radiation oncology, *Cancer Treat. Res.* 139, 41–62, 2008.

89. Flynn, R.T., Bowen, S.R., Bentzen, S.M. et al., Intensity-modulated x-ray (IMXT) versus proton (IMPT) therapy for theragnostic hypoxia-based dose painting, *Phys. Med. Biol.* 53, 4153–4167, 2008.

90. Lucignani, G., Nanoparticles for concurrent multimodality imaging and therapy: The dawn of new theragnostic synergies, *Eur. J. Nucl. Med. Mol. Imaging* 36, 869–874, 2009.

91. Subayey, V.I., Pisanic II T.R., and Jin, S., Magnetic nanoparticles for theragnostics, *Adv. Drug Deliv. Rev.* 61, 467–477, 2009.

92. Pene, F., Courtine, E., Cariou, A. et al., Toward theragnostics, *Crit. Care Med.* 37(1 Suppl), S50–S58, 2009.

93. Baron, A.T., Wilken, J.A., Haggstrom, D.E. et al., Clinical implementation of soluble EGFR (sEGFR) as a theragnostic serum biomarker of breast, lung and ovarian cancer, *IDrugs* 12, 302–308, 2009.

94. Langreth, R. and Waldholz, M., New era of personalized medicine: Targeting drugs for each unique genetic profile, *Oncologist* 4, 426–427, 1999.

95. Ginsburg, G.S. and McCarthy, J.J., Personalized medicine: Revolutionizing drug discovery and patient care, *Trends Biotechnol.* 19, 491–496, 2001.

96. Meyer, J.M. and Ginsburg, G.S., The path of personalized medicine, *Curr. Opin. Chem. Biol.* 6, 434–438, 2002.

97. Dean, C.E., Personalized medicine: Boon or budget-buster?, *Ann. Pharmacother.* 43, 958–962, 2009.

98. Jørgensen, J.T., New era of personalized medicine: A 10-year anniversary, *Oncologist* 14, 557–558, 2009.

99. Elledge, R.M., Clark, G.M., Hon, J. et al., Rapid in vitro assay for predicting response to fluorouracil in patients with metastatic breast cancer, *J. Clin. Oncol.* 13, 419–423, 1995.

100. Fruehauf, J.P., In vitro assay-assisted treatment selection for women with breast or ovarian cancer, *Endocr. Relat. Cancer* 9, 171–182, 2002.

101. Loizzi, V., Chan, J.K., Osann, K. et al., Survival outcomes in patients with recurrent ovarian cancer who were treated with chemoresistance assay-guided chemotherapy, *Am. J. Obstet. Gynecol.* 189, 1301–1307, 2003.

102. Kanasugi, M., Aoki, D., Suzuki, N. et al., Sensitivity to cisplatin determined by the histoculture drug response assay and clinical response of endometrial cancer, *Int. J. Gynecol. Cancer* 16, 409–415, 2006.

103. Lindhagen, E., Nygren, P., and Larsson, R., The fluorometric microculture cytotoxicity assay, *Nat. Protoc.* 3, 1364–1369, 2008.

104. Higashiyama, M., Oda, K., Okami, J. et al., In vitro-chemosensitivity test using the collagen gel droplet embedded culture drug test (CD-DST) for malignant pleural mesothelioma: Possibility of clinical application, *Ann. Thorac. Cardiovasc. Surg.* 14, 355–362, 2008.

105. Phillips, K.A., Closing the evidence gap in the use of emerging testing technologies in clinical practice, *JAMA* 300, 2542–2544, 2008.

106. Deverka, P.A., Pharmacogenomics, evidence, and the role of payers, *Public Health Genomics* 12, 149–157, 2009.

107. Ross, J.S., Slodkowska, E.A., Symmans, W.F. et al., The HER-2 receptor and breast cancer: Ten years of targeted anti-HER-2 therapy and personalized medicine, *Oncologist* 14, 320–368, 2009.

108. Frueh, F.W., Back to the future: Why randomized controlled trials cannot be the answer to pharmacogenomics and personalized medicine, *Pharmacogenomics* 10, 1077–1081, 2009.

109. Huckle, D., Point-of-care diagnostics: An advancing sector with nontechnical issues, *Expert Rev. Mol. Diagn.* 8, 679–688, 2008.
110. Mascini, M. and Tombelli, S., Biosensors for biomarkers in medical diagnostics, *Biomarkers* 13, 637–657, 2008.
111. Pfafflin, A. and Schleicher, E., Inflammation markers in point-of-care testing (POCT), *Anal. Bioanal. Chem.* 393, 1473–1480, 2008.
112. Warsinke, A., Point-of-care testing of proteins, *Anal. Bioanal. Chem.* 393, 1393–1405, 2009.
113. Tothill, I.E., Biosensors for cancer markers diagnosis, *Semin. Cell Dev. Biol.* 20, 55–62, 2009.
114. Conroy, P.J., Hearty, S., Leonard, P. et al., Antibody production, design and use for biosensor-based applications, *Semin. Cell Dev. Biol.* 20, 10–26, 2009.
115. Christenson, R.H. and Azzazy, H.M.E., Biomarkers of myocardial necrosis, in *Cardiovascular Biomarkers Pathophysiology and Disease Management*, D.A. Marrow (ed.), Humana Press, Totowa, NJ, Chapter 1, pp. 3–25, 2006.
116. Floriana, P.N., Christodoulides, N., Miller, C.S. et al., Use of saliva-based nano-biochip tests for acute myocardial infarction at the point of care: A feasibility study, *Clin. Chem.* 55(8), 1530–1538, 2009.
117. McDonnell, B., Hearty, S., Leonard, P., and O'Kennedy, R., Cardiac biomarkers and the case for point-of-care testing, *Clin. Biochem.* 42, 549–561, 2009.
118. Friess, U. and Stark, M., Cardiac markers: A clear cause of point-of-care testing, *Anal. Bioanal. Chem.* 393, 1453–1462, 2009.
119. Christenson, R.H. and Azzazy, H.M., Cardiac point of care testing: A focused review of current National Academy of Clinical Biochemistry guidelines and measurement platforms, *Clin. Biochem.* 42, 150–157, 2009.
120. Giannikis, E., Baum, H., Bertsch, T. et al., Multicenter evaluation of a new point-of-care test for the determination of CK-MB in whole blood, *Clin. Chem. Lab. Med.* 46, 630–638, 2008.
121. Gruson, D., Thys, F., Ketelsleger, J.M. et al., Multimarker panel in patients admitted to emergency department: A comparison with reference methods, *Clin. Biochem.* 42, 185–188, 2009.
122. McLean, A.S., Huang, S.J., and Salker, M., Bench-to-bedside review: The value of cardiac biomarkers in the intensive care patient, *Crit. Care* 12, 215, 2008.
123. Straface, A.L., Myers, J.H., Kirchick, H.J., and Black, K.E., A rapid point-of-care cardiac marker testing strategy facilitates the rapid diagnosis and management of chest pain patients in the emergency department, *Am. J. Clin. Pathol.* 129, 788–795, 2008.
124. Bharti, A., Ma, P.C., and Salgia, R., Biomarker discovery in lung cancer—Promises and challenges of clinical proteomics, *Mass Spectrom. Rev.* 26, 451–466, 2007.
125. Zetterberg, H., Rüetschi, U., Porelius, E. et al., Clinical proteomics in neurodegenerative disorders, *Acta Neurol. Scand.* 118, 1–11, 2008.
126. Mathivanan, S. and Pandey, A., Human proteinpedia as a resource for clinical proteomics, *Mol. Cell. Proteomics* 7, 2038–2047, 2008.
127. Decramer, S., Gonzalez de Peredo, A., and Breuil, B., Urine in clinical proteomics, *Mol. Cell. Proteomics* 7, 1850–1862, 2008.
128. Iwadate, Y., Clinical proteomics in cancer research-promises and limitations of current two-dimensional gel electrophoresis, *Curr. Med. Chem.* 15, 2393–2400, 2008.
129. Birmingham, K., What is translational research?, *Nat. Med.* 8, 647, 2003.
130. Mankoff, S.P., Brander, C., Ferrone, S., and Marincola, F.M., Lost in translation: Obstacles to translational medicine, *J. Transl. Med.* 2:14, 2004.
131. Feurerstein, G.Z. and Chavez, J., Translational medicine for stroke drug discovery: The pharmaceutical industry perspective, *Stroke* 40(3 Suppl), S121–S125, 2009.

132. Mendrick, D.L., Translational medicine: The discovery of bridging biomarkers using pharmacogenomics, *Pharmacogenomics* 7, 943–947, 2006.
133. VanMeter, A., Signore, M., Pierobon, M. et al., Reverse-phase protein microarrays: Application to biomarker discovery and translational medicine, *Expert Rev. Mol. Diagn.* 7, 625–633, 2007.
134. Wang, X., Keith Jr. J.C., Struthers, A.D. et al., Assessment of arterial stiffness, a translational medicine biomarker system for evaluation of vascular risk, *Cardiovasc. Ther.* 26, 214–223, 2008.
135. Keun, H.C., Biomarker discovery for drug development and translational medicine using metabolomics, *Ernst Schering Found. Symp. Proc.* 2007(4), 79–98, 2007.
136. Perrone, A., Molecular imaging technologies and translational medicine, *J. Nucl. Med.* 40, 25N, 2008.
137. *Illustrated Guide to Diagnostic Tests*, Springhouse Corporation, Springhouse, PA, 1994.
138. Tierney Jr. L.M., McPhee, S.J., and Papadaleis, M.A. (eds.), *2004 Lange Current Medical Diagnosis and Treatment*, Lange/McGraw-Hill, New York, 2004.
139. Loyda, H.-J., Putting new assays to the test, *IVD Technol.* 10(2), 38–42, March 2004.
140. Gospodarowicz, M.K., Hensen, D.E., Hutter, R.V.P., O'Sullivan, B., Sobin, L.H., and Wittekind, Ch. (eds.), *Prognostic Factors in Cancer*, 2nd edn., Wiley-Liss, New York, 2001.
141. Hacker, J. and Heesemann, J., Molecular diagnosis and epidemiology, in *Molecular Infectious Biology: Interactions between Microorganisms and Cells*, J. Hacker and J. Heesemann (eds.), Wiley-Liss/Spectrum, Heidelberg, Germany, Chapter 18, p. 195, 2002.
142. Reischl, U. (ed.), *Molecular Diagnosis of Infectious Disease*, Humana Press, Totowa, NJ, 1998.
143. Zieger, K., High throughput molecular diagnostics in bladder cancer—On the brink of clinical utility, *Mol. Oncol.* 1, 384–394, 2008.
144. Rohde, R.E., Falleur, D.M., and Kostroun, P., Molecular diagnostics clinical laboratory science course design: Making it real, *Clin. Lab. Sci.* 22, 9–15. 2009.
145. Muldrew, K.L., Molecular diagnostics of infectious diseases, *Curr. Opin. Pediatr.* 21, 102–111, 2009.
146. Shen, Y. and Wu, B.L., Microarray-based genomic DNA profiling technologies in clinical molecular diagnostics, *Clin. Chem.* 55, 659–669, 2009.
147. Kratz, A., Ferraro, M., Sluss, P.M., and Lewandrowski, K.B., Laboratory reference values, *N. Engl. J. Med.* 351, 1548, 2004.
148. Jones, H.B., On a new substance occurring in the urine of a patient with mollities ossium, *Philos. Trans. R. Soc. Lond.* 138, 55–62, 1848.
149. Roulston, J.E. and Leonard, R.C.F., *Serological Tumour Markers: An Introduction*, Churchill Livingstone, Edinburgh, U.K., 1993.
150. Sell, S. (ed.), *Serological Tumor Markers*, Humana Press, Totowa, NJ, 1992.
151. Luce, B.R. and Brown, R.E., The use of technology assessment by hospitals, health maintenance organizations, and third-party payers in the United States, *Int. J. Technol. Assess.* 11, 79, 1995.
152. Zarkowsky, H., Managed care organization's assessment of reimbursement for new technology, procedures, and drugs, *Arch. Pathol. Lab. Med.* 123, 677, 1999.
153. Zolg, J.W. and Langen, H., How industry is approaching the search for new diagnostic markers and biomarkers, *Mol. Cell. Proteomics* 3, 345, 2004.
154. Horton, G.L., Jortani, S.A., Ritchie Jr. J.C. et al., Proteomics: A new diagnostic frontier, *Clin. Chem.* 52, 1218–1222, 2006.
155. Lippi, G., Plebani, M., and Guidi, G.C., The paradox in translational medicine, *Clin. Chem.* 53, 1553, 2007.
156. Kockan, G., Bourgain, C., Fassina, A. et al., The role of breast FNAS in diagnosis and clinical management: A survey of current practice, *Cytopathology* 19, 271–278, 2008.

157. Moist, L.M., Foley, R.N., and Barrett, B.J., Clinical practice guidelines for evidence-based use of erythropoietic-stimulating agents, *Kidney Int. Suppl.* (110), S12–S18, 2008.

158. Gupta, R., Dastane, A.M., McKenna Jr. R. et al., The predictive value of epidermal growth factor receptor tests in patient with pulmonary adenocarcinoma: Review of current "best evidence" with meta-analysis, *Hum. Pathol.* 40, 356–365, 2008.

159. Bergmann, P., Body, J.J., Boonen, S. et al., Evidence-based guidelines for the use of biochemical markers of bone turnover in the selection and monitoring of bisphosphonate treatment in osteoporosis: A consensus document of the Belgium Bone Club, *Int. J. Clin. Pract.* 63, 19–26, 2009.

160. Taylor, R.S. and Elston, J., The use of surrogate outcomes in model-based cost-effectiveness analyses: A survey of UK health technology assessment reports, *Health Technol. Assess.* 13, 1–50, 2009.

161. Brent, D.A. and Mallouf, F.T., Pediatric depression: Is there evidence to improve evidence-based treatments?, *J. Child Psychol. Psychiatry* 50, 143–152, 2009.

162. Felker, G.M., Pang, P.S., Adams, K.F., et al., Clinical trials of pharmacological therapies in acute heart failure syndromes. Lessons learned and directions forward, *Circ. Heart Fail.* 3, 314–325, 2010.

163. Lyman, G.H., Comparative effectiveness research in oncology: The need for clarity, transparency and vision, *Cancer Invest.* 27, 593–593, 2009.

164. Wagner, J.A., Williams, S.A., and Webster, C.J., Biomarkers and surrogate end points for fit-for-purpose development and regulatory evaluation of new drugs, *Clin. Pharmacol. Ther.* 81, 104–107, 2007.

165. Venteuolo, C.E. and Levy, M.M., Biomarkers: Diagnosis and risk assessment in sepsis, *Clin. Chest Med.* 29, 591–603, 2008.

166. Lundblad, R.L., Approach to assay validation for the development of biopharmaceuticals, *Biotechnol. Appl. Biochem.* 34, 195, 2001.

167. Lundblad, R.L. and Wagner, P., Ruminations on the issue of assay validation for biopharmaceuticals, *Preclinica* 2, 1, 2004.

168. Colburn, W.A., Surrogate markers and clinical pharmacology, *J. Clin. Pharmacol.* 35, 441, 1995.

169. Bielokova, B. and Martin, R., Development of biomarkers in multiple sclerosis, *Brain* 127, 1463, 2004.

170. Colburn, W.A., Selecting and validating biological markers for drug development, *J. Clin. Pharmacol.* 27, 355, 1997.

171. Patterson, S., Selecting targets for therapeutic validation through differential protein expression using chromatography using chromatography-mass spectrometry, *Bioinformatics* 18(Suppl 2), 181, 2002.

172. Liu, S.C., Sauter, E.R., Clapper, M.L. et al., Markers of cell proliferation in normal epithelia and dysplastic leukoplakias of the oral cavity, *Cancer Epidemiol. Biomarkers Prev.* 7, 597, 1998.

173. Ahmed, M.I., Abd-Emalelib, F., Ziada, N.A., and Khalifa, A., Evaluation of some tissue and serum biomarkers in prostatic carcinoma among Egyptian males, *Clin. Biochem.* 32, 439, 1999.

174. Yasui, Y., Pepe, M., Thompson, M.L. et al., A data-analytic strategy for protein biomarker discovery: Profiling of high-dimensional proteomic data for cancer detection, *Biostatistics* 4, 449, 2003.

175. Eissa, S., Swellam, M., el-Mosallamy, H. et al., Diagnostic value of urinary molecular markers in bladder cancer, *Anticancer Res.* 23, 4347, 2003.

176. Ortolá, J., Castiñeiras, M.J., and Fuentes-Arderiu, X., Biological variation data applied to the selection of serum lipid rations used as risk markers of coronary heart disease, *Clin. Chem.* 38, 56–59, 1992.

177. Jensen, J.E., Kollerup, G., Sørensen, H.A., and Sørensen, O.H., Intraindividual variability in bone markers in the urine, *Scand. J. Clin. Lab. Invest.* 227, 29–34, 1007.

178. Beck-Jensen, J.E., Kollerup, G., Sørensen, H.A. et al., A single measurement of biochemical markers of bone turnover has limited utility in the individual person, *Scand. J. Clin. Lab. Invest.* 57, 351–359, 1997.

179. Kilpatrick, E.S., Maylor, P.W., and Keevil, B.G., Biological variation of glycated hemoglobin. Implications for diabetes screening and monitoring, *Diabetes Care* 21, 261–264, 1998.

180. Tuxen, M.K., Sölétormos, G., Rustin, G.J. et al., Biological variation and analytical imprecision of CA 125 in patients with ovarian cancer, *Scand. J. Clin. Lab. Invest.* 60, 713–721, 2000.

181. Møller, H.J., Petersen, P.H., Rejnmark, L., and Moestrup, S.K., Biological variation of soluble CD163, *Scand. J. Clin. Lab. Invest.* 63, 15–21, 2003.

182. Erden, G., Barazi, A.O., Tezcan, G., and Yildirimkaya, M.M., Biological variation and reference change values of CA 19-9, CEA, AFP in serum of healthy individuals, *Scand. J. Clin. Lab. Invest.* 68, 212–218, 2008.

183. Wu, A.H., Lu, Q.A., Todd, J. et al., Short- and long-term biological variation in cardiac troponin I measured with a high-sensitivity assay: Implications for clinical practice, *Clin. Chem.* 55, 52–58, 2008.

184. Nguyen, T.V., Nelson, A.E., Howe, C.J. et al., Within-subject variability and analytical imprecision of insulin like growth factor axis and collagen markers: Implications for clinical diagnosis and doping tests, *Clin. Chem.* 54, 1268–1276, 2008.

185. Wu, S.L., Hancock, W.S., Goodrich, G.G., and Kunitake, S.T., An approach to the proteomic analysis of a breast cancer cell line (SKBR-3), *Proteomics* 3, 1037–1046, 2003.

186. Rainey, T., Lesko, M., Sacho, R. et al., Predicting outcome after severe traumatic brain injury using the serum S100B biomarker: Results using a single (24h) time-point, *Resuscitation* 80, 341–345, 2009.

187. Wang, Y., Jacobs, E.J., McCullough, M.L. et al., Comparing methods for accounting for seasonal variability in a biomarker when only a single sample is available: Insights from simulations based on serum 25-hydroxyvitamin D, *Am. J. Epidemiol.* 170, 89–94, 2009.

188. Gupta, A.K., Brenner, D.E., and Turgeon, D.K., Early detection of colon cancer: New tests on the horizon, *Mol. Diagn. Ther.* 12, 77–85, 2008.

189. Dolci, A., Dominici, R., Cardinale, D. et al., Biochemical markers for prediction of chemotherapy-induced cardiotoxicity: Systematic review of the literature and recommendations for use, *Am. J. Clin. Pathol.* 130, 688–695, 2008.

190. Warner, E., The role of magnetic resonance imaging in screening women at high risk of breast cancer, *Top. Magn. Reson. Imaging* 19, 163–169, 2008.

191. Visser, T., Pillay, J., Koenderman, L., and Leenen, L.P., Post injury immune monitoring: Can multiple organ failure be predicted?, *Curr. Opin. Crit. Care* 14, 666–672, 2008.

192. Loitsch, S.M., Shastri, Y., and Stein, J. Stool test for colorectal cancer screening—It's time to move!, *Clin. Lab.* 54, 473–484, 2008.

193. Fukutomi, Y., Moriwaki, H., Nagase, S. et al., Metachronous colon tumors: Risk factors and rationale for the surveillance colonoscopy after initial polypectomy, *J. Cancer Res. Clin. Oncol.* 128, 569–574, 2002.

194. Menges, M., Gärtner, B., Georg, T. et al., Cost-benefit analysis of screening colonoscopy in 40- to 50-year-old first-degree relatives of patients with colorectal cancer, *Int. J. Colorectal Dis.* 21, 596–601, 2006.

195. Kolesar, J.M., Assessing therapeutically developed assays, *Manag. Care* 1(Suppl 7), 9–12, 2008.

196. Yong, J.H., Schuh, S., Rashidi, R. et al., A cost effectiveness analysis of omitting radiography in diagnosis of acute bronchiolitis, *Pediatr. Pulmonol.* 44, 122–127, 2009.

197. Pickhardt, P.J., Hassan, C., Laghi, A., and Kim, D.H., CT colonography to screen for colorectal cancer and aortic aneurysm in the medicare population: Cost-effectiveness analysis, *AJR Am. J. Roentgenol.* 192, 1332–1340, 2009.

198. de Kok, I.M., Polder, J.J., Habbema, J.D. et al., The impact of healthcare costs in the last year of life and in all life years gained on the cost-effectiveness of cancer screening, *Br. J. Cancer* 100, 1240–1244, 2009.

199. Pignone, M.P. and Lewis, C.L, Using quality improvement techniques to increase colon cancer screening, *Am. J. Med.* 122, 419–420, 2009.

200. Szucs, T.D. and Dedes, K.J., Is cancer prevention ever going to be profitable?, *Recent Results Cancer Res.* 181, 41–47, 2009.

201. Colombo, M., Screening and diagnosis of hepatocellular carcinoma, *Liver Int.* 29(Suppl 1), 143–147, 2009.

202. Gemmel, C., Eickhoff, A., Helmstädter, L., and Riemann, J.F., Pancreatic cancer screening: State of the art, *Expert Rev. Gastroenterol. Hepatol.* 3, 89–96, 2009.

203. Imamura, T. and Yasunaga, H., Economics evaluation of prostate cancer screening with prostate-specific antigen, *Int. J. Urol.* 15, 285–288, 2008.

204. MacKillop, W.J., The importance of prognosis in cancer medicine, in *Prognostic Factors in Cancer*, 2nd edn., M.K. Gospodarowicz, D.E. Hensen, R.V.P. Hutter, B. O'Sullivan, L.H. Sobin, and Ch. Wittekind (eds.), Wiley-Liss, New York, Chapter 1, pp. 3–16, 2001.

205. Hartmann, M. and Gundermann, C., Cost-benefit considerations regarding early detection of malignant diseases, *Onkologie* 14, 164, 2008.

206. Lansdorp-Vogelaar, I., van Ballegooijen, M., Zauber, A.G. et al., At what costs will screening with CT colonography be competitive? A cost-effectiveness approach, *Int. J. Cancer* 124, 1161–1168, 2009.

207. McIntosh, M.W. and Urban, N., A parametric empirical Bayes method for cancer screening using longitudinal observations of a biomarker, *Biostatistics* 4, 27–40, 2003.

208. Gornall, A.H., May, L.A., and Mulley Jr. A.G., *Primary Care Medicine*, J.B. Lippincott, Philadelphia, PA, Chapter 3, 1987.

209. Mulley Jr. A.G., The selection and interpretation of diagnostic tests, in *Primary Care Medicine*, A.H. Gornall, L.A. May, and A.G. Mulley Jr. (eds.), J.B. Lippincott, Philadelphia, PA, Chapter 2, 1987.

210. Schwartz, J.S., Clinical decision-making in cardiology, in *Braunwald's Heart Disease*, 3rd edn., P. Libby and E. Braunwald (eds.), Saunders/Elsevier, Philadelphia, PA, Chapter 3, pp. 27–34, 2008.

211. Conrad, D.H., Goyette, J., and Thomas, P.S., Proteomics as a method for early detection of cancer: A review of proteomics, exhaled breath condensate, and lung cancer screening, *J. Gen. Intern. Med.* 23(Suppl 1), 78–84, 2008.

212. Walsh, C.S. and Karlan, B.Y., Contemporary progress in ovarian cancer screening, *Curr. Oncol. Rep.* 9, 485–493, 2007.

213. Ziober, B.L., Maux, M.G., Falis, E.M. et al., Lab-on-a-chip for oral cancer screening and diagnosis, *Head Neck* 30, 111–121, 2008.

214. Jubb, A.M., Quirke, P., and Oates, A.J., DNA methylation, a biomarker for colorectal cancer: Implications for screening and pathological utility, *Ann. N. Y. Acad. Sci.* 983, 251–267, 2003.

215. Limburg, P.J., Devens, M.E., Harrington, J.J. et al., Prospective evaluation of fecal calprotectin as a screening biomarker for colorectal neoplasia, *Am. J. Gastroenterol.* 98, 2299–2305, 2003.

216. Lazar, I.M., Trisiripisal, P., and Sarvaiya, H.A., Microfluidic liquid chromatography system for proteomic applications and biomarker screening, *Anal. Chem.* 78, 5513–5524, 2006.

217. Linkov, F., Lisovich, A., Yurkovetsky, Z. et al., Early detection of head and neck cancer: Development of a novel screening tool using multiplexed immunobead-based biomarker profiling, *Cancer Epidemiol. Biomarkers Prev.* 16, 102–107, 2007.

218. Boulet, G.A., Horvath, C.A., Berghmans, S., and Bogers, J., Human papillomavirus in cervical cancer screening: Important role as biomarker, *Cancer Epidemiol. Biomarkers Prev.* 17, 810–817, 2008.

219. Yi, J.K., Chang, J.W., Han, W. et al., Autoantibody to tumor antigen, alpha-2-HS glycoprotein: A novel biomarker of breast cancer screening and diagnosis, *Cancer Epidemiol. Biomarkers Prev.* 18, 1357–1364, 2009.

220. McNeal, C.J., Wilson, D.P., Christou, D. et al., The use of surrogate vascular markers in youth at risk for premature cardiovascular disease, *J. Pediatr. Endocrinol. Metab.* 22, 195–211, 2009.

221. Xie, Y., Todd, N.W., Liu, Z. et al., Altered miRNA expression in sputum for diagnosis of non-small cell lung cancer, *Lung Cancer* 67, 170–176, 2010.

222. Favilli, S., Frenos, S., Lasagni, D. et al., The use of B-type natriuretic peptide in paediatric patients: A review of literature, *J. Cardiovasc. Med.* 10, 298–302, 2009.

223. Krizkova, S., Fabrik, I., Adam, V. et al., Metallothionein—A promising tool for cancer diagnostics, *Bratisl. Lek. Listy* 110, 93–97, 2009.

224. Wang, M., Long, R.E., Comunale, M.A. et al., Novel fucosylated biomarkers for the early detection of hepatocellular carcinoma, *Cancer Epidemiol. Biomarkers Prev.* 18, 1914–1921, 2009.

225. Kucur, M., Karadaq, B., Isman, F.K. et al., Plasma hyaluronidase activity as an indicated of atherosclerosis in patients with coronary artery disease, *Bratisl. Lek. Listy* 110, 21–26, 2009.

226. Karadag, B., Kucur, M., Isman, F.K. et al., Serum chitotriosidase activity in patients with coronary artery disease, *Circ. J.* 72, 71–75, 2008.

227. Tang, W.H., Katz, R., Brennan, M.L. et al., Usefulness of myeloperoxidase levels in healthy elderly subjects to predict risk of developing heart failure, *Am. J. Cardiol.* 103, 1269–1274, 2009.

228. Lu, S., Chiu, Y.S., Smith, A.P. et al., Biomarkers correlate with colon cancer and risks: A preliminary study, *Dis. Colon Rectum* 52, 715–724, 2009.

229. Charles, P.E., Kus, E., Aho, C. et al., Serum procalcitonin for the early recognition of nosocomial infection in the critically ill patients: A preliminary report, *BMC Infect. Dis.* 9, 49, 2009.

230. Zhang, J.Y., Looi, K.S., and Tan, E.M., Identification of tumor-associated antigens as diagnostic and predictive biomarkers in cancer, *Methods Mol. Biol.* 520, 1–10, 2009.

231. Jensen, M.B., Chacon, M.R., Sattin, J.A. et al., Potential biomarkers for the diagnosis of stroke, *Expert Rev. Cardiovasc. Ther.* 7, 389–393, 2009.

232. Schenk, J.M., Riboli, E., Chatterjee, N. et al., Serum retinol and prostate cancer risk: A nested case-control study in the prostate, lung, colorectal, and ovarian cancer screening trial, *Cancer Epidemiol. Biomarkers Prev.* 18, 1227–1231, 2009.

233. Menon, U., Gentry-Maharaj, A., Hallett, R. et al., Sensitivity and specificity of multimodal and ultrasound screening for ovarian cancer, and stage distribution of detected cancers: Results of the prevalence screen of the UK collaborative trial of ovarian cancer screening (UKCTOCS), *Lancet Oncol.* 10, 327–340, 2009.

234. Garrido-Delgado, R., Arce, L., Pérez-Marin, C.C., and Valcárcel, M., Use of ion mobility spectroscopy with an ultraviolet ionization source as vanguard screening system for the detection and determination of acetone in urine as a biomarker for cow and human diseases, *Talenta* 78, 863–868, 2009.

235. Gurbb III R.L., Black, A., Izmirlian, G. et al., Serum prostate-specific antigen hemodilution among obese men undergoing screening in the prostate, lung, colorectal, and ovarian cancer screen trial, *Cancer Epidemiol. Biomarkers Prev.* 18, 748–751, 2009.

236. Dey, P., Urinary markers of bladder carcinoma, *Clin. Chim. Acta* 340, 57–65, 2004.
237. Van Tilborg, A.A., Bangma, C.H., and Zwarthoff, E.C., Bladder cancer biomarkers and their role in surveillance and screening, *Int. J. Urol.* 16, 23–30, 2009.
238. Wright, R.J. and Stringer, S.A., Rapid testing strategies for HIV-1 serodiagnosis in high-prevalence African settings, *Am. J. Prev. Med.* 27, 42–48, 2004.
239. Baron, A.T., Boardman, C.H., Lafky, J.M. et al., Soluble epidermal growth factor receptor (SEG-FR) and cancer antigen 125 (CA125) as screening and diagnostic tests for epithelial ovarian cancer, *Cancer Epidemiol. Biomarkers Prev.* 14, 306–318, 2005.
240. Clarke-Pearson, D.L., Clinical practice. Screening for ovarian cancer, *N. Engl. J. Med.* 361, 170–177, 2009.
241. Rosenblatt, R., Jonmarker, S., Lewensohn, R. et al., Current status of prognostic immunohistochemical markers for urothelial bladder cancer, *Tumour Biol.* 29, 311–322, 2008.
242. Gould Rothberg, B.E., Bracken, M.B., and Rimm, D.L., Tissue biomarkers for prognosis in cutaneous melanoma: A systematic review and meta-analysis, *J. Natl. Cancer Inst.* 101, 452–474, 2009.
243. Gurda-Duda, A., Kuśnierz-Cabala, B., Nowak, W. et al., Assessment of the prognostic value of certain acute-phase proteins and procalcitonin in the prognosis of acute pancreatic, *Pancreas* 37, 449–453, 2008.
244. Donovan, M.J., Khan, F.M., Fernandez, G. et al., Personalized prediction of tumor response and cancer progression on prostate needle biopsy, *J. Urol.* 182, 125–132, 2009.
245. Shinozuka, K., Uzama, K., Fuhimi, K. et al., Downregulation of carcinoembryonic antigen-related cell adhesion molecule 1 in oral squamous cell carcinoma: Correlation with tumor progression and poor prognosis, *Oncology* ii, 387–397, 2009.
246. Gordon, G.J., Dong, L., Yeap, B.Y. et al., Four-gene expression ratio test for survival in patients undergoing surgery for mesothelioma, *J. Natl. Cancer Inst.* 101, 678–686, 2009.
247. Castelli, G.P., Pognani, C., Cita, M., and Paladini, R., Procalcitonin as a prognostic and diagnostic tool for septic complications after major trauma, *Crit. Care Med.* 37, 1845–1849, 2009.
248. Fleshner, N.E. and Lawrentschuk, N., Risk of developing prostate cancer in the future: Overview of prognostic biomarkers, *Urology* 73(5 Suppl), S21–S27, 2009.
249. Unoki, M., Kelly, J.D., Neal, D.E. et al., UHRF1 is a novel molecular marker for diagnosis and the prognosis of bladder cancer, *Br. J. Cancer* 101, 98–105, 2009.
250. Mikkelsen, M.E., Miltiades, A.N., Galeski, D.F. et al., Serum lactate is associated with mortality in severe sepsis independent of organ failure and shock, *Crit. Care Med.* 37, 1670–1677, 2009.
251. Whiteley, W., Chong, W.L., Sengupta, A. et al., Blood markers for the prognosis of ischemic stroke: A systematic review, *Stroke* 40, e380–e389, 2009.
252. Takikita, M., Altekruse, S., Lynch, C.F. et al., Association between selected biomarkers and prognosis in a population-based pancreatic cancer tissue microarray, *Cancer Res.* 69, 2950–2955, 2009.
253. Claus, R.A., Bockmeyer, C.L., Budde, U. et al., Variations in the ratio between von Willebrand factor and its cleaving protease during systemic inflammation and association with severity and prognosis of organ failure, *Thromb. Haemost.* 101, 239–247, 2009.
254. Kobayashi, M., Ikeda, K., Kawamura, Y. et al., High serum des-gamma-carboxy prothrombin level predicts poor prognosis after radiofrequency ablation of hepatocellular carcinoma, *Cancer* 115, 571–580, 2009.
255. Yu, J., Cheng, Y.Y., Tao, Q. et al., Methylation of protocadherin 10, a novel tumor suppressor, is associated with poor prognosis in patients with gastric cancer, *Gastroenterology* 136, 640–651, 2009.
256. Smits, K.M., Cleven, A.H., Weijenberg, M.P. et al., Pharmacoepigenomics in colorectal cancer: A step forward in predicting prognosis and treatment response, *Pharmacogenomics* 9, 1903–1916, 2008.

257. Chen, C.N., Chang, C.C., Su, T.E. et al., Identification of calreticulin as a prognosis marker and angiogenic regulator in human gastric cancer, *Ann. Surg. Oncol.* 16, 524–533, 2009.

258. Zlobec, I., Baker, K., Tarracciano, L. et al., Two-marker protein profile predicts poor prognosis in patients with early rectal cancer, *Br. J. Cancer* 99, 1712–1717, 2008.

259. Egler, R.A., Burlingame, S.M., Nuchtern, J.G., and Russell, H.V., Interleukin-6 and soluble interleukin-6 receptor levels as markers in disease extent and prognosis in neuroblastoma, *Clin. Cancer Res.* 14, 7028–7034, 2008.

260. Oikonomopoulou, K., Li, L., Zheng, Y. et al., Prediction of ovarian cancer prognosis and response to chemotherapy by a serum-based multiparametric biomarker panel, *Br. J. Cancer* 99, 1103–1113, 2008.

261. Boos, C.J., Balakrishnan, B., Blann, A.D., and Lip, G.Y., The relationship of circulating endothelial cells to plasma indices of endothelial damage/dysfunction and apoptosis in acute coronary syndrome: Implications for prognosis, *J. Thromb. Haemost.* 6, 1841–1850, 2008.

262. Boussekey, N., Van Grunderbeeck, N., and Leroy, O., CRP: A new prognosis marker in community-acquired pneumonia?, *Am. J. Med.* 121, e21, e23, 2008.

263. Haeuptle, J., Zaborsky, R., Fiumefreddo, R. et al., Prognostic value of procalcitonin in *Legionella pneumonia*, *Eur. J. Clin. Microbiol. Infect. Dis.* 28, 55–60, 2009.

264. Chalmers, J.D., Singanayagam, A., and Hill, A.T., C-reactive protein is an independent predictor of severity in community-acquired pneumonia, *Am. J. Med.* 121, 219–225, 2008.

265. Shimizu, K., Ogura, H., Goto, M. et al., Symbiotics decrease the incidence of septic complications in patients with severe SIRS: A preliminary report, *Dig. Dis. Sci.* 54, 1071–1078, 2008.

266. Seki, M., Watanabe, A., Mikase, K. et al., Revision of the severity rating and classification of hospital-acquired pneumonia in the Japanese respiratory society guidelines, *Respirology* 13, 880–885, 2008.

267. Ay, H., Arsava, E.M., Rosand, J. et al., Severity of leukoaraiosis and susceptibility to infarct growth in acute stroke, *Stroke* 39, 1409–1413, 2008.

268. Egan, L.J., Derijks, L.J., and Hommes, D.W., Pharmacogenomics in inflammatory bowel disease, *Clin. Gastroenterol. Hepatol.* 4, 21–28, 2006.

269. Anderson, J.E., Hansen, L.L., Mooren, F.C. et al., Methods and biomarkers for the diagnosis and prognosis of cancer and other diseases: Towards personalized medicine, *Drug Resist. Updat.* 9, 198–210, 2006.

270. Subbiah, M.T., Nutrigenetics and nutraceuticals: The next wave riding on personalized medicine, *Transl. Res.* 149, 55–61, 2007.

271. Tebbutt, S.J., James, A., and Paré, P.D., Single-nucleotide polymorphisms and lung disease: Clinical implications, *Chest* 131, 1216–1223, 2007.

272. Vandenbroeck, K. and Matute, C., Pharmacogenomics of the response to IFN-beta in multiple sclerosis: Ramifications from the first genome-wide screen, *Pharmacogenomics* 9, 639–645, 2008.

273. Heidecker, B. and Hare, J.M., The use of transcriptomic biomarkers for personalized medicine, *Heart Fail. Rev.* 12, 1–11, 2007.

274. Trevino, V., Falciani, F., and Barrere-Saldaña, H.A., DNA microarrays: A powerful genomic tool for biomedical and clinical research, *Mol. Med.* 13, 527–541, 2007.

275. Farragher, S.M., Tanney, S., Kennedy, R.D., and Harkin, P.D., RNA expression analysis from formalin fixed paraffin embedded tissues, *Histochem. Cell Biol.* 130, 435–445, 2008.

276. De Leon, J., Pharmacogenomics: The promise of personalized medicine for CNS disorders, *Neuropsychopharmacology* 34, 159–172, 2009.

277. Kojima, Y., Sasaki, S., Hayashi, Y. et al., Subtypes of α_1-adrenoceptors in BPH: Future prospects for personalized medicine, *Nat. Clin. Pract. Urol.* 6, 44–53, 2009.

278. Dowsett, M. and Dunbier, A.K., Emerging biomarkers and new understanding of traditional markers in personalized therapy for breast cancer, *Clin. Cancer Res.* 14, 8019–8026, 2008.

279. Cai, W., Niu, G., and Chen, X., Imaging of integrins as biomarkers for tumor angiogenesis, *Curr. Pharm. Des.* 14, 2943–2973, 2008.

280. Schrohl, A.S., Würtz, S., Kohn, E. et al., Banking of biological fluids for studies of disease-associated protein biomarkers, *Mol. Cell. Proteomics* 7, 2061–2066, 2008.

281. Lee, C.K., Lord, S.J., Coats, A.S., and Simes, R.J., Molecular biomarkers to individualise treatment: Assessing the evidence, *Med. J. Aust.* 190, 631–636, 2009.

282. Olopade, O.I., Grushko, T.A., Nanda, R., and Huo, D., Advances in breast cancer: Pathways to personalized medicine, *Clin. Cancer Res.* 14, 7988–7999, 2008.

283. Madreker, S.J. and Sargeant, D.J., Clinical trial designs for predictive biomarker validation: One size does not fit all, *J. Biopharm. Stat.* 19, 530–543, 2009.

284. Phan, J.H., Moffitt, R.A., Stokes, T.H. et al., Convergence of biomarkers, bioinformatics and nanotechnology for individualized cancer treatment, *Trends Biotechnol.* 27, 350–358, 2009.

285. Santos, E.S., Blaya, M., and Raez, L.E., Gene expression profiling and non-small-cell lung cancer: Where are we now?, *Clin. Lung Cancer* 10, 168–173, 2009.

286. Sawyers, C.L., The cancer biomarker problem, *Nature* 452, 548–552, 2008.

287. Macdermid, J.C., Gross, A.R., Galea, V. et al., Developing biologically-based assessment tools for physical therapy management of neck pain, *J. Orthop. Sports Phys. Ther.* 39, 388–399, 2009.

288. Phan, J.H., Moffitt, R.A., Stokes, T.H. et al., Convergence of biomarkers, bioinformatics and nanotechnology for individualized cancer treatment, *Trends Biotechnol.* 27, 350–358, 2009.

289. Mohr, S. and Liew, C.C., The peripheral-blood transcriptome: New insights into disease and risk assessment, *Trends Mol. Med.* 13, 422–432, 2007.

290. Garrison Jr. L.P. and Austin, M.J., Linking pharmacogenetic-based diagnostics and drugs for personalized medicine, *Health Aff. (Millwood)*, 25, 1281–1290, 2006.

291. van der Greef, J., Hankemeier, T., and McBurney, R.N., Metabolomics-based systems biology and personalized medicine: Moving toward $n = 1$ clinical trials, *Pharmacogenomics* 7, 1087–1094, 2006.

292. Danser, H., Batenburg, W.W., van den Meiracker, A.H., and Danilov, S.M., ACE phenotyping as a first step toward personalized medicine for ACE inhibitors. Why does ACE genotyping not predict the therapeutic efficacy of ACE inhibition?, *Pharmacol. Ther.* 113, 607–618, 2007.

293. Alexopoulos, G.S., Murphy, C.F., Gunning-Dixon, F.M. et al., Microstructural white matter abnormalities and remission of geriatric depression, *Am. J. Psychiatry* 165, 238–244, 2008.

294. Kumar, A. and Ajilore, O., Magnetic resonance imaging and late-life depression: Potential biomarkers in the era of personalized medicine, *Am. J. Psychiatry* 165, 166–168, 2008.

295. Moroi, S.E. and Heckenlively, J.R., Progress toward personalized medicine for age-related macular degeneration, *Ophthalmology* 115, 925–926, 2008.

296. Klein, M.L., Francis, P.J., Rosner, B. et al., CFH and LOC387715/ARMS2 genotypes and treatment with antioxidants and zinc for age-related macular degeneration, *Ophthalmology* 115, 1019–1025, 2008.

297. Eckelman, W.C., Reba, R.C., and Kelloff, G.J., Targeted imaging: An important biomarker for understanding disease progression in the era of personalized medicine, *Drug. Discov. Today* 13, 748–759, 2008.

298. Shai, R.M., Reichardt, J.K., and Chen, T.C., Pharmacogenomics of brain cancer and personalized medicine in malignant gliomas, *Future Oncol.* 4, 525–534, 2008.

299. Baek, S., Moon, H., Ahn, H. et al., Identifying high-dimensional biomarkers for personalized medicine via variable importance ranking, *J. Biopharm. Stat.* 853–868, 2008.

300. Vosslamber, S., van Baarsen, L.G., and Verweij, C.L., Pharmacogenomics of IFN-β in multiple sclerosis: Towards a personalized medicine approach, *Pharmacogenomics* 10, 97–108, 2009.

301. Cooke, G.E., Pharmacogenetics of multigene disease: Heart disease as an example, *Vascul. Pharmacol.* 44, 66–74, 2006.
302. Flores, C., Ma, S.F., Maresso, K. et al., Genomics for acute lung injury, *Semin. Respir. Crit. Care Med.* 27, 389–395, 2006.
303. Nagasaki, K. and Miki, Y., Molecular prediction of the therapeutic response to neoadjuvant chemotherapy in breast cancer, *Breast Cancer* 15, 117–120, 2008.
304. Bhushan, S., McLeod, H., and Walko, C.M., Role of pharmacogenetics as predictive biomarkers of response and/or toxicity in the treatment of colorectal cancer, *Clin. Colorectal Cancer* 8, 15–21, 2009.
305. Shin, J., Kayser, S.R., and Langee, T.Y., Pharmacogenetics: From discovery to patient care, *Am. J. Health Syst. Pharm.* 66, 625–637, 2009.
306. Walko, C.M. and McLeod, H., Pharmacogenomic progress in individualized dosing of drugs for cancer patients, *Nat. Clin. Pract. Oncol.* 6, 153–162, 2009.
307. Zhong, S. and Romkes, M., Pharmacogenomics, *Methods Mol. Biol.* 520, 231–245, 2009.
308. Trull, A.K. (ed.), *Biomarkers of Disease: An Evidence-Based Approach*, Cambridge University Press, New York, 2002.
309. Ming Chu, T. (ed.), *Biochemical Markers for Cancer*, Marcel Dekker, New York, 1982.
310. Anton-Guirgis, H. and Lynch, H.T. (eds.), *Biomarkers, Genetics, and Cancer*, Van Nostrand Reinhold, New York, 1985.
311. Miller, A.D., Bantsch, H., Bofetta, P., Dragsted, L., and Vaino, H. (eds.), *Biomarkers in Cancer Prevention*, IARC/Oxford University Press, Oxford, U.K., 2001.
312. Gasparini, G. and Hayes, D.F. (eds.), *Biomarkers in Breast Cancer Molecular Diagnostics for Predicting and Monitoring Therapeutic Effect*, Humana, Totowa, NJ, 2006.
313. Hamdan, M.H., *Cancer Biomarkers Analytical Techniques for Discovery*, Wiley-Interscience, Hoboken, NJ, 2007.
314. Sanchez-Carbayo, M., Recent advances in bladder cancer diagnosis, *Clin. Biochem.* 37, 562, 2004.
315. Mobley, J.A., Lam, Y.W., Lau, K.M. et al., Monitoring the serological proteome: The latest modality in prostate cancer detection, *J. Urol.* 172, 331, 2004.
316. Skates, S.J., Menon, U., McDonald, N. et al., Calculation of the risk of ovarian cancer from serial CA-125 values for preclinical detection in postmenopausal women, *J. Clin. Oncol.* 2(Suppl), 206S, 2003.
317. Jett, J.R. and Midthun, D.E., Screening for lung cancer: Current status and future direction, Thomas A. Neff lecture, *Chest* 125(Suppl), 158S, 2004.
318. Baydar, T., Yuksel, O., and Sahin, T.T., Neopterin as a prognostic biomarker in intensive care unit patients, *J. Crit. Care* 24, 318–321, 2009.
319. Petäjä, L., Salmenperä, M., Pulkki, K., and Pettila, V., Biochemical injury markers and mortality after coronary artery bypass grafting: A systematic review, *Ann. Thorac. Surg.* 87, 1981–1992, 2009.
320. Wang, Y., Ripa, R.S., Johansen, J.S. et al., YKL-40, a new biomarker in patients with acute coronary syndrome or stable coronary artery disease, *Scand. Cardiovasc. J.* 42, 295–302, 2008.
321. Lin, W.T., Tsai, C.C., Chen, C.Y. et al., Proteomic analysis of peritoneal dialysate fluid in patients with dialysis-related peritonitis, *Ren. Fail.* 30, 772–777, 2008.
322. Kallenberg, C.G., The last classification of vasculitis, *Clin. Rev. Allergy Immunol.* 35, 5–10, 2008.
323. Yin Ji, X., Kang, M.R., Choi, J.S. et al., Levels of intra- and extracellular heat shock 60 in Kawasaki disease patients treated with intravenous immunoglobulin, *Clin. Immunol.* 124, 304–310, 2007.
324. Jerums, G., Premaratne, E., Panagiotopoulos, S. et al., New and old markers of progression of diabetic nephropathy, *Diabetes Res. Clin. Pract.* 82(Suppl 1), S30–S37, 2008.

325. Futrakul, N., Sridama, V., and Futrakul, P., Microalbuminuira—A biomarker of renal microvascular disease, *Ren. Fail.* 31, 140–143, 2009.
326. Buszewki, B., Kesy, M., Ligor, T. et al., Human exhaled air analytics: Biomarkers of diseases, *Biomed. Chromatogr.* 21, 553–566, 2007.
327. Schulz, B.L., Laroy, W., and Callewaert, N., Clinical laboratory testing in human medicine based on the detection of glycoconjugates, *Curr. Mol. Med.* 7, 397–416, 2007.
328. Wang, X.M., Yao, T.W., Nadvi, N.A. et al., Fibroblast activation protein and chronic liver disease, *Front. Biosci.* 13, 3168–3180, 2008.
329. Rosa, H. and Parise, E.R., Is there a place of serum laminin determination in patients with liver disease and cancer?, *World J. Gastroenterol.* 14, 3628–2632, 2008.
330. Mehta, A. and Block, T.M., Fucosylated glycoproteins markers of liver disease, *Dis. Markers* 25, 259–265, 2008.
331. Janackova, S. and Sforza, E., Neurobiology of sleep fragmentation: Cortical and autonomic markers of sleep disorders, *Curr. Pharm. Des.* 14, 3474–3480, 2008.
332. Goertsches, R.H., Hecker, M., and Zettl, U.K., Monitoring of multiple sclerosis immunotherapy: From single candidates to biomarker networks, *J. Neurol.* 255(Suppl 6), 48–57, 2008.
333. Halperin, I., Morelli, M., Korczyn, A.D. et al., Biomarkers for evaluation of clinical efficacy of multipotential neuroprotective drugs for Alzheimer's and Parkinson's diseases, *Neurotherapeutics* 6, 128–140, 2009.
334. Rascol, O., "Disease-modification" trials in Parkinson disease: Target populations, endpoints and study design, *Neurology* 72(7 Suppl), S51–S58, 2009.
335. Leoni, V., Oxysterols as markers of neurological disease—A review, *Scand. J. Clin. Lab.* 69, 22–25, 2009.
336. Cho, H. and Kim, J.-H., Multiplication of neutrophil and monocyte counts (MNM) as an easily obtainable tumour marker for cervical cancer, *Biomarkers* 14, 161–170, 2009.
337. Ferri, N., Paoletti. R., and Corsini, A., Biomarkers for atherosclerosis: Pathophysiological role and pharmacological modulation, *Curr. Opin. Lipidol.* 17, 495–501, 2006.
338. Liang, S.L. and Chan, D.W., Enzymes and related proteins as cancer biomarkers: A proteomic approach, *Clin. Chim. Acta* 381, 93–97, 2007.
339. Malinowski, D.P., Multiple biomarkers in molecular oncology. II. Molecular diagnostics applications in breast cancer management, *Expert Rev. Mol. Diagn.* 7, 269–280, 2007.
340. Bentzen, S.M., Buffa, F.M., and Wilson, G.D., Multiple biomarker tissue microarrays: Bioinformatics and practical approaches, *Cancer Metastasis Rev.* 27, 481–494, 2008.
341. Li, X., Conklin, L., and Alex, P., New serological biomarkers of inflammatory bowel disease, *World J. Gastroenterol.* 14, 5115–5124, 2008.
342. Waiker, S.S. and Bonventre, J.V., Biomarkers for the diagnosis of acute kidney injury, *Nephron. Clin. Pract.* 109, c192–c197, 2008.
343. Feldmann, M., Brennan, F.M., Elliott, M. et al., TNF alpha as a therapeutic target in rheumatoid arthritis, *Circ. Shock.* 43, 179–184, 1994.
344. Feldmann, M. and Maini, R.N., Anti-TNFα therapy of rheumatoid arthritis: What have we learned?, *Annu. Rev. Immunol.* 19, 163–196, 2001.
345. Clark, I.A., How TNF was recognized as a key mechanism of disease, *Cytokine Growth Factor Rev.* 18, 335–343, 2007.
346. Gutierrez-Roelens, I. and Lauwerys, B.R., Genetic susceptibility to autoimmune disorders: Clues from gene association and gene expression studies, *Curr. Mol. Med.* 8, 551–561, 2008.
347. Feitsma, A.L., van der Helm-van Mil, A.H., Huizinga, T.W. et al., Protection against rheumatoid arthritis by HLA: Nature and nurture, *Ann. Rheum. Dis.* 67(Suppl 3), iii61–iii63, 2008.
348. Pratt, A.G., Isaacs, J.D., and Mattey, D.L., Current concepts in the pathogenesis of early rheumatoid arthritis, *Best Pract. Res. Clin. Rheumatol.* 23, 37–48, 2009.

349. Landewé, R., Predictivie markers in rapidly progressing rheumatoid arthritis, *J. Rheumatol. Suppl.* 80, 8–15, 2007.

350. Smolen, J.S., Aletaha, D., Grisar, J. et al., The need for prognosticators in rheumatoid arthritis. Biological and clinical markers: Where are we now?, *Arthritis Res. Ther.* 10, 208, 2008.

351. van den Broek, T., Tesser, J.R., and Albani, S., The evolution of biomarkers in rheumatoid arthritis: From clinical research to clinical care, *Expert Opin. Biol. Ther.* 8, 1773–1785, 2008.

352. Kiely, P., Williams, R., Walsh, C., and Young, A., Contemporary patterns of care and disease activity in early rheumatoid arthritis: The ERAN cohort, *Rheumatology* 48, 57–60, 2009.

353. Abramson, S.B., Attur, M., and Yazici, Y., Prospects for disease modification in osteoarthritis, *Nat. Clin. Pract. Rheumatol.* 2, 304–312, 2006.

354. Lundquist, L., Abatacept: A novel therapy approved for the treatment of patients with rheumatoid arthritis, *Adv. Ther.* 24, 333–345, 2007.

355. Sibilia, J., Gottenberg, J.E., and Mariette, X., Rituximab: A new therapeutic alternative in rheumatoid arthritis, *Joint Bone Spine* 75, 526–532, 2008.

356. Ranganathan, P., An update on pharmacogenomics in rheumatoid arthritis with a focus on TNF-blocking agents, *Curr. Opin. Mol. Ther.* 10, 562–567, 2008.

357. Ernst, E., Frankincense: Systematic review, *BMJ* 337, a2813, 2008.

358. Vliet Vlienland, T.P. and Pattison, D., Non-drug therapies in early rheumatoid arthritis, *Best Pract. Res. Clin. Rheumatol.* 23, 103–116, 2009.

359. Brent, L.H., Inflammatory arthritis: An overview for primary care physicians, *Postgrad. Med.* 121, 148–162, 2009.

360. Rubbert-Roth, A. and Finckh, A., Treatment options in patients with rheumatoid arthritis failing initial TNF inhibitor therapy: A critical review, *Arthritis Res.* 11(Suppl 1), S1, 2009.

361. Bansback, N., Marra, C.A., Finckh, A., and Anis, A., The economics of treatment in early rheumatoid arthritis, *Best Pract. Res. Clin. Rheumatol.* 23, 83–92, 2009.

362. Rosseau, J.C., Sangdell, L.C., Demas, P.D., and Garnero, P., Development and clinical application in arthritis of a new immunoassay for serum type IIa procollagen NH_2 propeptide, *Methods Mol. Med.* 101, 25–37, 2004.

363. Ingelgnoli F., Del Papa, N., Comina, D.P. et al., Autoantibodies to chromatin: Prevalence and clinical significance in juvenile rheumatoid arthritis, *Clin. Exp. Rheumatol.* 22, 499–501, 2004.

364. Low, J.M., Chaven, A.K., Kietz, D.A. et al., Determination of anti-cyclic citrullinated peptide antibodies in the sera of patients with juvenile idiopathic arthritis, *J. Rheumatol.* 31, 1829–1833, 2004.

365. Lindqvist, E., Eberhardt, K., Bendtzen, K. et al., Prognostic laboratory markers of joint damage in rheumatoid arthritis, *Ann. Rheum. Dis.* 64, 196–201, 2005.

366. Ates, A., Kihikli, G., Turgay, M., and Durnan, N., The levels of serum-soluble Fas in patients with rheumatoid arthritis and systemic sclerosis, *Clin. Rheumatol.* 23, 421–425, 2004.

367. Busso, N., Wagtmann, N., Herling, C. et al., Circulating CD26 is negatively associated with inflammation in human and experimental arthritis, *Am. J. Pathol.* 166, 433–442, 2005.

368. Deberg, M., Labasse, A., Christgua, S. et al., New serum biochemical markers (Coll 2-1 and Coll 2-1 NO_2) for studying oxidative-related type II collagen network degradation in patients with osteoarthritis and rheumatoid arthritis, *Osteoarthritis Cartilage* 13, 258–265, 2005.

369. Hueber, W., Utz, P.J., Steinman, L., and Robinson, W.H., Autoantibody profiling for the study and treatment of autoimmune disease, *Arthritis Res.* 4, 290–295, 2005.

370. Xiang, Y., Sekine, T., Nakamura, H. et al., Identification of triose phosphate isomerase as an autoantigen in patients with osteoarthritis, *Arthritis Rheum.* 50, 1511–1521, 2004.

371. Rantapää-Dahlqvist, S., Diagnostic and prognostic significance of autoantibodies in early rheumatoid arthritis, *Scand. J. Rheumatol.* 34, 83–96, 2005.

372. Marshall, K.W., Zhang, H., Yager, T.D. et al., Blood-based biomarkers for detecting mild osteoarthritis in the human knee, *Osteoarthritis Cartilage* 13, 861–871, 2005.

373. Butbul-Aviel, Y., Koren, A., Halvey, R., and Sakran, W., Procalcitonin as a diagnostic aid in osteomyelitis and septic arthritis, *Pediatr. Emerg. Care* 21, 828–832, 2005.

374. Hügle, T., Schuetz, P., Mueller, B. et al., Serum procalcitonin for discrimination between septic and non-septic arthritis, *Clin. Exp. Rheumatol.* 26, 453–456, 2008.

375. Maury, E.E. and Flores, R.H., Acute monarthritis: Diagnosis and management, *Prim. Care* 33, 779–793, 2006.

376. Allantaz, F., Chaussabel, D., Stichweb, D. et al., Blood leukocyte microarrays to diagnose systemic onset juvenile idiopathic arthritis and follow the response to IL-1 blockage, *J. Exp. Med.* 204, 2131–2144, 2007.

377. Wild, N., Karl, J., Grunert, V.P. et al., Diagnosis of rheumatoid arthritis: Multivariate analysis of biomarkers, *Biomarkers* 13, 88–105, 2008.

378. Tedesco, A., D'Agostino, D., Soriente, I. et al., A new strategy for the early diagnosis of rheumatoid arthritis: A combined approach, *Autoimmun. Rev.* 8, 233–237, 2009.

379. Matteson, E.L., Lowe, V.J., Prendergast, F.G. et al., Assessment of disease activity in rheumatoid arthritis using a novel folate targeted radiopharmaceutical FolateScan™, *Clin. Exp. Rheumatol.* 27, 253–259, 2009.

380. Sinz, A., Bantscheff, M., Nikkat, S. et al., Mass spectrometric proteome analyses of synovial fluids and plasmas from patients suffering from rheumatoid arthritis and comparison to reactive arthritis or osteoarthritis, *Electrophoresis* 23, 3445, 2002.

381. Drynda, S., Ringel, B., Kekow, M. et al., Proteome analysis reveals disease-associated marker proteins to differentiate RA patients from other inflammatory joint diseases with the potential to monitor anti-TNFalpha therapy, *Pathol. Res. Pract.* 200, 165, 2004.

382. Bliddal, H., Boesen, M., Christensen, R. et al., Imaging as a follow-up tool in clinical trials and clinical practice, *Best Pract. Res. Clin. Rheumatol.* 22, 1109–1126, 2008.

383. Kubassova, O., Boesen, M., Peloschek, P. et al., Quantifying disease activity and damage by imaging in rheumatoid arthritis and osteoarthritis, *Ann. N.Y. Acad. Sci.* 1154, 207–238, 2009.

384. Merchant, M. and Weinberger, S.R., Recent advancements in surface-enriched laser desorption ionization time-of-flight-mass spectrometry, *Electrophoresis* 21, 1164, 2000.

385. Tang, N., Tornatore, P., and Weinberger, S.R., Current developments in SELDI affinity technology, *Mass Spectrom. Rev.* 23, 34, 2004.

386. Uchida, T., Fukawa, A., Uchida, M., Fujito, K., and Saito, K., Application of a novel protein biochip technology for detection and identification of rheumatoid arthritis biomarkers in synovial fluid, *J. Proteome Res.* 1, 495, 2002.

387. Berntzen, H.B., Olmez, U., Fagerhol, M.K., and Monthe, E., The leukocyte protein L1 in plasma and synovial fluid from patients with rheumatoid arthritis and osteoarthritis, *Scand. J. Rheumatol.* 20, 74, 1991.

388. zur Weish, A.S., Foell, D., Frosch, M. et al., Myeloid related proteins MRP8'MRP14 may predict disease flares in juvenile idiopathic arthritis, *Clin. Exp. Rheumatol.* 22, 368, 2004.

389. Brun, J.G., Haga, H.J., Bøe, E. et al., Calprotectin in patients with rheumatoid arthritis: Relation to clinical and laboratory variable of disease activity, *J. Rheumatol.* 19, 859, 1992.

390. Hammer, H.B., Kvien, T.K., Glennås, A., and Melby, K., A longitudinal study of calprotectin as an inflammatory marker in patients with reactive arthritis, *Clin. Exp. Rheumatol.* 13, 59, 1995.

391. Yui, S., Nakatoni, Y., and Mikami, M., Calprotectin (S100A8/S100A9), an inflammatory protein complex from neutrophils with a broad apoptosis-inducing activity, *Biol. Pharm. Bull.* 26, 754, 2003.

392. Passey, R.J., Xu, K., Hume, D.A., and Gecsy, C.L., S100A8; emerging functions and regulation, *J. Leukoc. Biol.* 66, 549, 1999.
393. Korndörfer, I.P., Brueckner, F., and Skerra, A., The crystal structure of the human (S100A8/S100A9)2 heterotetramer, calprotectin, illustrates how conformational changes of interacting alpha-helices can determine specific association of two EF-hand proteins, *J. Mol. Biol.* 370, 887–898, 2007.
394. Fagerhol, M.K., Dale, I., and Andersson, T., Release and quantitation of a leukocyte derived protein L1, *Scand. J. Haematol.* 24, 393, 1980.
395. Vermeine, S., Van Assche, G., and Rutgeerts, P., C-reactive protein as a marker for inflammatory bowel disease, *Inflamm. Bowel Dis.* 10, 661, 2004.
396. Bruzzese, E., Rais, V., Gaudiello, G. et al., Intestinal inflammation in a frequent feature of cystic fibrosis and is reduced by probiotic administration, *Aliment. Pharmacol. Ther.* 20, 813, 2004.
397. Gravatt, M.G., Novy, M.J., Rosenfeld, R.G. et al., Diagnosis of intra-amniotic infection by proteomic profiling and identification of novel biomarkers, *JAMA* 292, 462, 2004.
398. Marenholz, I., Lovering, R.C., and Heizmann, C.W., An update for the S100 nomenclature, *Biochim. Biophys. Acta* 1763, 1282–1283, 2006.
399. Hammer, H.B., Haavardsholm, E.A., and Kviein, T.K., Calprotectin (a major leucocyte protein) is associated with the levels of anti-CCP and rheumatoid factor in a longitudinal study of patients with very early rheumatoid arthritis, *Scand. J. Rheumatol.* 37, 179–182, 2008.
400. Chen, Y.S., Yan, W., Geczy, C.L. et al., Serum levels of soluble receptor for advanced glycation end products and of S100 proteins are associated with inflammatory, autoantibody, and classical risk markers of joint and vascular damage in rheumatoid arthritis, *Arthritis Res. Ther.* 11, R39, 2009.
401. Ott, H.W., Lindner, H., Sarg, B. et al., Calgranulins in cystic fluid and serum from patients with ovarian carcinomas, *Cancer Res.* 63, 7507–7514, 2003.
402. Stulik, J., Osterreicher, J., Koupilova, K. et al., The analysis of S100A9 and S100A8 expression in matched sets of macroscopically normal colon mucosa and colorectal carcinoma: The S100A9 and S100A8 positive cells underlie and invade tumor mass, *Electrophoresis* 30, 1047–1054, 1999.
403. Melle, C., Ernst, G., Schimmel, B. et al., A technical triade for proteomic identification and characterization of cancer biomarkers, *Cancer Res.* 64, 4099–4104, 2004.
404. Lindahl, M., Irander, K., Tagesson, C., and Stahlbom, B., Nasal lavage fluid and proteomics as means to identify the effects of the irritating epoxy chemical dimethylbenzylamine, *Biomarkers* 9, 56–70, 2004.
405. Damms, A. and Bischoff, S.C., Validation and clinical significance of a new calprotectin rapid test for the diagnosis of gastrointestinal diseases, *Int. J. Colorectal Dis.* 23, 985–992, 2008.
406. Quail, M.A., Russell, R.K., Van Limbergen, J.E. et al., Fecal calprotectin complements routine laboratory investigations in diagnosing childhood inflammatory bowel disease, *Inflamm. Bowel Dis.* 15, 756–759, 2009.
407. Ho, G.T., Lee, H.M., Brydon, G. et al., Fecal calprotectin predicts the clinical course of actute severe ulceratie colitis, *Am. J. Gastroenterol.* 104, 673–678, 2009.
408. Gisbert, J.P., Bermejo, F., Pérez-Calle, J.L. et al., Fecal calprotectin and lactoferrin for the prediction of inflammatory bowel disease relapse, *Inflamm. Bowel Dis.* 15, 1190–1198, 2009.
409. Lamb, C.A., Mohiuddin, M.K., Gicquel, J. et al., Faecal calprotectin or lactoferrin can identify postoperative recurrence in Crohn's disease,. *Br. J. Surg.* 96, 663–674, 2009.
410. Bouma, G., Lam-Tse, W.K., Wierenga-Wolf, A.F., Drexhage, H.A., and Versnel, M.A., Increased serum levels of MRP-8/14 in type 1 diabetes induce an increased expression of CD11b and an enhanced adhesion of circulating monocytes to fibronectin, *Diabetes* 53, 1979–1986, 2004.

411. Lundy, F.T., Chalk, R., Lamey, P.J., Shaw, C., and Linden, G.J., Quantitative analysis of MRP-8 in gingival crevicular fluid in periodontal health and disease using microbore HPLC, *J. Clin. Periodontol.* 28, 1172–1177, 2001.

412. Anderson, H.C. and McCarty, M., The occurrence in the rabbit of an acute phase protein analogous to human C reactive protein, *J. Exp. Med.* 93, 25–36, 1951.

413. Kushner, I. and Rzewnicki, D., Acute phase response, in *Inflammation Basic Principles and Clinical Correlates*, 3rd edn., J.I. Gallin, R. Snyderman, D.T. Fecron, B.F. Haynes, and C. Nathan (eds.), Lippincott, Williams & Wilkins, Philadelphia, PA, Chapter 20, pp. 317–329, 1999.

414. Kaplan, M.A. and Lebretti, A., The role of C-reactive protein in allergic inflammation; relationship between the acute phase response and the antibody titer, *J. Allergy* 27, 450–460, 1956.

415. Cooper, E.H. and Ward, A.M., Acute phase reactant protein as aids to monitoring disease, *Invest. Cell Pathol.* 2, 293–301, 1979.

416. Pepys, M.B., C-reactive protein. A review of it structure and function, *Eur. J. Rheumatol. Inflamm.* 5, 386–397, 1982.

417. Kushner, I. and Mackiewicz, A., The acute phase response: An overview, in *Acute Phase Proteins Molecular Biology, Biochemistry, and Clinical Applications*, A. Mackiewicz, I. Kushner, and H. Baumann (eds.), CRC Press, Boca Raton, FL, 1993.

418. Morera, A.L., Henry, M., Garcia-Hernández, A., and Fernandez-López, L., Acute phase proteins as biological markers of negative psychopathology in paranoid schizophrenia, *Actas Esp. Psiquiatr.* 35, 249–252, 2007.

419. Fan, X., Pristach, C., Liu, E.Y. et al., Elevated serum levels of C-reactive protein are associated with more serve psychopathology in a subgroup of patients with schizophrenia, *Psychiatry Res.* 149, 267–271. 2007.

420. Waterson, C., Lanevsch, A., Horner, J., and Louden, C., A comparative analysis of acute-phase proteins as inflammatory biomarkers in preclinical toxicology studies: Implications for preclinical to clinical translation, *Toxicol. Pathol.* 37, 28–33, 2009.

421. Suntharalingam, G., Perry, M.R., Ward, S. et al., Cytokine storm in a phase I trial of the anti-CD28 monoclonal antibody TGN1412, *N. Engl. J. Med.* 355, 1018–1028, 2006.

422. Gribble, E.J., Sivakumar, P.V., Ponce, R.A. et al., Toxicity as a result of immunostimulation by biologics, *Expert Opin. Drug Metab. Toxicol.* 3, 209–234, 2007.

423. St. Clair, E.W., The calm after the cytokine storm: Lessons from the TGN1412 trial, *J. Clin. Invest.* 118, 1344–1347, 2008.

424. Muller, P.Y. and Brennan, F.R., Safety assessment and dose selection for first-clinical trials with immunomodulatory monoclonal antibodies, *Clin. Pharmacol. Ther.* 85, 247–258, 2009.

425. Emery, P., McInnes, I.B., van Vollenhoven, R., and Kraan, M.C., Clinical identification and treatment of a rapidly progressing disease state in patients with rheumatoid arthritis, *Rheumatology* 47, 392–398, 2008.

426. Feuchtenberger, M., Kneitz, C., Rooo, P. et al., Sustained remission after combination therapy with rituximab and etanercept in two patients with rheumatoid arthritis after TNF failure: Case report, *Open Rheumatol. J.* 3, 9–13, 2009.

427. Sano, H., Arai, K., Murai, T. et al., Tight control is important in patients with rheumatoid arthritis treated with an anti-tumor necrosis factor biological agent: Prospective study of 91 cases who used a biological agent for more than 1 year, *Mod. Rheumatol.* 19, 390–394, 2009.

428. May, A. and Wang, T.J., Evaluating the role of biomarkers for cardiovascular risk prediction: Focus on CRP, BNP and urinary microalbumin, *Expert Rev. Mol. Diagn.* 7, 793–804, 2007.

429. Casas, J.P., Shan, T., Hingorani, A.D. et al., C-reactive protein and coronary heart disease: A critical review, *J. Intern. Med.* 264, 295–314, 2008.

430. Shah, T., Casas, J.P., Cooper, J.A. et al., Critical appraisal of CRP measurement for the prediction of coronary heart disease events: New data and systematic review of 31 prospective cohorts, *Int. J. Epidemiol.* 38, 217–231, 2009.
431. Herzum, I. and Renz, H., Inflammatory markers in SIRS, sepsis and septic shock, *Curr. Med. Chem.* 15, 581–587, 2008.
432. Attreya, R. and Neurath, M.F., Involvement of the IL-6 in the pathogenesis of inflammatory bowel disease and colon cancer, *Clin. Rev. Allergy Immunol.* 28, 187–196, 2005.
433. Fain, J.N., Release of interleukins and other inflammatory cytokines by human adipose tissue is enhanced in obesity and primarily due to the nonfat cells, *Vitam. Horm.* 74, 443–477, 2006.
434. de Ferranti, S.D. and Rifai, N., C-reactive protein: A nontraditional serum marker of cardiovascular risk, *Cardiovasc. Pathol.* 16, 14–21, 2007.
435. Hingorani, A.D., Shat, T., Casas, J.P. et al., C-reactive protein and coronary heart disease: Predictive test or therapeutic target?, *Clin. Chem.* 55, 239–355, 2009.
436. Shen, J. and Ordovas, J.M., Impact of genetic and environmental factors on hsCRP concentrations and response to therapeutic agents, *Clin. Chem.* 55, 256–264, 2009.
437. Lamina, S., Okoye, C.G., and Dagogo, T.T., Managing erectile dysfunction in hypertension: The effects of a continuous training program on biomarker of inflammation, *BJU Int.* 103, 1218–1221, 2009.
438. Fung, E.T., Wilson, A.M., Zhang, F. et al., A biomarker panel for peripheral arterial disease, *Vasc. Med.* 13, 217–224, 2008.
439. Packard, R.R. and Libby, P., Inflammation in atherosclerosis: From vascular biology to biomarker discovery and risk prediction, *Clin. Chem.* 54, 24–38, 2008.
440. Apple, F.S., Pearce, L.A., Chung, A. et al., Multiple biomarkers use for detection of adverse events in patients presenting with symptoms suggestive of acute coronary syndrome, *Clin. Chem.* 53, 874–881, 2007.
441. Weisman, M.H., Durez, P., Hallegua, D. et al., Reduction of inflammatory biomarker response by abatacept in treatment of rheumatoid arthritis, *J. Rheumatol.* 33, 2162–2166, 2006.
442. Willcox, B.J., Abbott, R.D., Yano, K. et al., C-reactive protein, cardiovascular disease and stroke: New roles for an old biomarker, *Expert Rev. Neurother.* 4, 507–518, 2004.
443. Andersson, J., Johansson, L., Ladenvall, P. et al., C-reactive protein is a determinant of first-ever stroke: Prospective nested care-referent study, *Cerebrosvasc. Dis.* 27, 544–551, 2009.
444. Idicula, T.T., Brogger, J., Naess, H. et al., Admission C-reactive protein after acute ischemic stroke is associated with stroke severity and mortality: The 'Bergen stroke study', *BMC Neurol.* 9, 18, 2009.
445. Casamassima, A., Picciariello, M., Quaranta, M. et al., C-reactive protein: A biomarker of survival in patients with metastatic renal cell carcinoma treated with subcutaneous interleukin-2 based immunotherapy, *J. Urol.* 173, 52–55, 2005.
446. Sowers, M., Jannausch, M., Stein, E. et al., C-reactive protein as a biomarker of emergent osteoarthritis, *Osteoarthritis Cartilage* 10, 595–601, 2002.
447. Berliner, S., Zeltser, D., Shapira, I. et al., A simple biomarker to exclude the presence of low grade inflammation in apparently healthy individuals, *J. Cardiovasc. Risk* 9, 281–286, 2002.
448. Eroglu, S., Elif Sade, L., Yildirir, A. et al., Serum levels of C-reactive protein and uric acid in patients with cardiac syndrome X, *Acta Cardiol.* 64, 207–211, 2009.
449. Otake, T., Uezono, K., Takahashi, R. et al., C-reactive protein and colorectal adenomas: Self Defense Forces Health Study, *Cancer Sci.* 100, 709–714, 2009.
450. Nguyen, X.M., Lane, J., Smith, B.R., and Nguyen, N.T., Changes in inflammatory biomarkers across weight classes in a representative US population: A link between obesity and inflammation, *J. Gastrointest. Surg.* 13, 1205–1212, 2009.

451. Snaedal, S., Heimbürger, O., Qureshi, A.R. et al., Comorbidity and acute clinical events as determinants of C-reactive protein variation in hemodialysis patients: Implications for patient survival, *Am. J. Kidney Dis.* 53, 1024–1033, 2009.

452. Müller, B., Schuetz, P., and Trampuz, A., Circulating biomarkers as surrogates for bloodstream infections, *Int. J. Antimicrob. Agents* 30(Suppl 1), S16–S23, 2007.

453. Mageed, R.A., Børretzen, M., Moyes, S.P. et al., Rheumatoid factor autoantibodies in health and disease, *Ann. N. Y. Acad. Sci.* 815, 296–311, 1997.

454. Sutton, B., Corper, A., Bonagura, V., and Taussig, M., The structure and origin of rheumatoid factors, *Immunol. Today* 21, 177–183, 2000.

455. Epstein, W., Johnson, A., and Ragan, C., Observations on a precipitin reaction between serum of patients with rheumatoid arthritis and a preparation (Cohn fraction II), *Proc. Soc. Exp. Biol. Med.* 91, 235–237, 1956.

456. Epstein, W. and Ragan, C., Serological reactions in rheumatoid arthritis, *Am. J. Med.* 20, 487–489, 1956.

457. Edelman, G.M., Kunkel, H.G., and Franklin, E.C., Interaction of the rheumatoid factor with antigen-antibody complexes and aggregated gamma globulin, *J. Exp. Med.* 108, 105–120, 1958.

458. Franklin, E.C., Kunkel, H.G., and Ward, J.R., Clinical studies of seven patients with rheumatoid arthritis and uniquely large amounts of rheumatoid factor, *Arthritis Rheum.* 1, 400–409, 1958.

459. Kunkel, H.G., Franklin, E.D., and Muller-Eberhard, H.J., Studies on the isolation and characterization of the "rheumatoid factor", *J. Clin. Invest.* 38, 424–434, 1959.

460. Lospalluto, J. and Ziff, M., Chromatographic studies of the rheumatoid factor, *J. Exp. Med.* 110, 169–186, 1959.

461. Ribas, J.L., da Rosa Utiyama, S.R., Nisihara, R.M. et al., High prevalence of rheumatoid factor associated with clinical manifestations of rheumatic disease in Kaingang and Guarini Indians from Southern Brazil, *Rheumatol. Int.* 29, 427–430, 2009.

462. Attar, S.M., Bunting, P.S., Smith, C.D., and Karsh, J., Comparison of the anti-cyclic citrullinated peptide and rheumatoid factor in rheumatoid arthritis at an arthritis center, *Saudi Med. J.* 30, 446–447, 2009.

463. Reneses, S., González-Escribano, M.F., Fernández-Suárez, A. et al., The value of HLA-DRB1 shared epitope,-308 tumor necrosis factor-α, gene promoter polymorphism, Rheumatoid factor, anti-citrullinated peptide antibodies, and early erosion for predicting radiological outcome in recent-onset rheumatoid arthritis, *J. Rheumatol.* 36, 1143–1149, 2009.

464. Sokka, T. and Pincus, T., Erythrocyte sedimentation rate, C-reactive protein, or Rheumatoid factor are normal at presentation in 35%–45% of patients with rheumatoid arthritis seen between 1980 and 2004: Analysis from Finland and the United States, *J. Rheumatol.* 36, 1143–1149, 2009.

465. Lundberg, I.E. and Nader, G.A., Molecular effects of exercise in patients with inflammatory rheumatic disease, *Nat. Clin. Pract. Rheumatol.* 4, 597–604, 2008.

466. Scott, D.L., What have we learnt about the development and progression of early RA from RCTs?, *Best Pract. Res. Clin. Rheumatol.* 23, 13–24, 2009.

467. Machado, P., Santos, A., Pereira, C. et al., Increased prevalence of allergic sensitization in rheumatoid arthritis patients treated with anit-TNFα, *Joint Bone Spine* 76, 508–513, 2009.

468. Ranzolin, A., Brenol, J.C., Bredemeier, M. et al., Association of concomitant fibromyalgia with worse disease activity score in 28 joints, health assessment questionnaire, and short form 36 scores in patients with rheumatoid arthritis, *Arthritis Rheum.* 61, 794–800, 2009.

469. Saha, N. and Banerjee, B., Correlation of erythrocyte sedimentation rate with serum composition in the tropics, *Trop. Geogr. Med.* 23, 30–34, 1971.

470. Fearon, K.C., Barber, M.D., Falconer, J.S. et al., Pancreatic cancer as a model: Inflammatory mediators, acute-phase response, and cancer, *World J. Surg.* 23, 584–588, 1999.

471. Willemze, A., Ioan-Facsinay, A., and El-Gabalawy, H., Anti-citrullinated protein antibody response associated with synovial immune deposits in a patient with suspected early rheumatoid arthritis, *J. Rheumatol.* 35, 2282–2284, 2008.

472. Okumrua, N., Haneishi, A., Teresawa, M.M. et al., Citrullinated fibrinogen shows defects in FPA and FPB release and fibrin polymerization catalyzed by thrombin, *Clin. Chim. Acta* 401, 119–123, 2009.

473. Uysal, H., Bockermann, R., Nandakumar, K.S. et al., Structure and pathogenicity of antibodies specific for citrullinated collagen type II in experimental arthritis, *J. Exp. Med.* 206, 449–462, 2009.

474. Goëb, V., Tron, F., Gilbert, D. et al., Candidate autoantigens identified by mass spectrometry in early rheumatoid arthritis are chaperones and citrullinated glycolytic enzymes, *Arthritis Res. Ther.* 11, R38, 2009.

475. Szodoray, P., Szabó, Z., Kapitány, A. et al., Anti-citrullinated protein/peptide autoantibodies in association with genetic and environmental factors as indicators of disease outcome in rheumatoid arthritis, *Autoimmun. Rev.* 9, 140–143, 2010.

476. Schellekens, G.A., de Jong, B.A.W., van den Hoogen, F.H.J. et al., Citrulline is an essential constituent of antigenic determinants recognized by rheumatoid arthritis-specific autoantibodies, *J. Clin. Invest.* 101, 273–281, 1998.

477. Scheellekens, G.A., Visser, H., de Jong, B.A.W. et al., The diagnostic properties of rheumatoid arthritis antibodies recognizing a cyclic citrullinated peptide, *Arthritis Rheumat.* 43, 155–163, 2000.

478. Bas, S., Perneger, T.V., Seitz, M. et al., Diagnostic tests for rheumatoid arthritis: Comparison of anti-cyclic citrullinated peptide antibodies, anti-keratin antibodies and IgM rheumatoid factors, *Rheumatology* 41, 809–814, 2002.

479. Rojas-Serrano, J., Burgos-Vargas, R., Lino Pérez, L. et al., Very recent onset arthritis: The value of initial rheumatologist evaluation and anti-cyclic citrullinated peptide antibodies in the diagnosis of rheumatoid arthritis, *Clin. Rheumatol.* 28, 1135–1139, 2009.

480. Varadé, J., Loza-Santamaria, E., Fernández-Arquero, M. et al., Shared epitope and anti-cyclic citrullinated peptide antibodies: Relationship with age at onset and duration of disease in rheumatoid arthritis, *J. Rheumatol.* 36, 1085–1086, 2009.

481. Babey, M., Aeberli, D., Vajtai, I. et al., Acute onset of polyarthralgia and high anti-cyclic citrullinated peptide antibodies in a case of idiopathic granulomatous hypophysitis, *J. Rheumatol.* 36, 204–207, 2009.

482. Labrador-Horrillo, M., Martinez, M.A., Selva-O-Callaghan, A. et al., Anti-cyclic citrullinated peptide and anti-keratin antibodies in patients with idiopathic inflammatory myopathy, *Rheumatology* 48, 676–679, 2009.

483. Zing, Y.V., Zhang, Q.B., Zhou, J.G. et al., The detecting and clinical value of anti-cyclic citrullinated peptide antibodies in patients with systemic lupus erythematosus, *Lupus* 18, 713–717, 2009.

484. Fearon, W.R., The carbamido diacetyl reaction: A test for citrulline, *Biochem. J.* 33, 902–907, 1939.

485. Rogers, G.E., Occurrence of citrulline in proteins, *Nature* 194, 1149–1151, 1962.

486. György, B., Tóth, E., Tarcsa, E. et al., Citrullination: A post translational modification in health and disease, *Int. J. Biochem. Cell Biol.* 38, 1662–1677, 2006.

487. Vossenaar, E.R., Zendman, A.J., van Venrooij, W.J. et al., PAD, a growing family of citrullinating enzymes: Genes, features and involvement in disease, *Bioessays* 25, 1106–1118, 2003.

488. Shmerling, R.H., Testing for anti-cyclic citrullinated peptide antibodies, *Arch. Intern. Med.* 169, 9–14, 2009.

489. Dejaco, C., Duftner, C., Klotz, W. et al., Third generation anti-cyclic citrullinated peptide antibodies do not predict anti-TNFα treatment response in rheumatoid arthritis, *Rheumatol. Int.* 30, 451–454, 2010.

490. Saxne, T., Månsson, B., and Heinegård, D., Biomarkers for cartilage and bone in rheumatoid arthritis, in *Rheumatoid Arthritis*, 2nd edn., G.S. Firestein, G.S. Pavayi, and F.A. Wollheim (eds.), Oxford University Press, Oxford, U.K., 2006.
491. Singer, F.R. and Eyre, D.R., Using biochemical markers of bone turnover in clinical practice, *Cleve. Clin. J. Med.* 75, 739–750, 2008.
492. Rosseau, J.C. and Delmas, P.D., Biological markers in osteoarthritis, *Nat. Clin. Pract. Rheumatol.* 3, 346–356, 2007.
493. Garnero, P., Use of biochemical markers to study and follow patients with osteoarthritis, *Curr. Rheumatol. Rep.* 8, 37–44, 2006.
494. Rannou, F., Francois, M., Corvol, M.T., and Berenbaum, F., Cartilage breakdown in rheumatoid arthritis, *Joint Bone Spine* 73, 29–36, 2006.
495. de Seny, D., Fillet, M., Meuwis, M.A. et al., Discovery of new rheumatoid arthritis biomarkers using the surface-enhanced laser desorption/ionization time-of-flight mass spectrometry proteinchip approach, *Arthritis Rheum.* 52, 3801–3812, 2005.
496. Dotzlaw, H., Schulz, M., Eggert, M., and Neeck, G., A pattern of protein expression in peripheral blood mononuclear cells distinguishes rheumatoid arthritis patients from healthy individuals, *Biochim. Biophys. Acta* 1696, 121–129, 2004.
497. Guo, D., Tan, W., Wang, F. et al., Proteomic analysis of human articular cartilage: Identification of differentially expressed proteins in knee osteoarthritis, *Joint Bone Spine* 75, 439–444, 2008.
498. Sekigawa, I., Yanagida, M., Iwabuchi, K. et al., Protein biomarker analysis by mass spectrometry in patients with rheumatoid arthritis receiving anti-tumor necrosis factor-alpha therapy, *Clin. Exp. Rheumatol.* 26, 261–267, 2008.
499. Zheng, X., Wu, S.L., Hincapie, M., and Hancock, W.S., Study of the human plasma proteome of rheumatoid arthritis, *J. Chromatogr. A* 1216, 3538–3545, 2009.
500. Chang, X., Cui, Y., Zong, M. et al., Identification of proteins with increased expression in rheumatoid arthritis synovial tissues, *J. Rheumatol.* 36, 872–880, 2009.
501. Hueber, W., Tomooka, B.H., Batliwalla, F. et al., Blood autoantibody and cytokine profiles predict response to anti-tumor necrosis factor therapy in rheumatoid arthritis, *Arthritis Res. Ther.* 11, R76, 2009.
502. Loos, B.G., Systemic markers of inflammation in periodontitis, *J. Periodontol.* 76(11 Suppl), 2106–2115, 2005.
503. Wang, T.Y., Al-Jaroudi, W.A., and Newby, C.K., Markers of cardiac ischemia and inflammation, *Cardiol. Clin.* 23, 491–501, 2005.
504. Elkind, M.S.V., Inflammation, atherosclerosis, and stroke, *Neurologist* 12, 140–148, 2006.
505. Maggio, M., Basaria, S., Ceda, G.P. et al., The relationship between testosterone and molecular markers of inflammation in older men, *J. Endocrinol. Invest.* 28(11 Suppl proc), 116–119, 2005.
506. Konikoff, M.R. and Denson, L.A., Role of fecal calprotectin as a biomarker of intestinal inflammation in inflammatory bowel disease, *Inflamm. Bowel Dis.* 12, 524–536, 2006.
507. Ray, K.K., Biomarkers, C-reactive proteins and statins in acute coronary syndromes, *Fundam. Clin. Pharmacol.* 21(Suppl 2), 31–33, 2007.
508. Gortney, J.S. and Sanders, R.M., Impact of C-reactive protein on treatment of patients with cardiovascular disease. *Am. J. Health. Sys. Pharm.* 64, 2009–2016, 2007.
509. Gottsäter, A., Flondell-Site, D., Kölbel, T., and Lindblad, B., Associations between statin treatment and markers of inflammation, vasoconstriction, and coagulation in patients with abdominal aortic aneurysm, *Vasc. Endovasc. Surg.* 42, 567–573, 2008.
510. Elkind, M.S., Inflammatory markers and stroke, *Curr. Cardiol. Rep.* 11, 12–20, 2009.
511. Nøjgaard, C., Høst, N.B., Christensen, I.J. et al., Serum levels of YKL-40 increases in patients with acute myocardial infarction, *Coron. Artery Dis.* 19, 257–263, 2008.

512. Rathcke, C.N., Raymond, I., Kistorp, C. et al., Low grade inflammation as measure by YKL-40: Association with an increased overall and cardiovascular mortality rate in an elderly population, *Int. J. Cardiol.*, in press, 2009.

513. Roslind, A. and Johansen, J.S., YKL-40: A novel marker shared by chronic inflammation and oncogenic transformation, *Methods Mol. Biol.* 511, 159–184, 2009.

514. Lemiere, C., Induced sputum and exhaled nitric oxide as noninvasive markers of airway inflammation from work exposure, *Curr. Opin. Allergy Clin. Immunol.* 7, 133–137, 2007.

515. Minderhoud, I.M., Sansom, M., and Oldenburg, B., What predicts mucosal inflammation in Crohn's disease patients?, *Inflamm. Bowel Dis.* 13, 1567–1572, 2007.

516. Virani, S.S., Polsani, V.R., and Nambi, V., Novel markers of inflammation in atherosclerosis, *Curr. Atheroscler. Rep.* 10, 164–170, 2008.

517. Salas-Salvadó, J., Casas-Agustench, P., Murphy, M.M. et al., The effect of nuts on inflammation, *Asia Pac. J. Clin. Nutr.* 17(Suppl 1), 333–336, 2008.

518. Son, J., Pang, G., McFallin, J.L. et al., Surveying the damage: The challenge of developing nucleic acid biomarkers of inflammation, *Mol. Biosyst.* 4, 902–908, 2008.

519. Forsythe, L.K., Wallace, J.M., and Livingstone, M.B., Obesity and inflammation: The effects of weight loss, *Nutr. Res. Rev.* 21, 117–133, 2008.

520. McMillan, D.C., Systemic inflammation, nutritional status and survival in patients with cancer, *Curr. Opin. Clin. Nutr. Metab. Care* 12, 223–226, 2009.

521. Steinvil, A., Shirom, A., Melamed, S. et al., Relation of educational level to inflammation-sensitive biomarker level, *Am. J. Cardiol.* 102, 1034–1039, 2008.

522. Wan, J., Sun, W., Li, X. et al., Inflammation inhibitors were remarkably up-regulated in plasma of sever respiratory syndrome patients at progressive phase, *Proteomics* 6, 2886–2894, 2006.

523. Liu, T., Qian, W.J., Gritsenko, M.A. et al., High dynamic range characterization of the trauma patient plasma proteome, *Mol. Cell. Proteomics* 5, 1899–1913, 2006.

524. de Torre, C., Ying, S.X., Munson, P.J. et al., Proteomic analysis of inflammatory biomarkers in bronchoalveolar lavage, *Proteomics* 6, 3949–3957, 2006.

525. Lull, M.E., Freeman, W.M., Myers, J.L. et al., Plasma proteomics: A noninvasive window on pathology and pediatric cardiac surgery, *ASAIO J.* 52, 562–566, 2006.

526. Goligorsky, M.S., Clinical assessment of endothelial dysfunction: Combine and rule, *Curr. Opin. Nephrol. Hypertens.* 15, 617–624, 2006.

527. Galli, F., Protein damage and inflammation in uraemia and dialysis patients, *Nephrol. Dial. Transplant.* 22(Suppl 5), v20–v36, 2007.

528. He, Q.Y., Yang, H., Wong, B.C., and Chiu, J.F., Serological proteomics of gastritis, degradation of apolipoprotein A-I and alpha1-antitrypsin is a common response to inflammation irrespective of *Helicobacter pylori* infection, *Dig. Dis. Sci.* 53, 3112–3118, 2008.

529. Kojima, K., Asmellash, S., Klug, C.A. et al., Applying proteomic-based biomarker tools for the accurate diagnosis of pancreatic cancer, *J. Gastrointest. Surg.* 12, 1683–1690, 2008.

530. Reif, D.M., Motsinger-Reif, A.A., McKinney, B.A. et al., Integrated analysis of genetic and proteomic data identifies biomarkers associated with adverse events following smallpox vaccination, *Genes Immun.* 10, 112–119, 2009.

531. Heinecke, J.W., The HDL proteome: A marker—and perhaps mediator—of coronary artery disease, *J. Lipid Res.* 50(Suppl), S167–S171, 2009.

532. Buhimschi, C.S., Bhandari, V., Han, Y.W. et al., Using proteomics in perinatal and neonatal sepsis: Hopes and challenges for the future, *Curr. Opin. Infect. Dis.* 22, 235–243, 2009.

533. Ardoin, S.P., Shanahan, J.C., and Pisetsky, D.S., The role of microparticles in inflammation and thrombosis, *Scand. J. Immunol.* 66, 159–165, 2007.

534 Corrado, E., Rizzo, M., Muratori, I. et al., Older age and markers of inflammation are strong predictors of clinical events in women with asymptomatic carotid lesions, *Menopause* 15, 212–214, 2008.

535. Spagnoli, L.G., Bonanno, E., Sangiorgi, G., and Mauriello, A., Role of inflammation in atherosclerosis, *J. Nucl. Med.* 48, 1800–1815, 2007.

536. Hannawi, S., Haluska, B., Marwick, T.H., and Thomas, R., Atherosclerotic disease is increased in recent-onset rheumatoid arthritis: A critical role of inflammation, *Arthritis Res. Ther.* 9, R116, 2007.
537. Hansson, G.K., Robertson, A.K., and Söderberg-Nauclér, C., Inflammation and atherosclerosis, *Annu. Rev. Pathol.* 1, 297–329, 2006.
538. Akalin, A., Alatas, O., and Colak, O., Relation of plasma homocysteine levels to atherosclerotic vascular disease and inflammation markers in type 2 diabetic patients, *Eur. J. Endocrinol.* 158, 47–52, 2008.
539. Aronson, D., Avizohar, O., Levy, Y. et al., Factor analysis of risk variables associated with low-grade inflammation, *Atherosclerosis* 200, 206–212, 2008.
540. Hamirani, Y.S., Pandey, S., Rivera, J.J. et al., Markers of inflammation and coronary artery calcification: A systematic review, *Atherosclerosis* 201, 1–7, 2008.
541. Krupinski, J., Font, A., Luque, A. et al., Angiogenesis and inflammation in carotid atherosclerosis, *Front. Biosci.* 13, 6472–6482, 2008.
542. Engström, G., Hedblad, B., Tydén, P., and Lindgärde, F., Inflammation-sensitive plasma proteins are associated with increased incidence of heart failure: A population-based cohort study, *Atherosclerosis* 202, 617–622, 2009.
543. Libby, P., Role of inflammation in atherosclerosis associated with rheumatoid arthritis, *Am. J. Med.* 121(10 Suppl 1), S21–S31, 2008.
544. Rizzo, M., Corrado, E., Coppola, G. et al., Markers of inflammation are strong predictors of subclinical and clinical atherosclerosis in women with hypertension, *Coron. Artery Dis.* 20, 15–20, 2009.
545. Rathcke, C.N., Persson, F., Tarnow, L. et al., YKL-40, a marker of inflammation and endothelial dysfunction, is elevated in patients with type 1 diabetes and increases with levels of albuminuria, *Diabetes Care* 32, 323–328, 2009.
546. Bash, L.D., Erlinger, T.P., Coresh, J. et al., Inflammation, hemostasis, and the risk of kidney function decline in the atherosclerosis risk in communities (ARIC) study, *Am. J. Kidney Dis.* 53, 596–605, 2009.
547. He, K., Liu, K., Daviglus, M.L. et al., Associations of dietary long-chain *n*-3 polyunsaturated fatty acids and fish with biomarkers of inflammation and endothelial activation (from the multi-ethnic study of atherosclerosis [MESA]), *Am. J. Cardiol.* 103, 1238–1243, 2009.
548. Lehrke, M., Millington, S.C., Leferova, M. et al., CXCL16 is a marker of inflammation, atherosclerosis, and acute coronary syndromes in humans, *J. Am. Coll. Cardiol.* 49, 442–449, 2007.
549. Petrovsky, N., Winter, W.E., and Schatz, P.A., Type 1 diabetes: Immunology and genetics, in *Clinical Diabetes Translating Research into Practice*, V.A. Fonseca (ed.), Saunders/Elsevier, Philadelphia, PA, Chapter 2, pp. 6–20, 2006.
550. Winter, W.E., Harris, N., and Schatz, P.A., Type 1 diabetes islet autoantibody markers, *Diabetes Technol. Ther.* 4, 817–839, 3003.
551. Shehadeh, N., Pollack, S., Wildbaum, G. et al., Selective autoantibody production against CCL3 is associated with human type 1 diabetes mellitus and serves as a novel biomarker for its diagnosis, *J. Immunol.* 182, 8104–8109, 2009.
552. Roep, B.O., Immune markers of disease and therapeutic intervention in type 1 diabetes, *Novartis Found. Symp.* 292, 159–171, 2008.
553. Deveraj, S., Chandalia, M., and Jialal, I., Diabetes and inflammation, in *Clinical Diabetes Translating Research into Practice*, V.A. Fonseca (eds.), Saunders/Elsevier, Philadelphia, PA, Chapter 8, pp. 89–99, 2006.
554. Chavez, B.E. and Henry, R.R., Type 2 diabetes: Insulin resistance, beta cell dysfunction, and other metabolic and hormonal abnormalities, in *Clinical Diabetes Translating Research into Practice*, V.A. Fonseca (ed.), Saunders/Elsevier, Philadelphia, PA, Chapter 3, pp. 21–34, 2006.

555. Inzucchi, S.E., Classification and diagnosis of diabetes mellitus, in *Ellenberg & Rifkin's Diabetes Mellitus*, 6th edn., D. Porte Jr., R.S. Sherwin, and A. Boron (eds.), McGraw-Hill Medical, New York, Chapter 18, pp. 265–300, 2003.

556. Koch, A., Gressner, O.A., Sanson, E. et al., Serum resistin levels in critically ill patients are associated with inflammation, organ dysfunction and metabolism, and may predict survival of non-septic patients, *Crit. Care* 13, R95, 2009.

557. Morgenthaler, N.G., Struck, J., Jochberger, S., and Dünser, M.W., Copeptin: Clinical use of a new biomarker, *Trends Endocrinol. Metab.* 19, 43–49, 2008.

558. Rao, P.V., Reddy, A.P., Lu, X. et al., Proteomic identification of salivary biomarkers of type-2 diabetes, *J. Proteome Res.* 8, 239–245, 2009.

559. Farah, R., Shurtz-Swirski, R., and Lapin, O., Intensification of oxidative stress and inflammation in type 2 diabetes despite antihyperglycemic treatment, *Cardiovasc. Diabetol.* 7, 20, 2008.

560. Bertz, L., Barani, J., Gottsäter, A. et al., Are there differences in inflammatory bio-markers between diabetic and non-diabetic patients with critical limb ischemia?, *Int. Angiol.* 25, 370–377, 2006.

561. Kralisch, S., Ziegelmeier, M., Bachman, A. et al., Serum levels of the atherosclerosis biomarker sTWEAK are decrease in type 2 diabetes and end-stage renal disease, *Atherosclerosis* 199, 440–444, 2008.

562. Rathcke, C. N. and Vestergaard, H., YKL-40—an emerging biomarker in cardiovascular disease and diabetes, *Cardiovasc. Diabetol.* 8, 61, 2009.

563. Mantzoros, C.S. (ed.), *Obesity and Diabetes*, Humana Press, Totowa, NJ, 2006.

564. Anderson, R.E. (ed.), *Obesity Etiology Assessment Treatment and Prevention*, Human Kinetics Press, Champaign, IL, 2003.

565. Crespo, C.J. and Smit, E., Prevalence of overweight and obesity in the United States, in *Obesity and Diabetes*, C.S. Mantzoros (ed.), Humana Press, Totowa, NJ, Chapter 1, pp. 3–15, 2006.

566. Woodward-Lopez, G., Ritchie, L.D., Gerstein, D.E., and Crawford, P.B., *Obesity Dietary and Development Influences*, CRC Press/Taylor & Francis, Boca Raton, FL, Chapter 1, 2008.

567. Finkelstein, E.A., Trogdon, J.G., Cohen, J.W., and Dietz, W., Annual medical spending attributable to obesity: Payer- and service-specific estimates, *Health Aff.* 28, w822–w831, 2009.

568. Meigs, J.B., The role of obesity in insulin resistance. Epidemiological and metabolic aspects, in *The Metabolic Syndrome Epidemiology, Clinical Treatment, and Underlying Mechanisms*, B.C. Hansen and G.A., Bray (eds.), Humana Press, Totowa, NJ, Chapter 3, pp. 37–55, 2005.

569. Chaiken, R.L. and Banerji, M.A., Metabolic syndrome, in *Obesity and Diabetes*, C.S. Mantzoros (ed.), Chapter 9, pp. 155–168, 2006.

570. Hansen, B.C. and Bray, G.A. (eds.), *The Metabolic Syndrome Epidemiology, Clinical Treatment, and Underlying Mechanisms*, Humana Press, Totowa, NJ, 2005.

571. Zeitler, P.S. and Nadeau, K.J. (eds.), *Insulin Resistance Childhood Precursors and Adult Disease*, Humana Press, Totowa, NJ, 2008.

572. Levine, T.B. and Levine, A.B., *Metabolic Syndrome and Cardiovascular Disease*, Saunders/Elsevier, Philadelphia, PA, 2006.

573. Inadera, H., The usefulness of circulating adipokine levels for the assessment of obesity-related health problems, *Int. J. Med. Sci.* 5, 248–262, 2008.

574. de Mello, V.D., Kolehmainen, M., Pulkkinen, L. et al., Downregulation of genes involved in NFκB activation in peripheral blood mononuclear cells after weight loss is associated with the improvement of insulin sensitivity in individuals with the metabolic syndrome: The GENOBIN study, *Diabetologia* 51, 2060–2067, 2008.

575. Saltevo, J., Vanhala, M., Kautianinen, R. et al., Gender differences in C-reactive protein, interleukin-1 receptor antagonist and adiponectin levels in the metabolic syndrome: A population-based study, *Diabet. Med.* 25, 747–750, 2008.

576. Faintuch, J., Marques, P.C., Bortolotto, L.A. et al., Systemic inflammation and cardiovascular risk factors: Are morbidly obese subjects different?, *Obes. Surg.* 18, 854–862, 2008.

577. Hung, J., McQuillan, B.M., Thompson, P.L., and Beilby, J.P., Circulating adiponectin levels associate with inflammatory markers, insulin resistance and metabolic syndrome independent of obesity, *Int. J. Obes.* 32, 772–779, 2008.

578. Odrowaz-Sypniewska, G., Markers of pro-inflammatory and pro-thrombotic state in the diagnosis of metabolic syndrome, *Adv. Med. Sci.* 52, 246–250, 2007.

579. Badellino, K.O., Wolfe, M.L., Reilly, M.P., and Rader, D.J., Endothelial lipase is increased in vivo by inflammation in humans, *Circulation* 117, 678–685, 2008.

580. Lee, S., Bacha, F., Gungor, N., and Arslanian, S., Comparison of different definitions of pediatric metabolic syndrome: Relation to abdominal adiposity, insulin resistance, adiponectin, and inflammatory biomarkers, *J. Pediatr.* 152, 177–184, 2008.

581. Kraja, A.T., Province, M.A., Arnett, D. et al., Do inflammation and procoagulation biomarkers contribute to the metabolic syndrome cluster?, *Nutr. Metab.* 4, 28, 2007.

582. Chapidze, G., Dolidze, N., Enquobahrie, D.A. et al., Metabolic syndrome and C-reactive protein among cardiology patients, *Arch. Med. Res.* 38, 783–788, 2007.

583. Norberg, M., Stenlund, H., Lindahl, B. et al., Components of metabolic syndrome predicting diabetes: No role of inflammation or dyslipidemia, *Obesity* 15, 1875–1885, 2007.

584. Axelsson, J., Møller, H.J., Witasp, A. et al., Changes in fat mass correlate with changes in soluble sCD163, a marker of mature macrophages, in patients with CKD, *Am. J. Kidney Dis.* 48, 916–925, 2006.

585. Koh, K.K., Han, S.H., and Quon, M.J., Inflammatory markers and the metabolic syndrome, *J. Am. Coll. Cardiol.* 46, 1978–1985, 2005.

586. Stenson, W.F. and Korzenik, J., Inflammatory bowel disease, in *Textbook of Gastroenterology*, 4th edn., T.Yamada, D.H. Alpers, N. Kaplowitz, L. Laine, C. Owyang, and D.W. Powell (eds.), Lippincott, Williams, & Wilkins, Philadelphia, PA., Chapter 83, pp. 1699–1759, 2003.

587. Sutherland, A.D., Gearry, R.B., and Frizelle, F.A., Review of fecal biomarkers in inflammatory bowel disease, *Dis. Colon Rectum* 51, 1283–1291, 2008.

588. Jones, J., Loftus Jr. E.V., Panaccione, R. et al., Relationships between disease activity and serum and fecal biomarkers in patients with Crohn's disease, *Clin. Gastroenterol. Hepatol.* 6, 1218–1224, 2008.

589. Joishy, M., Davies, I., Ahmed, M. et al., Fecal calprotectin and lactoferrin as noninvasive markers of pediatric inflammatory bowel disease, *J. Pediatr. Gastroenterol. Nutr.* 48, 48–54, 2009.

590. Hayakawa, T., Jin, C.X., Ko, S.B. et al., Lactoferrin in gastrointestinal disease, *Intern. Med.* 48, 1251–1254, 2009.

591. Williams, H.R., Cox, I.J., Walker, D.G. et al., Characterization of inflammatory bowel disease with urinary metabolic profiling, *Am. J. Gastroenterol.* 104, 1435–1444, 2009.

592. Seow, C.H., Stempak, J.M., Xu, W. et al., Novel anti-glycan antibodies related to inflammatory bowel disease diagnosis and phenotypes, *Am. J. Gastroenterol.* 104, 1426–1434, 2009.

593. Mokrowiecka, A., Daniel, P., Siomka, M. et al., Clinical utility of serological markers in inflammatory bowel disease, *Hepatogastroenterology* 56, 162–166, 2009.

594. Haas, S.L., Abbatista, M., Brade, J. et al., Interleukin-18 serum levels in inflammatory bowel diseases: Correlation with disease activity and inflammatory markers, *Swiss Med. Wkly.* 139, 140–145, 2009.

595. Zahn, A., Giese, T., Karner, M. et al., Transcript levels of different cytokines and chemokines correlates with clinical and endoscopic activity in ulcerative colitis, *BMC Gastroenterol.* 9, 13, 2009.

596. Osada, T., Ohkusa, T., Okayasu, I. et al., Correlations among total colonoscopic findings, clinical symptoms, and laboratory markers in ulcerative colitis, *J. Gastroenterol. Hepatol.* 23(Suppl 2), S262–S267, 2008.

597. Kanmura, S., Uto, H., Numata, M. et al., Human neutrophil peptides 1–3 are useful biomarkers in patients with active ulcerative colitis, *Inflamm. Bowel Dis.* 15, 909–917, 2009.

598. Arsenescu, R., Bruno, M.E., Rogier, E.W. et al., Signature biomarkers in Crohn's disease: Toward a molecular classification, *Mucosal Immunol.* 1, 399–411, 2008.

599. Solberg, I.C., Lygren, I., Cvancarova, M. et al., Predictive value of serologic markers in a population-based Norwegian cohort with inflammatory bowel disease, *Inflamm. Bowel Dis.* 15, 406–414, 2009.

600. Lehrk, M., Konrad, A., Schachinger, V. et al., CXCL16 is a surrogate marker of inflammatory bowel disease, *Scand. J. Gastroenterol.* 43, 283–288, 2008.

601. Wagner, M., Peterson, C.G., Ridefelt, P. et al., Fecal markers of inflammation used as surrogate markers for treatment outcome in relapsing inflammatory bowel disease, *World J. Gastroenterol.* 14, 5584–5589, 2008.

602. Grip, O., Janciauskiene, S., and Bredberg, A., Use of atorvastatin as an inflammatory treatment in Crohn's disease, *Br. J. Pharmacol.* 155, 1085–1092, 2008.

603. Vincent, J.L., Sepsis: The magnitude of the problem, in *The Sepsis Text*, J.-L. Vincent, J. Carlet, and S.M. Opal (eds.), Kluwer Academic, Boston, MA, Chapter 1, pp. 1–9, 2002.

604. Cruz, K. and Dellinger, R.P., Diagnosis and source of sepsis: The utility of clinical findings, in *The Sepsis Text*, J.-L. Vincent, J. Carlet, and S.M. Opal (eds.), Kluwer Academic, Boston, MA, Chapter 2, pp. 11–28, 2002.

605. Redl, H., Spittler, A., and Strohmaier, W., Markers of sepsis, in *The Sepsis Text*, J.-L. Vincent, J. Carlet, and S.M. Opal (eds.), Kluwer Academic, Boston, MA, Chapter 4, pp. 47–66, 2002.

606. Sponholz, C., Sakr, Y., Reinhart, K., and Brunkhorst, F., Diagnostic value and prognostic implications of serum procalcitonin after cardiac surgery: A systematic review of the literature, *Crit. Care* 10, R145, 2006.

607. Cohen, J., Diagnosing sepsis: Does the microbiology matter?, *Crit. Care* 12, 145, 2008.

608. Gao, H., Evans, T.W., and Finney, S.J., Bench-to-bedside: Sepsis, severe sepsis and septic shock—Does the nature of the infecting organism matter?, *Crit. Care* 12, 213, 2008.

609. Reier-Nilsen, T., Farstad, T., Nakstad, B. et al., Comparison of broad range 16S rDNA PCR and conventional blood culture for diagnosis of sepsis in the newborn: A case control study, *BMC Pediatr.* 9, 5, 2009.

610. Reny, J.-L., Vuagnat, A., Ract, C. et al., Diagnosis and follow-up of infections in intensive care patients: Value of C-reactive protein compared with other clinical and biological variables, *Crit. Care Med.* 30, 529–535, 2002.

611. Saito, K., Wagatsuma, T., and Toyama, H., Sepsis is characterized by the increases in percentages of circulating $CD4^+CD25^+$ regulatory T cells and plasma levels of soluble CD25, *Tohoku J. Exp. Med.* 216, 61–68, 2008.

612. Koussoulas, V., Tzivras, M., Karagianni, V. et al., Monocytes in systemic inflammatory response syndrome: Differences between sepsis and acute pancreatitis, *World J. Gastroenterol.* 7, 6711–6714, 2006.

613. Prucha, M., Herold, I., Zazula, R. et al., Significance of lipopolysaccharide-binding protein (an acute phase protein) in monitoring critically ill patients, *Crit. Care* 7, R154–R159, 2003.

614. Morlion, B.J., Torwesten, E., Kuhn, K.S. et al., Cysteinyl-leukotriene generation as a biomarker for survival of sepsis in critically ill children, *Crit. Care Med.* 28, 3655–3658, 2000.

615. Assicot, M., Gendrel, D., Carsin, H. et al., High serum procalcitonin concentrations in patients with sepsis and infection, *Lancet* 341, 515–518, 1993.

616. Billeter, A., Turina, M., Seifert, B. et al., Early serum procalcitonin, interleukin-6, and 24-hour lactate clearance: Useful indicators of septic infections in severely traumatized patients, *World J. Surg.* 33, 558–566, 2009.
617. Maier, M., Wutzler, S., Lehnet, M. et al., Serum procalcitonin levels in patients with multiple injuries including visceral trauma, *J. Trauma* 66, 243–249, 2009.
618. Moller, H.J., Moestrup, S.K., Weis, N. et al., Macrophage serum markers in pneumococcal bacteremia: Prediction of survival by soluble CD163, *Crit. Care Med.* 34, 2561–2566, 2006.
619. Gaïni, S., Pedersen, S.S., Koldkjaer, O.G. et al., New immunological serum markers in bacteraemia: Anti-inflammatory soluble CD163, but not proinflammatory high mobility group-box 1 protein, is related to prognosis, *Clin. Exp. Immunol.* 151, 423–431, 2008.
620. Nakamura, T., Fujiwara, N., Sato, E. et al., Effect of Polymyxin B-immobilized fiber hemoperfusion on serum high mobility group box-1 protein levels and oxidative stress in patients with acute respiratory distress syndrome, *ASAIO J.* 55, 395–399, 2009.
621. Komen, N., de Bruin, R.W.F., Kleinrensink, G.J. et al., Anastomotic leakage, the search for a reliable biomarker. A review of the literature, *Colorectal Dis.* 10, 109–117, 2008.
622. Hollenbach, B., Morgenthaler, N.G., Struck, J. et al., New assay for the measurement of selenoprotein P as a sepsis biomarker from serum, *J. Trace Elem. Med. Biol.* 22, 24–32, 2008.
623. Hesselvik, J.F., Blombäck, M., Brodin, B. et al., Coagulation, fibrinolysis, and kallikrein systems in sepsis: Relation to outcome, *Crit. Care Med.* 17, 724–733, 1989.
624. Shorr, A.F., Nelson, D.R., Wyncoll, D.L. et al., Protein C: A potential biomarker in severe sepsis and a possible tool for monitoring treatment with drotrecogin alfa (activated), *Crit. Care* 12, R45, 2008.
625. Bagshaw, S.M., Brophy, P.D., Cruz, D., and Ronco, C., Fluid balance as a biomarker: Impact of fluid overload on outcome in critically ill patients with acute kidney injury, *Crit. Care* 12, 169, 2008.
626. Baum, A., Pohl, M., Kreusch, S. et al., Searching biomarker candidates in serum using multidimensional native chromatography. II Method evaluation with Alport syndrome and severe inflammation, *J. Chromatogr. B Analyt. Technol. Biomed. Life Sci.* 876, 31–40, 2008.
627. Harvey, S.B., Zhang, Y., Wilson-Grady, J. et al., *O*-Glycoside biomarker of apolipoprotein C3: Responsiveness to obesity, bariatric surgery, and therapy with metformin, to chronic or severe liver disease and to mortality in severe sepsis and graft vs host disease, *J. Proteome Res.* 8, 603–612, 2009.
628. Hattori, N., Oda, S., Sadhiro, T. et al., YKL-40 identified by proteomic analysis as a biomarker of sepsis, *Shock* 32, 393–400, 2009.
629. Mailliard-Lefebvre, H., Boulanger, E., Darous, M. et al., Soluble receptor for advanced glycation end products: A new biomarker in diagnosis and prognosis of chronic inflammatory diseases, *Rheumatology* 48, 1190–1196, 2009.
630. Liu, Y., Sanoff, H.K., Cho, H. et al., Expression of p16 in peripheral blood T-cells is a biomarker of human aging, *Aging Cell* 8, 439–448, 2009.
631. Wu, X., Keith Jr. J.C., Struthers, A.D., and Feuerstein, G.Z., Assessment of arterial stiffness, a translational medicine biomarker system for evaluation of vascular risk, *Cardiovasc. Ther.* 26, 214–223, 2008.
632. Malkin, I., Kalichman, L., and Kobyliansky, E., Heritability of a skeletal biomarker of biological aging, *Biogerontology* 8, 627–637, 2007.
633. Zubenko, G.S., Zubenko, W.N., Maher, B.S., and Wolf, N.S., Reduced age-related cataracts among elderly persons who reach age 90 with preserved cognition: A biomarker of successful aging?, *J. Gerontol. A Biol. Sci. Med. Sci.* 62, 500–506, 2007.
634. Simm, A., Wagner, J., Gurinsky, T. et al., Advanced glycation endproducts: A biomarker for age as an outcome predictor after cardiac surgery?, *Exp. Gerontol.* 42, 668–675, 2007.

635. Terry, D.F., Wyszynski, D.F., Nolan, V.G. et al., Serum heat shock protein 70 level as a biomarker of exception longevity, *Mech. Ageing Dev.* 127, 862–868, 2006.

636. DiMarco, T. and Giulivi, C., Current analytical methods for the detection of dityrosine, a biomarker of oxidative stress, in biological samples, *Mass Spectrom. Rev.* 26, 108–120, 2007.

637. Eppenberger-Castori, S., Moore Jr. D.H., Thor, A.D. et al., Age-associated biomarker profiles of human breast cancer, *Int. J. Biochem. Cell Biol.* 34, 1318–1330, 2002.

638. Chan, S.S., Monteiro, H.P., Deucher, G.P. et al., Functional activity of blood polymorphonuclear leukocytes as an oxidative stress biomarker in human subjects, *Free Radic. Biol. Med.* 24, 1411–1418, 1998.

639. Requena, J.R., Ahmed, M.U., Fountain, C.W. et al., Carboxymethylethanolamine, a biomarker of phospholipid modification during the maillard reaction in vivo, *J. Biol. Chem.* 272, 17473–17479, 1997.

640. Bortz IV W.M. and Bortz II W.M., How fast do we age? Exercise performance over time as a biomarker, *J. Gerontol. A Biol. Sci. Med. Sci.* 51, M223–M225, 1996.

641. Vanhoeren, V., Laroy, W., Libert, C., and Chen, C., N-glycan profiling in the study of human aging, *Biogerontology* 9, 351–356, 2008.

642. Baynes, J.W., The role of AGEs in aging: Causation or correlation, *Exp. Gerontol.* 36, 1527–1537, 2001.

643. Meli, M., Frey, J., and Perier, C., Native protein glycoxidation and aging, *J. Nutr. Health Aging*, 7, 263–266, 2003.

644. Grillo, M.A. and Colombatto, S., Advanced glycation end-products (AGEs): Involvement in aging and in neurodegenerative diseases, *Amino Acids* 35, 39–36, 2008.

645. Osawa, M., Hayashi, T., Nomura, H. et al., Nitric oxide (NO) is a new clinical biomarker of survival in the elderly patients and its efficacy may be equal to albumin, *Nitric Oxide* 16, 157–163, 2007.

646. Davuluri, G., Espina, V., Petricoin III E.F. et al., Activated VEGF receptor shed into the vitreous with potential for predicting the treatment time and monitoring response, *Arch. Ophthalmol.* 127, 613–621, 2009.

647. Tolppanen, A.M., Nevalainen, T., Kolehmainen, M. et al., Single nucleotide polymorphisms of the tenomodulin gene (TNMD) in age-related macular degeneration, *Mol. Vis.* 15, 762–770, 2009.

648. Klein, R., Knudtson, M.D., Lee, K.E., and Klein, B.E., Serum cystatin C level, kidney disease markers, and incidence of age-related macular degeneration: The Beaver Dam Eye Study, *Arch. Opthalmol.* 127, 193–199, 2009.

649. Funatsu, H., Noma, H., Minura, T. et al., Association of vitreous inflammatory factors with diabetic macular edema, *Ophthalmology* 116, 73–79, 2009.

650. Ahlers, C., Golbaz, I., Einwallner, E. et al., Identification of optical density ratios in subretinal fluid as a clinical relevant biomarker in exudative macular disease, *Invest. Ophthalmol. Vis. Sci.* 50, 3417–3424, 2009.

651. Pawlak, A.M., Glenn, J.V., Beattie, J.R. et al., Advanced glycation as a basis for understanding retinal aging and noninvasive risk prediction, *Ann. N. Y. Acad. Sci.* 1126, 59–65, 2008.

652. Zaliuniene, D., Paunksnis, A., Gustiene, O. et al., Pre- and postoperative C-reactive protein levels in patients with cataract and age-related macular degeneration, *Eur. J. Ophthalmol.* 17, 919–927, 2007.

653. Boekhoorn, S.S., Vingerling, J.R., Witteman, J.C. et al., C-reactive protein level and risk of aging macula disorder: The Rotterdam Study, *Arch. Ophthalmol.* 125, 1396–1401, 2007.

654. Joachim, S.C., Bruns, K., Lackner, K.J. et al., Analysis of IgG antibody patterns against retinal antigens and antibodies to α-crystallin, GFAP, and α-enolase in sera of patients with "wet" age-related macular degeneration, *Graefes Arch. Clin. Exp. Ophthalmol.* 245, 619–626, 2007.

655. Sakuri, S., Yamamoto, Y., Tamei, H. et al., Development of an ELISA for esRAGE and its application to type 1 diabetic patients, *Diabetes Res. Clin. Pract.* 73, 158–165, 2006.

656. La Heij, E.C., van de Waarenburg, M.P.H., Blaauwgeers, H.G.T. et al., Levels of basic fibroblast growth factor, glutamine synthetase, and interleukin-6 in subretinal fluid from patients with retinal detachment, *Am. J. Ophthalmol.* 132, 544–550, 2001.
657. Perez, V.L., Papaliodis, G.N., Chu, D. et al., Elevated levels of interleukin 6 in the vitreous fluid of patients with pars planitis and posterior uveitis: The Massachusetts Eye & Ear experience and review of previous studies, *Ocul. Immunol. Inflamm.* 12, 193–202, 2004.
658. Cummings, J.L., Clinical evaluation as a biomarker for Alzheimer's disease, *J. Alzheimers Dis.* 8, 327–337, 2005.
659. Dassan, P., Keir, G., and Brown, M.M., Criteria for a clinically informative serum biomarker in acute ischemic stroke: A review of S100B, *Cerebrovasc. Dis.* 27, 295–302, 2009.
660. Yin, G.N., Lee, H.W., Cho, J.Y., and Suk, K., Neuronal pentraxin receptor in cerebrospinal fluid as a potential biomarker for neurodegenerative diseases, *Brain Res.* 1265, 158–170, 2009.
661. Blekher, T., Weaver, M., Rupp, J. et al., Multiple step pattern as a biomarker in Parkinson disease, *Parkinsonism Relat. Disord.* 15, 506–510, 2009.
662. Borroni, B., Malinverno, M., Gardoni, F. et al., Tau forms in CSF as a reliable biomarker for progressive supranuclear palsy, *Neurology* 71, 1796–1803, 2008.
663. Shi, M., Caudle, W.M., and Zhang, J., Biomarker discovery in neurodegenerative diseases: A proteomic approach, *Neurobiol. Dis.* 35, 157–164, 2009.
664. Schwarzschild, M.A., Schwid, S.R., Marek, K. et al., Serum urate as a predictor of clinical and radiographic progression in Parkinson disease, *Arch. Neurol.* 65, 716–723, 2008.
665. Martin, W.R., Wieler, M., and Gee, M., Midbrain iron content in early Parkinson disease: A potential biomarker of disease status, *Neurology* 70, 1411–1417, 2008.
666. Bohnen, N.I., Kuwabara, H., Constantine, G.M. et al., Grooved pegboard test as a biomarker of nigrostriatal denervation in Parkinson's disease, *Neurosci. Lett.* 424, 185–189, 2007.
667. Zhang, J., Goodlett, D.R., and Montine, T.J., Proteomic biomarker discovery in cerebrospinal fluid for neurodegenerative diseases, *J. Alzheimers Dis.* 8, 377–386, 2005.
668. El-Agnaf, O.M., Salem, S.A., Paleologou, K.E. et al., Detection of oligomeric forms of α-synuclein protein in human plasma as a potential biomarker for Parkinson's disease, *FASEB J.* 20, 419–425, 2006.
669. Sato, S., Mizuno, Y., and Hattori, N., Urinary 8-hydroxydeoxyguanosine levels as a biomarker for progression of Parkinson disease, *Neurology* 64, 1081–1083, 2005.
670. Khan, M.G., *Heart Disease and Therapy. A Practical Approach*, 2nd edn., Humana Press, Totowa, NJ, Chapter 4, 2005.
671. Graeber, G.M., Creatine kinase (CK): Its use in the evaluation of perioperative myocardial infarction, *Surg. Clin. North Am.* 65, 539–551, 1985.
672. Apple, F.S., Diagnostic use of CK-MM and CK-MB isoforms for detecting myocardial infarction, *Clin. Lab. Med.* 9, 643–654, 1989.
673. Wu, A.H., Use of cardiac markers as assessed by outcomes analysis, *Clin. Biochem.* 30, 339–350, 1997.
674. Roberts, R., Rapid MB Ck subform assay and the early diagnosis of myocardial infarction, *Clin. Lab. Med.* 17, 669–683, 1997.
675. Plebani, M., Biochemical markers of cardiac damage: From efficiency to effectiveness, *Clin. Chim. Acta* 311, 3–7, 2001.
676. Fesmire, F.M., Improved identification of acute coronary syndromes with delta cardiac serum marker measurements during the emergency department evaluation of chest pain patients, *Cardiovasc. Toxicol.* 1, 117–123, 2001.
677. Jaffe, A.S., Use of biomarkers in the emergency department and chest pain unit, *Cardiol. Clin.* 23, 453–465, 2005.
678. Cohen, M., Diez, J., Fry, E. et al., Strategies for optimizing outcomes in the NSTE-ACS patient The CATH (cardiac catherization and antithrombotic therapy in the hospital) clinical consensus panel report, *J. Invasive Cardiol.* 18, 617–639, 2006.

679. Kagen, L.J., Myoglobin: Methods and diagnostic uses, *CRC Crit. Rev. Clin. Lab. Sci.* 9, 273–302, 1978.

680. Lott, J.A. and Stang, J.M., Differential diagnosis of patients with abnormal serum creatine kinase isoenzymes, *Clin. Lab. Med.* 9, 627–642, 1989.

681. Hamm, C.W. and Katus, H.A., New biochemical markers for myocardial cell injury, *Curr. Opin. Cardiol.* 10, 355–360, 1995.

682. Henderson, A.R., An overview and ranking of biochemical markers of cardiac disease. Strengths and limitations, *Clin. Lab. Med.* 17, 625–654, 1997.

683. Balk, E.M., Ioannidis, J.P., Salem, D. et al., Accuracy of biomarkers to diagnose acute cardiac ischemia in the emergency department: A meta-analysis, *Emerg. Med.* 37, 478–494, 2001.

684. Daves, M., Trevisan, D., and Cemin, R., Different collection tubes in cardiac biomarkers detection, *J. Clin. Lab. Anal.* 22, 391–394, 2008.

685. Lippi, G., Schena, F., Montagnana, M., Salvagno, G.L., and Guidi, G.C., Influence of acute physical exercise on emerging muscular biomarkers, *Clin. Chem. Lab. Med.* 46, 1313–1318, 2008.

686. Futterman, L.G. and Lemberg, L., SGOT, LDH, HBD, CPK, CK-MS, MB1:MB2, cTnT, cTnI, *Am. J. Crit. Care* 6, 333–338, 1997.

687. Kost, G.J., Kirk, J.D., and Omand, K., A strategy for the use of cardiac injury markers (troponin I and T, creatine kinase-MB mass and isoforms, and myoglobin) in the diagnosis of acute myocardial infarction, *Arch. Pathol. Lab. Med.* 122, 245–251, 1998.

688. Lindahl, B., Cardiac troponin for risk assessment and management of non-ST-elevation acute coronary syndrome, in *Cardiovascular Biomarkers Pathophysiology and Disease Management*, D.A. Morrow (ed.), Humana Press, Totowa, NJ, Chapter 5, pp. 79–92, 2006.

689. Neizel, M., Futterer, S., Steen, H. et al., Predicting microvascular obstruction with cardiac troponin T after acute myocardial infarction: A correlative study with contrast-enhanced magnetic resonance imaging, *Clin. Res. Cardiol.* 98, 555–562, 2009.

690. Zahid, M., Good, C.G., Singla, I., and Sonel, A.F., Clinical significance of borderline elevated troponin I across different assays in patients with suspected acute coronary syndrome, *Am. J. Cardiol.* 104, 164–168, 2009.

691. Javed, U., Aftab, W., Ambrose, J.A. et al., Frequency of elevated troponin I and diagnosis of acute myocardial infarction, *Am. J. Cardiol.* 104, 9–13, 2009.

692. Chia, S., Senatore, F., Raffel, O.C. et al., Utility of cardiac biomarkers in predicting infarct size, left ventricular function, and clinical outcome after primary percutaneous coronary intervention for ST-segment elevation myocardial infarction, *JACC Cardiovasc. Interv.* 1, 415–423, 2008.

693. Hallén, J. and Atar, D., In ST-elevation myocardial infarction patients receiving primary percutaneous coronary intervention, admission cardiac troponin T and peak cardiac troponin T values differ in their prognostic properties, *Am. J. Cardiol.* 103, 1331, 2009.

694. Eggers, K.M., Lagerqvist, B., Venge, P. et al., Prognostic value of biomarkers during and after non-ST-segment elevation acute coronary syndrome, *J. Am. Coll. Cardiol.* 54, 357–364, 2009.

695. Hall, C., Essential biochemistry and physiology of (NT-pro)BNP, *Eur. J. Heart Fail.* 6, 257–260, 2004.

696. Hall, C., NT-ProBNP: The mechanism behind the marker, *J. Card. Fail.* 11(Suppl), S81–S83, 2005.

697. Hoffmann, U., Borggrefe, M., and Brueckmann, M., New horizons: NT-proBNP for risk stratification in patients with shock in the intensive care unit, *Crit. Care.* 10, 134, 2006.

698. Novo, G., Amoroso, G.R., Fazio, G. et al., Biomarkers in heart failure, *Front. Biosci.* 14, 2484–2493, 2009.

699. Madhok, V., Falk, G., and Rogers, A., The accuracy of symptoms, signs and diagnostic tests in the diagnosis of left ventricular dysfunction in primary care: A diagnostic accuracy systematic review, *BMC Fam. Pract.* 9, 56, 2008.

700. Mueller, C., Cost-effectiveness of B-type natriuretic peptide testing, *Congest. Heart Fail.* 14(4 Suppl 1), 35–37, 2008.
701. Maisel, A., Mueller, C., Adams Jr. K., et al., State of the art: Using natriuretic peptide levels in clinical practice, *Eur. J. Heart Fail.* 10, 824–839, 2008.
702. Cea, L.B., Natriuretic peptide family: New aspects, *Curr. Med. Chem. Cardiovasc. Hematol. Agents* 3, 87–98, 2005.
703. Sobel, B.E. and Shell, W.E., Serum enzyme determinations in the diagnosis and assessment of myocardial infarction, *Circulation* 45, 471–482, 1972.
704. Lott, J.A. and Stang, J.M., Serum enzymes and isoenzymes in the diagnosis and differential diagnosis of myocardial infarction and necrosis, *Clin. Chem.* 26, 1241–1250, 1980.
705. Brancaccio, P., Maffulli, N., Buonauro, R., and Limongelli, F.M., Serum enzyme monitoring in sports medicine, *Clin. Sports Med.* 27, 1–18, 2008.
706. Patschan, D., Witzke, O., Dührsen, U. et al., Acute myocardial infarction in thrombotic microangiopathies—Clinical characteristics, risk factors and outcome, *Nephrol. Dial. Transplant.* 21, 1549–1554, 2006.
707. Ege, T., Us, M.H., Cikirikcioglu, M. et al., Analysis of C-reactive protein and biochemical parameters in pericardial fluid, *Yonsei Med. J.* 47, 372–376, 2006.
708. Jong, G.P., Wang, Y.F., Tsai, F.J. et al., Immunoglobulin E and matrix metalloproteinase-9 in patients with different stages of coronary artery disease, *Chin. J. Physiol.* 50, 277–282, 2007.
709. Pesek, K., Buković, D., Pesek, T. et al., Risk factor analysis and diagnosis of coronary heart disease in patients with hypercholesterolemia from Croatian Zagorje Country, *Coll. Antropol.* 32, 369–374, 2008.
710. Hjortshøj, S., Otterstad, J.E., Lindahl, B. et al., Biochemical diagnosis of myocardial infarction evolves towards ESC/ACC consensus experiences from the Nordic countries, *Scand. Cardiovasc. J.* 39, 159–166, 2005.
711. Mangi, A.A., Christison-Lagay, E.R., Torchiana, D.F. et al., Gastrointestinal complications in patients undergoing heart operation: An analysis of 8709 consecutive cardiac surgical patients, *Ann. Surg.* 241, 895–901, 2005.
712. Shin, D.H., Kwon, Y.I., Choi, S.I. et al., Accidental ten times overdose administration of recombinant human erythropoietin (rh-EPO) up to 318,000 units a day in acute myocardial infarction: Report of two cases, *Basic Clin. Pharmacol. Toxicol.* 98, 222–224, 2006.
713. Papadopoulos, C.E., Karvounis, H.I., Gourasas, I.T. et al., Evidence of ischemic preconditioning in patients experiencing first non-ST-segment elevation myocardial infarction (NSTEMI), *Int. J. Cardiol.* 92, 209–217, 2003.
714. Johnston, C.C. and Bolton, E.C., Cardiac enzymes, *Ann. Emerg. Med.* 11, 27–35, 1982.
715. Valle, H.A., Riesgo, L.G., Bel, M.S. et al., Clinical assessment of heart-type fatty acid binding protein in early diagnosis of acute coronary syndrome, *Eur. J. Emerg. Med.* 15, 140–144, 2008.
716. Liyan, C., Jie, Z., and Xiaozhou, H., Prognostic value of combination of heart-type fatty acid-binding protein and ischemia-modified albumin in patients with acute coronary syndromes and normal troponin T values, *J. Clin. Lab. Anal.* 23, 14–18, 2009.
717. Sbarouni, E., Georgiadou, P., Sklavainas, I. et al., Increases in serum concentration of human heart-type fatty acid-binding protein following elective coronary intervention, *Biomarkers* 14, 317–320, 2009.
718. Bathia, D.P., Carless, D.R., Viswanathan, K. et al., Serum 99th centile values for two heart-type fatty acid binding protein assays, *Ann. Clin. Biochem.* 46, 464–467, 2009.
719. Turedi, S., Gunduz, A., Mentese, A. et al., Investigation of the possibility of using ischemia-modified albumin as a novel and early prognostic marker in cardiac arrest patients after cardiopulmonary resuscitation, *Resuscitation* 80, 994–999, 2009.

720. Bhagavan, N.V., Ha, J.S., Park, J.H. et al., Utility of serum fatty acid concentrations as a marker for acute myocardial infarction and their potential role in the formation of ischemia-modified albumin: A pilot study, *Clin. Chem.* 55, 1588–1590, 2009.

721. Lippi, G., Montagnana, M., Salvagno, G.L., and Guidi, G.C., Standardization of ischemia-modified albumin testing: Adjustment for serum albumin, *Clin. Chem. Lab. Med.* 45, 261–262, 2007.

722. Sinha, M.K., Roy, D., Gaze, D.C. et al., Role of "ischemia modified albumin", a new biochemical marker of myocardial ischemia, in the early diagnosis of acute coronary syndromes, *Emerg. Med. J.* 21, 29–34, 2004.

723. Christenson, R.H., Duh, S.H., Sanhai, W.R. et al., Characterization of an albumin cobalt binding test for assessment of acute coronary syndrome patients: A multicenter study, *Clin. Chem.* 47, 464–470, 2001.

724. Vaananen, H.K., Syrjala, H., Rahkita, P. et al., Serum carbonic anhydrase III and myoglobin concentrations in acute myocardial infarction, *Clin. Chem.* 36, 635–638, 1990.

725. Brogan Jr. G.X., Vuori, J., Friedman, S. et al., Improved specificity of myoglobin plus carbonic anhydrase assay versus that of creatine kinase-MB for early diagnosis of acute myocardial infarction, *Ann. Emerg. Med.* 27, 22–28, 1996.

726. Beurele, J.R., Azzazy, H.M., Styba, G., Duh, S.H., and Christenson, R.H., Characteristics of myoglobin, carbonic anhydrase III and the myoglobin/carbonic anhydrase III ratio in trauma, exercise, and myocardial infarction patients, *Clin. Chim. Acta* 294, 115–128, 2000.

727. Vuotikka, P., Ylitalo, K., Vuori, J. et al., Serum myoglobin/carbonic anhydrase III ratio in the diagnosis of perioperative myocardial infarction during coronary bypass surgery, *Scand. Cardiovasc. J.* 37, 23–29, 2003.

728. Fu Q. and Van Eyk, J.E., Proteomics and heart disease: Identifying biomarkers of clinical utility, *Expert Rev. Proteomics* 3, 237–249, 2006.

729. de Roos, B., Geelen, A., Ross, K. et al., Identification of potential serum biomarkers of inflammation and lipid modulation that are altered by fish oil supplementation in healthy volunteers, *Proteomics* 8, 1965–1974, 2008.

730. Kuhn, E., Addona, T., Keshishian, H. et al., Developing multiplexed assays for troponin I and interleukin-33 in plasma by peptide immunoaffinity enrichment and targeted mass spectrometry, *Clin. Chem.* 55, 1108–1117, 2009.

731. Zannad, F., Roosignol, P., and Iraqi, W., Extracellular fibrotic markers in heart failure, *Heart Fail. Rev.*, in press, 2009.

732. O'Dell, S.D., Humphries, S.E., and Day, I.N., Rapid methods for population-scale analysis for gene polymorphisms: The ACE gene as an example, *Br. Heart J.* 73, 368–371, 1995.

733. Yndestad, A., Damås, J.K., Geir Eiken, H. et al., Increased gene expression of tumor necrosis factor superfamily ligands in peripheral blood mononuclear cells during chronic heart failure, *Cardiovasc. Res.* 54, 175–182, 2002.

734. Hilker, M., Långin, T., Hake, U. et al., Gene expression profiling of human stenotic aorto-coronary bypass grafts by cDN array analysis, *Eur. J. Cardiothorac. Surg.* 23, 620–625, 2003.

735. Horwitz, P.A., Tsai, E.J., Putt, M.E. et al., Detection of cardiac allograft rejection and response to immunosuppressive therapy with peripheral blood gene expression, *Circulation* 110, 3815–3821, 2004.

736. Kittleson, M.M. and Hare, J.M., Molecular signature analysis: Using the myocardial transcriptome as a biomarker in cardiovascular disease, *Trends Cardiovasc. Med.* 15, 130–138, 2005.

737. Beisvag, V., Lehre, P.K., Midelfart, H. et al., Aetiology-specific patterns in end-stage heart failure patients identified by functional annotation and classification of microarray data, *Eur. J. Heart Fail.* 8, 381–389, 2006.

738. Karason, K., Jernås, M., Hågg, D.A., and Svensson, P.A., Evaluation of CXCL9 and CXCL10 as circulating biomarkers of human cardiac allograft rejection, *BMC Cardiovasc. Disord.* 6, 29, 2006.

739. Mehra, M.R., Uber, P.A., Walther, D. et al., Gene expression profiles and B-type natri-
    uretic peptide elevation in heart transplantation: More than a hemodynamic marker,
    Circulation 114(1 Suppl), I21–I26, 2006.
740. Hägg, D.A., Jernås, M., Wiklund, O. et al., Expression profiling of macrophages from
    subjects with atherosclerosis to identify novel susceptibility genes, *Int. J. Mol. Med.* 21,
    697–704, 2008.
741. Heidecker, B., Kasper, E.K., Wittstein, I.S. et al., Transcriptomic biomarkers for indi-
    vidual risk assessment in new-onset heart failure, *Circulation* 118, 238–246, 2008.
742. Lara-Pezzi, E., Felkin, L.E., Birks, E.J. et al., Expression of follistatin-related genes is
    altered in heart failure, *Endocrinology* 149, 5822–5827, 2008.
743. Wei, Y.J., Huang, Y.X., Zhang, X.L. et al., Apolipoprotein D as a novel marker in human
    end-stage heart failure: A preliminary study, *Biomarkers* 13, 535–548, 2008.
744. D. Marrow (ed.), *Cardiovascular Biomarkers Pathophysiology and Disease Management*,
    Humana Press, Totowa, NJ, 2006.
745. Collison, P.O., Cardiac markers, *Br. J. Hosp. Med.* 70, M84–M87, 2009.
746. Setiadi, B.M., Lei, H., and Chang, J., Troponin not just a simple cardiac marker:
    Prognostic significance of cardiac troponin, *Chin. Med. J. (Engl.)* 122, 351–358,
    2009.
747. Emdin, M., Vittorini, S., Passino, C., and Clerico, A., Old and new biomarkers of heart
    failure, *Eur. J. Heart Fail.* 11, 331–335, 2009.
748. Adreassi, M.G., Gasttaldelli, A., Clerico, A. et al., Imaging and laboratory biomarkers in
    cardiovascular disease, *Curr. Pharm. Des.* 15, 1131–1141, 2009.
749. Froelicher, V.F. and Myers, J., *Exercise and the Heart*, 5th edn., Saunders/Elsevier,
    Philadelphia, PA, Chapter 7, 2006.
750. Morrow, D.A., A multimarker approach to evaluation of patients with acute coronary
    syndrome, in *Cardiovascular Biomarkers Pathophysiology and Disease Management*,
    D.A. Morrow (ed.), Humana, Totowa, NJ, Chapter 31, pp. 545–558, 2006.
751. Braunwald, E., Unstable angina. An etiologic approach to management, *Circulation* 98,
    2219–2222, 1998.
752. Lee, T.H. and Cannon, C.P., Approaches to the patient with chest pain, in *Braunwald's
    Heart Disease*, 7th edn., D.P. Zipes, P.Libby, R.O. Bonow, and E. Braunwald (eds.),
    Elsevier/Saunders, Philadelphia, PA, Chapter 45, pp. 1129–1139, 2005.
753. Ashton-Chess, J., Giral, M., Soulillou, J.P., and Brouard, S., Using biomarkers of toler-
    ance and rejection to identify high- and low-risk patients following kidney transplanta-
    tion, *Transplantation* 87(9 Suppl), S95–S99, 2009.
754. Ichimura, T., Hung, C.C., Yang, S.A. et al., Kidney injury molecule-1: A tissue and uri-
    nary biomarker for nephrotoxicant-induced renal injury, *Am. J. Physiol. Renal Physiol.*
    286, F552–F563, 2004.
755. Vaidya, V.S., Ramirez, V., Ichimura, T. et al., Urinary kidney injury molecule-1: A sensi-
    tive quantitative biomarker for early detection of kidney tubular injury, *Am. J. Physiol.
    Renal Physiol.* 290, F517–F529, 2006.
756. Zhang, P.L., Rothblum, L.I., Han, W.K. et al., Kidney injury molecule-1 expression
    in transplant biopsies is a sensitive measure of cell injury, *Kidney Int.* 73, 608–614,
    2008.
757. Perco, P. and Oberbauer, R., Kidney injury molecule-1 as a biomarker of acute kidney
    injury in renal transplant recipients, *Nat. Clin. Pract. Nephrol.* 4, 362–363, 2008.
758. Chaturvedi, S., Farmer, T., and Kapke, G.F., Assay validation for KIM-1: Human uri-
    nary dysfunction biomarker, *Int. J. Biol. Sci.* 5, 128–134, 2009.
759. Nguyen, M.T. and Devarajan, P., Biomarkers for the early detection of acute kidney
    injury, *Pediatr. Nephrol.* 23, 2151–2157, 2008.
760. Wagner, D.K., Harris, T., and Madan, J.H., Proteinuria as biomarker: Risk of subsequent
    morbidity and mortality, *Environ. Res.* 66, 160–172, 1994.

761. Lambers Heerspink, H.J., Brinkman, J.W., Bakker, S.J. et al., Update on microalbuminuria as a biomarker in renal and cardiovascular disease, *Curr. Opin. Nephrol. Hypertens.* 15, 631–636, 2006.
762. Danziger, J., Importance of low-grade albuminuria, *Mayo Clin. Proc.* 83, 806–812, 2008.
763. Wu, X., Jin, T., Wang, Z. et al., Urinary calcium as a biomarker of renal dysfunction in a general population exposed to cadmium, *J. Occup. Environ. Med.* 43, 898–904, 2001.
764. Pawlak, K., Pawlak, D., and Mysliwiec, M., Cu/Zn superoxide dismutase plasma levels as a new useful clinical biomarker of oxidative stress in patients with end-stage renal disease, *Clin. Biochem.* 38, 700–705, 2005.
765. Noordzij, M.J., Lefrandt, J.D., and Smit, A.J., Advanced glycation end products in renal failure: An overview, *J. Ren. Care.* 34, 207–212, 2008.
766. Washburn, K.K., Zappitelli, M., Arikan, K.K. et al., Urinary interleukin-18 is an acute kidney injury biomarker in critically ill children, *Nephrol. Dial. Transplant.* 23, 566–572, 2008.
767. Niwa, T., Biomarker discovery for kidney diseases by mass spectrometry, *J. Chromatogr. B* 870, 148–153, 2008.
768. Portilla, D., Dent, C., Sugaya, T. et al., Liver fatty acid-binding protein as a biomarker of acute kidney injury after cardiac surgery, *Kidney Int.* 73, 465–472, 2008.
769. Kato, H., Si, H., Hostetter, T., and Susztak, K., Smad1 as a biomarker for diabetic nephropathy, *Diabetes* 57, 1459–1460, 2008.
770. Mima, A.A.H., Matsubara, T., Abe, H. et al., Urinary Smad1 is a novel marker to predict later onset of mesangial matrix expansion in diabetic nephropathy, *Diabetes* 57, 1712–1722, 2008.
771. Bao, J., Tu. Z., Wang, J. et al., A novel accurate rapid ELISA for detection of urinary connective tissue growth factor, a biomarker of chronic allograft nephropathy, *Transplant Proc.* 40, 2361–2364, 2008.
772. Kobori, H., Alper Jr. A.B., Shenava, R. et al., Urinary angiotensinogen as a novel biomarker of the intrarenal renin-angiotensin system status in hypertensive patients, *Hypertension* 53, 344–350, 2009.
773. Zaritsky, J., Young, B., Wang, H.J. et al., Hepcidin—A potential novel biomarker for iron status in chronic kidney disease, *Clin. J. Am. Soc. Nephrol.* 4, 1051–1056, 2009.
774. Devarajan, P., Neutrophil gelatinase-associated lipocalin—An emerging troponin for kidney injury, *Nephrol. Dial. Transplant.* 23, 3737–3743, 2008.
775. Soni. S.S., Cruz, D., Bobek, I. et al., NGAL: A biomarker of acute kidney injury and other systemic conditions, *Int. Urol. Nephrol.* 42, 141–150, 2010.
776. de Jong, P.E. and Gansevoort, R.T., Screening techniques for detecting chronic kidney disease, *Curr. Opin. Nephrol. Hypertens.* 14, 567–572, 2005.
777. Trof, R.J., Di Maggio, F., Leemreis, J., and Groeneveld, A.B., Biomarkers of acute renal injury and renal failure, *Shock* 26, 245–253, 2006.
778. Caubet, C., Lacroix, C., Decramer, S. et al., Advances in urinary proteome analysis and biomarker discovery in pediatric renal disease, *Pediatr. Nephrol.* 25, 27–35, 2010.
779. Hajjar, V. and Schreiber Jr. M.J., Does measuring natriuretic peptides have a role in patients with chronic kidney disease?, *Cleve. Clin. J. Med.* 76, 476–478, 2009.
780. Khalifeh, N., Haider, D., and Hört, W.H., Natriuretic peptides in chronic kidney disease and during renal replacement therapy: An update, *J. Invest. Med.* 57, 33–39, 2009.
781. Dhar, S., Pressman, G.S., Subramanian, S. et al., Natriuretic peptides and heart failure in the patient with chronic kidney disease: A review of current evidence, *Postgrad. Med. J.* 85, 299–302, 2009.
782. deFilippi, C.R. and Christenson, R.H., B-type natriuretic peptide (BNP)/NT-proBNP and renal function: Is the controversy over?, *Clin. Chem.* 55, 1271–1273, 2009.
783. Zhang, G.Y., Hu, M., Wang, Y.M. et al., Foxp3 as a marker of tolerance induction versus rejection, *Curr. Opin. Organ Transplant.* 14, 40–45, 2009.

784. Herget-Rosenthal, S., Bökenkamp, A., and Hoffman, W., How to estimate GFR-serum creatinine, serum cystatin C or equations?, *Clin. Biochem.* 40, 153–161, 2007.
785. Rule, A.D., Understanding estimated glomerular filtration rate: Implications for identifying chronic kidney disease, *Curr. Opin. Nephrol. Hypertens.* 16, 242–249, 2007.
786. Soloman, R. and Segal, A., Defining acute kidney injury: What is the most appropriate metric?, *Nat. Clin. Pract. Nephrol.* 4, 208–215, 2008.
787. Taal, M.R. and Brenner, B.M., Renal risk scores: Progress and prospects, *Kidney Int.* 73, 1216–1219, 2008.
788. Stevens, L.A., Zhang, Y., and Schmid, C.H., Evaluating the performance of equations for estimating glomerular filtration rate, *J. Nephrol.* 21, 797–807, 2008.
789. Clamp, A., Danson, S., Nguyen, H. et al., Assessment of therapeutic response in patients with metastatic bone disease, *Lancet Oncol.* 5, 607–616, 2004.
790. Palma, M.A. and Body, J.J., Usefulness of bone formation markers in breast cancer, *Int. J. Biol. Markers* 20, 146–155, 2005.
791. Kehoe, T., Bone quality: A perspective from the food and drug administration, *Curr. Osteoporos. Rep.* 4, 76–79, 2006.
792. Majumdar, S., Magnetic resonance imaging and spectroscopy of the intervertebral disc, *NMR Biomed.* 19, 894–903, 2006.
793. Tankó, L.B., Karsdal, M.A., Christiansen, C., and Leeming, D.J., Biochemical approach to the detection and monitoring of metastatic bone disease: What do we know and what questions need answers?, *Cancer Metastasis Rev.* 25, 659–668, 2006.
794. Rousseau, J.C. and Delmas, P.D., Biological markers in osteoarthritis, *Nat. Clin. Pract. Rheumatol.* 3, 346–356, 2007.
795. Zimmerman, G., Müller, U., and Wentzensen, A., The value of laboratory and imaging studies in the evaluation of long-bone non-unions, *Injury* 38(Suppl 2), S33–S37, 2007.
796. Samuels, J., Krasnokutsky, S., and Abramson, S.B., Osteoarthritis: A tale of three tissues, *Bull. NYU Hosp. Jt. Dis.* 66, 244–250, 2008.
797. Raza, K. and Filer, A., Predicting the development of RA in patients with early undifferentiated arthritis, *Best Pract. Res. Clin. Rheumatol.* 23, 25–36, 2009.
798. Coe, F.L. and Favus, M.J. (eds.), *Disorders of Bone and Mineral Metabolism*, 2nd edn., Lippincott, Williams, & Wilkins, Philadelphia, PA, 2002.
799. Kosnett, M.J., Becker, C.E., Osterloh, J.D. et al., Factors influencing bone lead concentration in a suburban community assessed by noninvasive K x-ray fluorescence, *JAMA* 271, 197–203, 1994.
800. Kuiper, J.I., Verbeek, J.H., Frings-Dresen, M.H., and Ikkink, A.J., Keratan sulfate as a potential biomarker of loading of the intervertebral disc, *Spine* 23, 657–663, 1998.
801. Leeming, D.J., Delling, G., Koizumi, M. et al., Alpha CTX as a biomarker of skeletal invasion of breast cancer: Immunolocalization and the load dependency of urinary excretion, *Cancer Epidemiol. Biomarkers Prev.* 15, 1392–1995, 2006.
802. Sharif, M., Granell, R., Johansen, J. et al., Serum cartilage oligomeric matrix protein and other biomarker profiles in tibiofemoral and patellofemoral osteoarthritis of the knee, *Rheumatology* 45, 522–526, 2006.
803. Davis, C.R., Karl, J., Granell, R. et al., Can biochemical markers serve as surrogates for imaging in knee osteoarthritis?, *Arthritis Rheum.* 56, 4038–4047, 2007.
804. Nelson, E.C., Evans, C.F., Pan, C.X., and Lara Jr. P.N., Prostate cancer and markers of bone metabolism: Diagnostic, prognostic, and therapeutic implications, *World J. Urol.* 25, 393–399, 2007.
805. Bhattacharyya, S., Siegel, E.R., Achenbach, S.J. et al., Serum biomarker profile associated with high bone turnover and BMD in postmenopausal women, *J. Bone Miner. Res.* 23, 1106–1117, 2008.

806. Hannan, F.M., Athansou, N.A., Teh, J. et al., Oncogenic hypophosphataemic osteomalacia: Biomarker roles of fibroblast growth factor 23, 1,25-dihydroxyvitamin D3 and lymphatic vessel endothelial hyaluronan receptor 1, *Eur. J. Endocrinol.* 158, 265–271, 2008.

807. Dane, F., Tunk, H.M., Sevinc, A. et al., Markers of bone turnover in patients with lung cancer, *J. Nat. Med. Assoc*, 100, 425–428, 2008.

808. Fine, D.H., Markowitz, K., Furgang, D. et al., Macrophage inflammatory protein-1α: A salivary biomarker for bone loss in a longitudinal cohort study of children at risk for aggressive periodontal disease?, *J. Periodont.* 80, 106–113, 2009.

809. Wang, J., Lee, J., Burns, D. et al., "Fit-for-purpose" method validation and application of a biomarker (C-terminal telopeptides of type I collagen) in denosumab clinical studies, *AAPS J.* 11, 385–394, 2009.

810. DeCaprio, A.P. (ed.), *Toxicological Biomarkers*, Taylor & Francis, New York, 2006.

811. Anderson, D., Factors contributing to biomarker responses in exposed workers, *Mutat. Res.* 428, 197–202, 1998.

812. Shuker, D.E., Prevost, V., Friesen, M.D. et al., Urinary markers for measuring exposure to endogenous and exogenous alkylating agents and precursors, *Environ. Health Perspect.* 99, 33–37, 1993.

813. Chang, X., Jin, T., Nordberg, M., and Lei, L., Metallothionein I isoform mRNA expression in peripheral lymphocytes as a biomarker for occupational cadmium exposure, *Exp. Biol. Med.* 234, 666–672, 2009.

814. Wiwanitkit, V., Classification of occupations at risk for lead exposure using blood lead levels as a biomarker, *Toxicol. Environ. Chem.* 91, 75–78, 2009.

815. Sobus, J.R., Waidyanatha, S., McClean, M.D. et al., Urinary naphthalene and phenanthrene as biomarkers of occupational exposure to polycyclic aromatic hydrocarbons, *Occup. Environ. Med.* 66, 99–104, 2009.

816. Barbieri, A., Violante, F.S., Sabatini, L. et al., Urinary biomarkers and low-level environmental benzene concentration: Assessing occupational and general exposure, *Chemosphere* 74, 64–69, 2008.

817. Shahtaheri, S.J., Abdollahi, M., Golbabaei, F. et al., Monitoring of mandelic acid as a biomarker of environmental and occupational exposures to styrene, *Int. J. Environ. Res.* 2, 169–176, 2008.

818. Lindh, C.H., Littorin, M., Johannesson, G., and Joensson, B.A.G., Analysis of ethylenethiourea as a biomarker to human urine using liquid chromatography/triple quadrupole mass spectrometry, *Rapid Commun. Mass. Spectrom.* 22, 2573–2579, 2008.

819. Tinnerberg, H., Sennbro, C.J., and Jonsson, B.A.G., Aniline in hydrolyzed urine and plasma-possible biomarkers for phenylisocyanate exposure, *J. Occup. Environ. Hyg.* 5, 629–632, 2008.

820. Staruchova, M., Collins, A.R., Volkovova, K. et al., Occupational exposure to mineral fibres. Biomarkers of oxidative damage and antioxidant defence and associations with DNA damage repair, *Mutagenesis* 23, 249–260, 2008.

821. Sabatini, L., Barbieri, A., Mattioli, S. et al., Validation of an HPLC-MS/MS method for the simultaneous determination of phenylmercapturic acid, benzylmercapturic acid, and o-methylbenzyl mercapturic acid in urine as biomarkers of exposure of benzene, toluene, and xylenes, *J. Chromatogr. B: Analyt. Technol. Biomed. Life Sci.* 863, 115–122, 2008.

822. Inayat-Hussain, S.H., Lubis, S.H., Sakian, N.I.M. et al., Is plasma β-glucuronidase a novel human biomarker for monitoring anticholinesterase pesticides exposure? A Malaysian experience, *Toxicol. Appl. Pharmacol.* 219, 210–216, 2007.

823. Joo, W.-A., Sui, D., Lee, D.-Y. et al., Proteomic analysis of plasma proteins of workers exposed to benzene, *Mutat Res.: Genet. Toxicol. Environ. Mutagen.* 558, 35–44, 2004.

824. Witzmann, F.A., Bauer, M.D., Fieno, A.M. et al., Proteomic analysis of the renal effects of simulated occupational jet fuel exposure, *Electrophoresis* 21, 976–984, 2000.

825. Lee, Y.-S., Chen, P.-W., Tsai, P.-J. et al., Proteomics analysis revealed changes in rat bronchoalveolar lavage fluid proteins associated with oil mist exposure, *Proteomics* 6, 2236–2250, 2006.

826. Ahsan, N., Renault, J., and Komatsu, S., Recent developments in the application of proteomics to the analysis of plant responses to heavy metals, *Proteomics* 9, 2602–2621, 2009.

827. Sala, G.L., Ronzitti, G., Sasaki, M. et al., Proteomic analysis reveals multiple patterns of response in cells exposed to a toxin mixture, *Chem. Res. Toxicol.* 22, 1077–1085, 2009.

828. Hegedus, C.M., Skibola, C.F., Warner, M. et al., Decreased urinary beta-defensin-1 expression as a biomarker of response to arsenic, *Toxicol. Sci.* 106, 74–82, 2008.

829. Ahsan, N., Lee, D.-G., Alam, I. et al., Comparative proteomic analysis of differentially expressed proteins in rice roots reveals glutathione plays a central role during As stress, *Proteomics* 8, 3561–3576, 2008.

830. Merrick, B.A., The plasma proteome, adductome and idiosyncratic toxicity in toxico-proteomics research, *Brief. Funct. Genomic Proteomic* 7, 35–49, 2008.

831. Preston, A., Fodey, T., Douglas, A., and Elliott, C.T., Monoclonal antibody development for acrylamide-adducted human haemoglobin: A biomarker of dietary acrylamide exposure, *J. Immunol. Methods* 341, 19–29, 2009.

832. Petricoin, E.F., Rajapaske, V., Herman, E.H. et al., Toxicoproteomics: Serum proteomic pattern diagnostics for early detection of drug induced cardiac toxicities and cardiopro-tection, *Toxicol. Pathol.* 32(Suppl 1), 122–130, 2004.

833. Benningoff, A.D., Toxicoproteomics—The next step in the evolution of environmental biomarkers, *Toxicol. Sci.* 95, 1–4, 2007.

834 Grzegorczyk, M., Comparison of two different stochastic models for extracting pro-tein regulatory pathways using Bayesian networks, *J. Toxicol. Environ. Health A* 71, 827–834, 2008.

835. Santos, P.M., Simões, T., and Sá-Correia, I., Insights into yeast adaptive response to the agricultural fungicide mancozeb: A toxioproteomic approach, *Proteomics* 9, 657–670, 2009.

836. Gao, Y., Holland, R.D., and Yu, L.R., Quantitative proteomics for drug toxicity, *Brief Funct. Genomic. Proteomics* 8, 158–166, 2009.

837. Lee, J.W. and Hall, M., Method validation of protein biomarkers in support of drug develelopment or clinical diagnosis/prognosis, *J. Chromatogr. B* 877, 1259–1271, 2009.

838. Scaros, O. and Fisler, R., Biomarker technology roundup: From discovery to clinical applications, a broad set of tools is required to translated from the lab to the clinic, *Biotechniques* 38, 30–32, 2005.

839. Lindon, J.C., Holmes, E., and Nicholson, J.K., Metabolomics techniques and applica-tion to pharmaceutical research & development, *Pharm. Res.* 23, 1075–1088, 2006.

840. McBurney, R.N., Hines, W.M., Von Tungeln, L.S. et al., The liver toxicity biomarker study: Phase I design and preliminary results, *Toxicol. Pathol.* 37, 52–64, 2009.

841. Merrick, B.A. and Witzmann, F.A., The role of toxicoproteomics in assessing organ specific toxicity, *EXS* 99, 367–400, 2009.

842. Psaty, B.M., Weiss, N.S., Furberg, C.D. et al., Surrogate end points, health outcomes, and the drug-approval process for the treatment of risk factors for cardiovascular dis-ease, *JAMA* 282, 786–790, 1999.

843. Fleming, T.R., Surrogate endpoints and FDA's accelerated approval process, *Health Aff.* (Milwood) 24, 67–78, 2005.

844. Petkau, J., Reingold, S.C., Held, U. et al., Magnetic resonance imaging as a surrogate outcome for multiple sclerosis relapses, *Mult. Scler.* 14, 700–778, 2008.

845. Fleming, T.R. and DeMets, D.L., Surrogate end points in clinical trials: Are we being misled?, *Ann. Intern. Med.* 125, 605–613, 1996.

846. Peterson, R.C. and Trojanowski, J.Q., Use of Alzheimer disease biomarkers. Potentially yes for clinical trials but not yet for clinical practice, *JAMA* 302, 436–437, 2009.

847. Mattsson, N., Zetterberg, H., Hansson, O. et al., CSF biomarkers and incipient Alzheimer disease in patients with mild cognitive impairment, *JAMA* 302, 385–393, 2009.
848. Vellas, B., Andrieu, S., Sampaio, C. et al., Endpoints for trials in Alzheimer's disease: A European task force consensus, *Lancet Neurol.* 7. 436–450, 2008.
849. Prentice, R.L., Surrogate endpoints in clinical trials: Definition and operational criteria, *Stat. Med.* 8, 431–440, 1989.
850. Simon, R.M., Guidelines for the design of clinical studies for the development and validation of therapeutic relevant biomarkers and biomarker-based classification systems, in *Biomarkers in Breast Cancer*, G. Gasparini and D.F. Hayes (eds.), Humana Press, Totowa, NJ, Chapter 1, pp. 3–15, 2006.
851. Mandrekar, S.J. and Sargent, D.J., Clinical trial design for predictive biomarker validation: One size does not fit all, *J. Biopharm. Stat.* 19, 530–542, 2009.
852. Blair, E.D., Assessing the value-adding impact of diagnostic-type tests on drug development and marketing, *Mol. Diagn. Ther.* 12, 331–337, 2009.
853. Hinman, L., Spear, B., Tsuchihashi, Z. et al., Drug-diagnostic codevelopment strategies: FDA and industry dialogue at the 4th FDA/DIA/PhRMA/PWG/BIO pharmacogenomics workshop, *Pharmacogenomics* 10, 127–136, 2009.
854. Papadopoulos, N., Kinzler, K.W., and Vogelstein, B., The role of companion diagnostics in the development and use of mutation-targeted cancer therapies, *Nat. Biotechnol.* 24, 985–995, 2006.
855. Jones, D., Biomarker debate highlights retrospective challenge, *Nat. Rev. Drug Discov.* 8, 179–180, 2009.
856. Cross, J., DxS Ltd., *Pharmacogenomics* 9, 463–467, 2008.
857. Ashton-Chess, J., Guillet, M., and Huricz, A., TcLand Expression, *Personalized Med.* 6, 381–384, 2009.
858. Viljoen, A. and Wierzbicki, A.S., Towards companion diagnostics for the management of statins, *Expert Opin. Med. Diagn.* 3, 659–671, 2009.
859. Landais, P., Eresse, V., and Ghislain, J.-C., Evaluation and validation of diagnostic tests for guiding therapeutic decisions, *Therapie* 64, 187–201, 2009.

3 Development of Biomarkers for Oncology

The great interest in the development of biomarkers for oncology is driven, in part, by the presence of precancerous conditions and the frequent absence of physical signs prior to the development of malignant tumors. It is generally accepted that an early diagnosis of cancer is the most important factor in therapeutic outcome,[1–4] and that the development of validated biomarkers is important for early diagnosis.[5–9] A PubMed search for the terms "biomarker" and "cancer" yielded more that 160,000 citations, with search results extending back to 1963; since the first biomarker for cancer was described in 1848,[*,10] it is safe to assume that there are substantially more studies in the literature that are not contained in current databases.[†] Given the large number of studies, it is fair to say that the study of cancer biomarkers is a mature discipline. The reader is directed to various books on the subject of cancer biomarkers, which have been published since 1990,[11–16] for more information on development over the last 20 years. In addition, there is an informative chapter[17] on tumor markers in the most recent edition of one of the leading reference books on clinical chemistry.

While cancer biomarkers can be used in various phases of the disease process as discussed in Chapter 2, it is fair to say that the greatest interest in cancer biomarkers is for screening with somewhat less interest in prognostic and/or predictive applications. There is interest in diagnosis but the use of biomarkers is overshadowed by histological or cytological examination, which is referred to as a "gold standard" for diagnosis or prognosis[18–21] although others consider clinical outcome as the "gold standard."[22,23] The spectrum of biomarkers in oncology is broad, ranging from tumor size and other imaging biomarkers using technology such as positron emission tomography to the molecular biomarkers described below.[24–26]

There is a need for caution in this area of research. The need for a useful screening diagnostic for cancer is clear and hope is high. There have, however, been a number of "false alarms," which have required a good bit of unraveling.[27] The most recent issue of *Science* (this is being written on September 29, 2009) contains an article[28] of litigation concerning the validity of data for a prostate cancer test. The fact is that

---

* The reader is directed to Roulston, J.E. and Leonard, R.C.F., *Serological Tumour Markers: An Introduction*, Chapter 1, Churchill Livingstone, Edinburgh, U.K., 1993, for a useful discussion of the historical development of biomarkers for cancer.

† Not all journals, particularly some new Internet (nonprint), are indexed in PubMed, ISI, or SciFinder™ databases. Most books are not indexed in these databases. The author's experience suggests that even the most rigorous of searches misses 10%–20% of the literature—of course, as has been noted by various national leaders, you don't know what you don't know.

this is an area where there is great public hope and it is easy for companies (and scientists) be more concerned with the financial outcome than the provision of a service to the community. As noted below, it may not be possible to develop a screening test for a normal population. One notable exception is the Pap smear,[29,30] named after George Pananikolaou.

INTERMEDIATE BIOMARKERS AND THE PRECANCEROUS CONDITION

As with many chronic diseases, the development of cancer is a process starting from normal and proceeding through "precancerous" to stages to malignancy. The intermediate biomarker appears to be unique to oncology[31–35] and linked to precancerous conditions.[32,36,37] Kosmeder and Pezzuto[33] defined intermediate biomarkers of cancer as those phenotypic, genotypic, and molecular changes that characterized the multistage nature of carcinogenesis. Boffetta and Trichopoulos[35] have suggested that an intermediate biomarker measures early, nonpersistent biological events such as cellular toxicity (cytogenetic abnormalities), changes in DNA, RNA, and protein expression, and DNA repair taking place between exposure and cancer development. Intermediate biomarkers are important to molecular epidemiology.[38–43] Examples of intermediate biomarkers are listed in Table 3.1 while the use of biomarkers for precancerous conditions is listed in Table 3.2. It is acknowledged that the distinction between intermediate and precancerous biomarkers is somewhat artificial as it would seem that intermediate biomarkers measure the precancerous condition. While intermediate biomarkers/biomarkers of precancerous conditions tend to be directed toward cell morphology, there are DNA adducts (Figure 3.1) and oxidation products (Figure 3.2) of considerable interest.[58,59] It is also noted that DNA adducts overlap between exposure biomarkers and intermediate biomarkers.[58] Intermediate biomarkers are considered important for chemoprevention.[60,61]

TABLE 3.1
Some Examples of Intermediate Biomarkers

Biomarker(s)	Reference
Intermediate biomarkers for colorectal cancer includes expression of mucins, intermediate filaments and cytoskeletal proteins, and the structure and expression of a variety of genes associated with normal and abnormal cell development	36
Rectal cell proliferation as an intermediate biomarker in colorectal cancer (altered distribution of proliferating cells in colorectal crypts)	44
p53 as an intermediate biomarker in Barrett's esophagus	45
Definition of biomarkers of exposure, biomarkers of susceptibility, and biomarkers of effect. The use of intermediate biomarkers to determine cancer risk at very early stages	46
Hypomethylation of DNA as an intermediate biomarker in colorectal carcinogenesis	47
Nuclear matrix composition as an intermediate biomarker in squamous cell carcinoma	48
Use of quantitative reverse transcription-polymerase chain reaction (quantitative RT-PCR) to measure intermediate biomarkers in tissue biopsies from oral cancer patients	49

TABLE 3.2

Some Examples of Biomarkers for Precancerous Lesions

Biomarker(s)	References
Cellular proliferation (labeling index), apoptosis, polyamine metabolism, arachidonic acid metabolism, genetic alterations	50
NAD(P)H and collagen as biomarkers for epithelial precancerous lesions	51
Cytogenetic changes (aneusomy demonstrated by fluorescence in situ hybridization [FISH] analysis) as biomarker for preinvasive changes in the development of lung cancer	52–54
Proliferating cell nuclear antigen (PCNA), p53, and polyamines as biomarker for early skin tumorogenesis	55
Discussion of precancerous biomarkers for cervical cancer including human papillovirus DNA, E6/E7 mRNA, minichromosome maintenance protein, PCNA, MIB-1, cyclin E and p16INK4A	56
Use of FISH to detect chromosomal abnormalities in preinvasive ulcerative colitis neoplasia	57

CANCER AND INFLAMMATION

There is the complex relationship between tumors and inflammation.[62] There are chronic inflammatory conditions such as pancreatitis (pancreatic cancer), Crohn's disease (colon cancer), and lichen planus (oral cancer) that predispose to tumor development.[63–68] The inflammatory response might also support or suppress tumor growth.[69–71] Examples of inflammation biomarkers in oncology are listed in Table 3.3.

S100 protein was first described by Moore in 1965 as a fraction from brain, which was soluble in saturated ammonium sulfate.[82] S100 protein was subsequently demonstrated to be a family of small proteins[83,84] characterized by diverse function and the ability to bind calcium ions.[85,86] A search of PubMed yielded more than 3000 citations to the use of S100 proteins as biomarkers. One of these S100 biomarkers is calprotectin, originally described as leukocyte protein L1,[87–89] which is a heterodimer of S100A8 and S100A9[90,91] involved in the modulation of the inflammatory response.[92] The majority of S100 biomarker citations are for use in cancer, and some examples of serum S100 protein biomarkers are presented in Table 3.4. There has been considerable interest in the use of serum levels S100B protein (S100β; S100 beta; S100 calcium-binding protein B) as a diagnostic, predictive, and prognostic biomarker in melanoma.[101–122] Table 3.5 lists some examples of the expression of S100 proteins in oncology.

There has been particular interest in calprotectin as a fecal marker for colorectal cancer[130–134] but the specificity is complicated by presence in inflammatory digestive disease.[135–138] There is also interest in CRP as biomarker for increased cancer risk (Table 3.6).[151–153] While such studies do not suggest the direct involvement of CRP or calprotectin in the metastatic process, it is clear that inflammation predisposes to tumor development and tumor growth, and that biomarkers such as CRP can serve to measure risk and prognosis. An examination of the information in Table 3.6 suggests that the presence of CRP is not reflective of a specific tumor but a general part of the metastatic process.

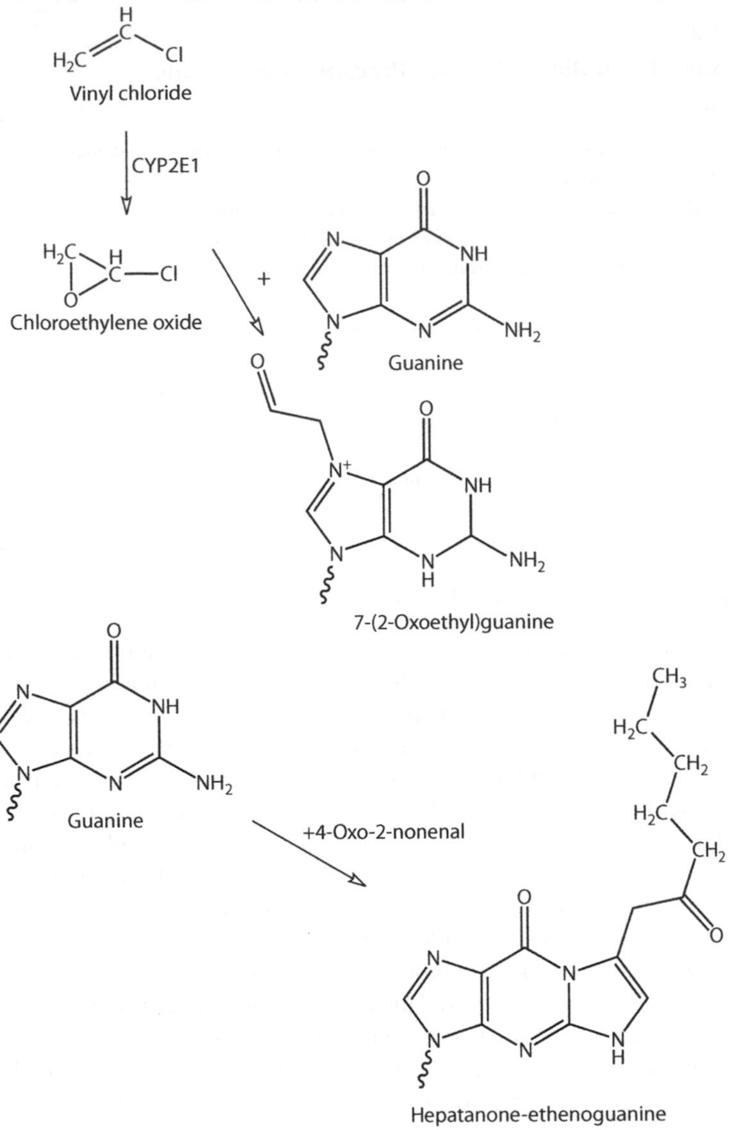

FIGURE 3.1 The formation of adducts of nucleic acids. The term adduct refers to the combination of two different materials to form a compound. While this could refer to the chemical interaction of any two chemicals to form a derivative compound, the term adduct appears to have a unique application within nucleic acid chemistry. Shown here are adducts of vinyl chloride with guanosine (Weyandt, J., Ellsworth, R.E., Hooke, J.A. et al., Environmental chemicals and breast cancer risk—A structural chemistry perspective, *Curr. Med. Chem.* 15, 2680–2701, 2008) and the reaction of 2-hydroxy-4-nonenal with guanosine (Blair, I.A., DNA adducts with lipid peroxidation products, *J. Biol. Chem.* 283, 15545–15549, 2008). See also Broyde, S., Wang, L., Zhang, L. et al., DNA adduct structure-function relationships: Comparing solution with polymerase structures, *Chem. Res. Toxicol.* 21, 45–52, 2008; Gallo, V., Khan, A., Gonzales, C. et al., Validation of biomarkers for the study of environmental carcinogens: A review, *Biomarkers* 13, 505–534, 2008.

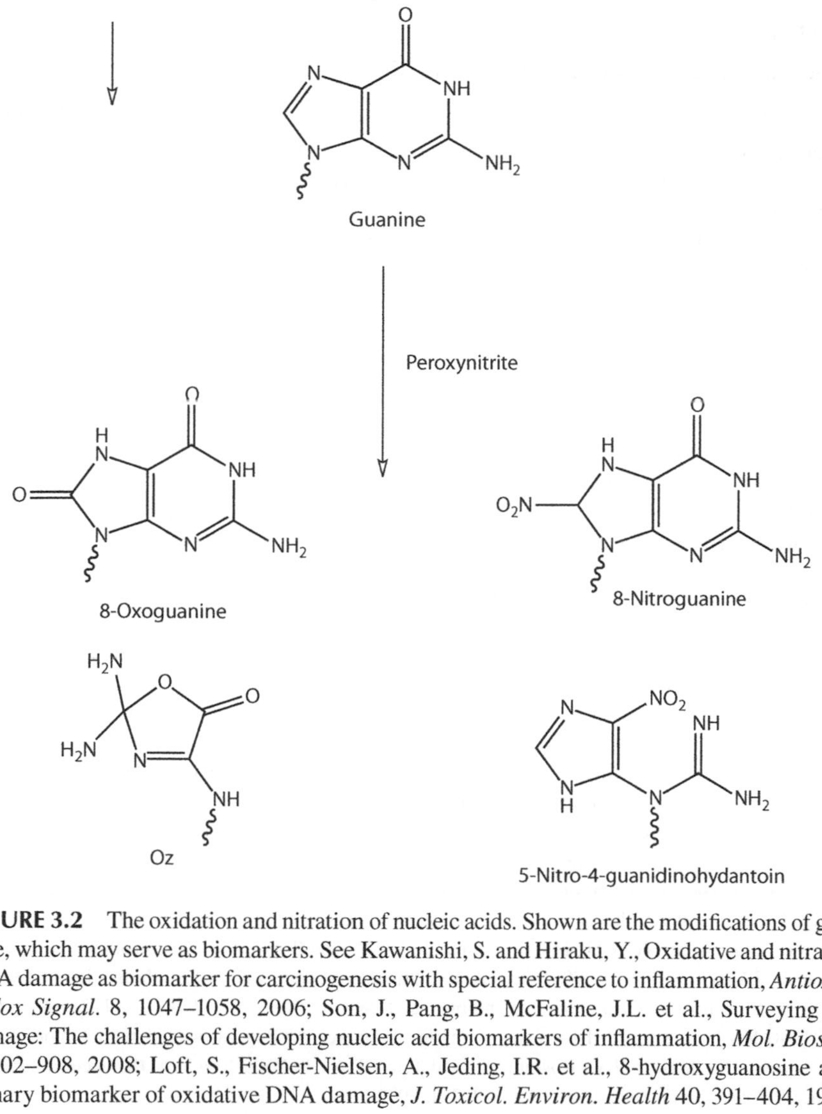

FIGURE 3.2   The oxidation and nitration of nucleic acids. Shown are the modifications of guanine, which may serve as biomarkers. See Kawanishi, S. and Hiraku, Y., Oxidative and nitrative
DNA damage as biomarker for carcinogenesis with special reference to inflammation, *Antioxid.
Redox Signal.* 8, 1047–1058, 2006; Son, J., Pang, B., McFaline, J.L. et al., Surveying the
damage: The challenges of developing nucleic acid biomarkers of inflammation, *Mol. Biosyst.*
4, 902–908, 2008; Loft, S., Fischer-Nielsen, A., Jeding, I.R. et al., 8-hydroxyguanosine as a
urinary biomarker of oxidative DNA damage, *J. Toxicol. Environ. Health* 40, 391–404, 1993.

A consideration of the material in the preceding tables suggest that (1) changes specific to a tumor are detected either by histochemistry or gene expression and (2) serological markers are nonspecific in that, as with CRP, they are markers of systemic response
rather than individual tumor type. Roulston and Leonard present an excellent discussion[154] of the basis of serological basis of tumor markers noting that there is little basis
for thinking that tumors will produce a biomarker not produced by non-neoplastic tissue. In a latter consideration, Roulston[155] restates that position and notes disease prevalence is the driving force for screening and disease prevalence in cancer is too low in the
general population to justify screening. This does not mean that the search for unique
biomarkers for screening and diagnosis should end but rather that greater thought should

TABLE 3.3

Some Inflammation Biomarkers in Oncology

Biomarker(s)	References
The Glasgow Prognostic Score (GPS),[a] which is derived from the level of CRP and albumin is a prognostic biomarker for advance lung and gastrointestinal cancer. An increase in γ-glutamyl transferase and alkaline phosphatase is also associated with the systemic inflammatory response in these advanced cancers	72
The formation of 8-nitroguanine[b] is an inflammation biomarker found in various tumors	73–76
IL-6 and VEGF are predictive biomarkers of inflammation in anthracycline-based chemotherapy in breast cancer	77
Serum levels of PSA is elevated in prostate inflammation (prostatitis) and can be a confounding factor in the use of PSA in prostate cancer	78
2-deoxy-2-^{18}F-glucose (FDG)[c] accumulation in inflammatory tissue may cause a false-positive during cancer screening	79
YKL-40 (HCgp39)[d] is a prognostic biomarker in cancer	80
Endothelial nitric oxide synthase (eNOS) is a biomarker for chronic inflammation and carcinogenesis in upper airways	81

[a] See Forrest, L.M., McMillan, D.C., McArdle, C.S. et al., Evaluation of cumulative prognostic scores based on the systemic inflammatory response in patients with inoperable non-small-cell lung cancer, *Br. J. Cancer* 89, 1028–1030, 2003.

[b] See Yermilov, V., Rubio, J., Becchi, M. et al., Formation of 8-nitroguanine by the reaction of guanine with peroxynitrite *in vitro*, *Carcinogenesis* 16, 2045–2050, 1995; Szabó, C. and Oshima, H., DNA damage induced by peroxynitrite: Subsequent biological effects, *Nitric Oxide* 1, 373–385, 1997.

[c] Positron emission tomography (PET) is used with radiotracers such as 2-deoxy-2-^{18}Fluoro-glucose for the identification of tumors. See Phelps, M.E., PET: The merging of biology and imaging into molecular imaging, *J. Nucl. Med.* 41, 661–681, 2000; Yasuda, S. and Ide, M., PET and cancer screening, *Ann. Nucl. Med.* 19, 167–177, 2005; Facey, K., Bradbury, I., Laking, G., and Payne, E., Overview of the clinical effectiveness of positron emission tomography imaging in selected cancers, *Heath Technol. Assess.* 11(44), xi-267, 2007.

[d] YKL-40 is a member of the mammalian chitinase family involved in inflammation, tissue remodeling, fibroses, and cancer (Bleau, G., Massicotte, F., Merlen, Y., and Boisvert, C., Mammalian chitinase-like proteins, *EXS* 87, 211–221, 1999; Johansen, J.S., Studies on serum YKL-40 as a biomarker in diseases with inflammation, tissue remodeling, fibroses and cancer, *Dan. Med. Bull.* 53, 172–209, 2006). YKL-40 has been suggested to involved in the tissue remodeling associated with tumor growth (Johansen, J.S., Jensen, B.V., Roslind, A., and Price, P.A., Is YKL-40 a new therapeutic target in cancer?, *Expert Opin. Ther. Targets* 11, 219–234, 2007).

be given to the further characterization of prognostic biomarker to distinguish between tumor markers and host markers.[156] Emphasis on predictive biomarkers is critical for the application of personalized medicine to cancer treatment.[157–160] The following sections will describe the development of biomarkers in various cancers.

HEAD AND NECK CANCER

The term "head and neck cancer" defines a group of squamous cell carcinomas accounting for 2.5% of all cancer (approximately 35,000 cases/year) in the United States,[161] which originate in the mucosal surfaces of the head and neck, identified by

TABLE 3.4
The Use of S100 Proteins as Biomarkers in Oncology

Biomarker Study	Reference
(Calgranulin) Calgranulin A (S100A8) and calgranulin B (S100A9) in cystic fibrosis for ovarian fluid	93
S100A6 as biomarker in human osteosarcoma	94
Expression of S100A2 and S100A4 as biomarker in Barrett's esophagus (Barrett's adenocarcinoma)	95
Immobilized anti-S100 protein antibody as assay matrix for measurement of serum S100 protein prostate cancer	96
S100A4, S100A8, and S100A11 are prognostic indicators in bladder cancer	97
Serum metastasin mRNA is a prognostic biomarker in breast cancer	98
Calprotectin as a prognostic biomarker in invasive ductal carcinoma of the breast	99
Increased expression of S100A4 (fibroblast-specific protein 1) as a prognostic biomarker for advanced-stage endometrial cancer	100

TABLE 3.5
The Use of S100 Protein Expression as Biomarker in Oncology

Biomarker Study	Reference
Expression of S100 proteins as a prognostic biomarker for gastrointestinal stromal tumors	123
Expression of S100A2 as a prognostic biomarker in pancreatic cancer	124
Expression of S100A6 protein as a differentiating diagnostic biomarker for cutaneous tumors; expression was observed in squamous cell carcinoma but not in basal cell carcinoma	125
Expression of S100A4 as a biomarker for lymph node metastasis	126
S100 proteins as biomarkers in squamous cell carcinoma (downregulation of S100A4, S100A6, S100A8, S100A14)	127
S100B (serum) is a prognostic biomarker for stage III melanoma but lack of statistical significance between serum S100B and expression in tumor specimen	119
S100A2 is a predictive biomarker for pancreatectomy in pancreatic cancer	128
Expression of S100A4 as a biomarker for colorectal cancer	129

location of origin[162–164] and are notable by poor outcome.[165,166] As with most tumors, the prognosis is good if the diagnosis occurs early in the disease process.[167,168] Notwithstanding the above arguments on the specificity of biomarkers, it is thought that biomarkers would be useful in the early recognition of head and neck cancer.[169–172] While progress is being made, a recent study suggests that event-free survival or perhaps locoregional control are surrogate end points for overall survival, which is still considered the gold standard for assessing the effectiveness of chemotherapy or radiation therapy.[23] Some examples of biomarkers for head and neck cancer are presented in Table 3.7. Wineland and Stack[185] have recently presented a discussion on the importance of biomarkers in the use of evidence-based medicine in optimizing the quality and cost of the treatment of head and neck cancer.

TABLE 3.6

The Use of C-Reactive Protein as an Oncology Biomarker

Biomarker Study	Reference
CRP is a prognostic biomarker for advanced pancreatic cancer	139
CRP is an independent prognostic factor in non-small-cell lung cancer	140
CRP (and albumin) is a predictive biomarker for radiotherapy in esophageal cancer	141
CRP is a biomarker for progression of epithelial ovarian cancer from stage I to stage IV. A new method, protein cleavage isotope dilution mass spectrometry, gives a higher value than an ELISA technique	142
CRP as a biomarker of risk for incident cancer	143
CRP is a risk biomarker for epithelial ovarian cancer; risk of mucinous tumors increased with high CRP levels	144
CRP is a predictive biomarker for mucositis in the radiation treatment of head and neck cancer	145
Patients with inoperable tumors have higher CRP levels. Patients with operable tumors and hypertension had higher CRP levels than normotensive patients. Anti-hypertensive drugs suppressed tumor-induced CRP production	146
High-sensitivity CRP is a biomarker for large colorectal adenomas but not to small adenomas. It is suggested that inflammation as measured by CRP is linked to the growth of colorectal adenomas	147
CRP is a prognostic biomarker for oral squamous cell carcinoma; combining tumor size with CRP increases predictive power of CRP	148
CRP is a biomarker for ovarian cancer. CRP is independent of CA125 and is an additional biomarker for the differential diagnosis of ovarian cancer	149
CRP is a biomarker with albumin, which is a part of the modified Glasgow Prognostic Score (mGPS), an inflammation-based prognostic score. The mGPS is a predictive biomarker for chemotherapy in unrectable colorectal cancer	150

OVARIAN CANCER

The search for biomarkers to be used for screening for ovarian cancer is a high-profile area in oncology research. Ovarian cancer represents approximately 25% of gynecological cancer but is responsible for approximately 50% of cancer deaths in women.[186,187] The high mortality rate reflects the late diagnosis of ovarian cancer. It is of interest that there is a difference in incidence rate based on the country of origin, age, and race.[188] Recent information from the World Health Organization[189] shows higher incidence/mortality in more developed regions than in less developed regions.

Tumor-specific antigens have been of interest for some time.[190] Carcinoembryonic antigen (CEA) was the first well-described tumor-specific antigen.[191] CEA was characterized in human colonic carcinoma by immunological tolerance and adsorption techniques by Gold and Freedman in 1965.[192] Gold and coworkers subsequently described a radioimmunoassay for CEA.[193] CEA has been useful for colorectal cancer but is more a general than specific tumor biomarker.[194–199] Work on the identification of tumor-specific antigens for ovarian cancer was proceeding at the same time[200]

TABLE 3.7
Some Biomarkers for Head and Neck Cancer

Biomarker(s)	References
Gene expression as predictive biomarker for docetaxel therapy in head and neck squamous cell cancer. cDNA microarray data is combined with Powerblot™ immunoblotting[a]	173
Serum amyloid A protein is a predictive biomarker in nasopharyngeal cancer; serum amyloid A protein increased in patients who relapsed during a trial with gemciabine/cisplatin[b]	174
Telomerase expression in peripheral blood mononuclear cells is a prognostic biomarker for head and neck cancer	175
DNA methylation in saliva is a biomarker for prognosis in head and neck cancer; abnormal methylation is detectable in saliva prior to clinical signs of relapse	176
Soluble CD44[c] in saliva or oral rinses is a biomarker for head and neck cancer[d]	177, 178
Serum zinc concentration is a prognostic biomarker for head and neck cancer	179
Serum soluble IL-2 receptor and MMP-9 are prognostic biomarkers for head and neck cancer	180
Use of phage display of tumor antigens for inclusion in a microarray platform for the identification of an autoantibody panel for the screening/early diagnosis of head and neck cancer	181
Two 14-3-3 proteins[e] [stratifin(14-3-3sigma) and YWHAZ (14-3-3zeta)] are suggested as prognostic biomarkers for head and neck cancer	182
Circulating endothelial progenitor cells (CD133+/KDR+) are biomarkers for head and neck cancer	183
HIF-1α[f] and CAIX are prognostic biomarkers in head and neck cancer	184

[a] Also known as high-throughput immunoblotting. See Kim, H.J. and Lotan, R., Identification of retinoid-modulated proteins in squamous carcinoma cells using high-throughput immunoblotting, *Cancer Res.* 64, 2439–2448, 2004; Zhou, Q. and Amar, S., Identification of proteins differentially expressed in human monocytes exposed to *Porphyromas gingivalis* and its purified components by high-throughput immunoblotting, *Infect. Immun.* 74, 1204–1214, 2006; Whyte, L., Huang, Y.Y., Torres, K., and Mehta, R.G., Molecular mechanisms of resveratrol action in lung cancer cells using dual protein and microarray analyses, *Cancer Res.* 67, 12007–12017, 2007.

[b] Serum amyloid protein A is an acute phase protein, which is synthesized in the liver and is a marker for inflammatory disease; serum amyloid protein A is also synthesized in tumor tissue and is thought to be directly involved in neoplastic disease (see Malle, E., Sodin-Semrl, S., and Kovacevic, A., Serum amyloid A: An acute-phase protein involved in tumour pathogenesis, *Cell. Mol. Life Sci.* 66, 9–26, 2009). Another study showed the production of serum amyloid protein A in uterine serous papillary cancer (Cocco, E., Bellone, S., El-Sahwi, K. et al., Serum amyloid A (SAA): A novel biomarker for uterine serous papillary cancer, *Br. J. Cancer* 101, 335–341, 2009).

[c] CD44 is a receptor with complex functions. It probably best known as a receptor for hyaluronan (Lesley, J. and Hyman, R., CD44 structure and function, *Front. Biosci.* 3, d616–d30, 1998) but is now recognized to function in inflammation and tumor growth (Naor, D., Nedvetzki, S., Golan, I. et al., CD44 in cancer, *Crit. Rev. Clin. Lab. Sci.* 39, 527–579, 2002; Cichy, J. and Puré, E., The liberation of CD44, *J. Cell Biol.* 161, 839–843, 2003). As with other cell surface receptors, isoforms exist including soluble forms derived from proteolysis (Cichy, J., Bals, R., Potempa, J. et al., Proteinase-mediated release of epithelial cell-associated CD44. Extracellular CD44 complexes with components of cellular matrices, *J. Biol. Chem.* 277, 44440–44447, 2002) or alternative splicing (Mayer, S., zur Hausen, A., Watermann, D.O.

(*continued*)

TABLE 3.7 (continued)
Some Biomarkers for Head and Neck Cancer

et al., Increased soluble CD44 concentrations are associated with larger tumor size and lymph node metastasis in breast cancer patients, *J. Cancer Res. Clin. Oncol.* 134, 1229–1235, 2008).

d CD44 was observed to be constitutively expressed in permanent head and neck squamous cell carcinoma cells lines; CD44 expressing potential stem cells were found in solid tumors from patients with head and neck squamous cell carcinoma (see Pries, R., Witrkopf, N., Trenkle, T. et al., Potential stem cell marker CD44 is constitutively expressed in permanent cell lines of head and neck cancer, *In Vivo* 22, 89–92, 2008).

e The 14-3-3 protein family are small acidic proteins serving a regulatory process, which conserved in eukaryotes (see Zannis-Hadjopoulos, M., Yahyaoui, W., and Callejo, M., 14-3-3 cruciform-binding proteins as regulators of eukaryotic DNA replication, *Trends Biochem. Sci.* 33, 44–50, 2008; Obsilová, V., Silhan, J., Boura, E. et al., 14-3-3 proteins: A family of versatile molecular regulators, *Physiol. Res.* 57(Suppl 3), S11–S21, 2008; Morrison, D.K., The 14-3-3 proteins: Integrators of diverse signaling cues that impact cell fate and cancer development, *Trends Cell Biol.* 19, 16–23, 2009). The nomenclature 14-3-3 was derived from the original systematic classification of nervous system proteins (Aitken, A., Collinge, D.B., van Heusden, B.P. et al., 14-3-3 proteins: A highly conserved, widespread family of eukaryotic proteins, *Trends Biochem. Sci.* 17, 498–501, 1992).

f HIF-1α is a general prognostic biomarker for cancer (see Tang, B., Qu, Y., Zhao, F. et al., In vitro effects of hypoxia-inducible factor 1α on the biological characteristics of the SiHA uterine cervix cancer cell line, *Int. J. Gynecol. Cancer* 19, 898–904, 2009; Naidu, R., Har, Y.C., and Taib, N.A., Associations between developing breast cancer, *Neoplasma* 56, 441–447, 2009; Monolescu, B., Oprea, E., and Busu, C., Natural compounds and the hypoxia-inducible factor (HIF) signaling pathway, *Biochimie* 91, 1347–1358, 2009).

because of the recognition of the need for an early diagnostic tool[201] although it was recognized that such antigens were not unique but rather increased in concentration in malignancy.[202]

The most extensively used biomarker for ovarian cancer is CA125 (Mucin 16; Muc16[203,204]), which was first described in 1981 by Bast and colleagues,[205] who described a mouse monoclonal antibody, OC125, which reacted with each of six epithelial ovarian cancer cell lines, with tissue from ovarian cancer patients but not with non-malignant tissues including fetal and adult ovary. There was evidence to support specificity for epithelial ovarian cancer as OC125 did not react with a pancreatic cancer cell line, a breast cancer cell line, or CEA. An immunoassay was subsequently developed for diagnostic use,[206,207] which has seen considerable use in monitoring ovarian cancer. Table 3.8 describes work on the development of CA125 as a biomarker for ovarian cancer.

It was recognized earlier that while the measurement of CA125 was useful in the monitoring the treatment of ovarian cancer, it was not possible to use it for early diagnosis because of lack of specificity[217]; the specificity of the assay for CA125 antigen assay has 83% (RIA) for ovarian tumors; there was also high reactivity (66%) with cervical tumors and uterine tumors (56%) with much less reactivity (20%) for colon cancer.

There are other studies that demonstrate lack of sensitivity of CA125 for the diagnosis of ovarian cancer. There is a report of elevation of CA125 in a patient with right

TABLE 3.8

Observations on the Use of CA125 as a Biomarker for Ovarian Cancer

Study Description	References
Failure of CA125 to detect early stage ovarian cancer	208, 209
Identification of a new biomarker complementary to CA125	210
Value of pattern recognition in diagnostics	211
Haptoglobin α-subunit complementary to CA125 in the diagnosis of ovarian cancer	212
Value of other biomarkers such as CA 19.9, CA 15.3, and TAG.72 in the diagnosis of ovarian cancer. CA 19.9 has a high sensitivity for the mucinous histotype. CA125 is also useful for endometrial cancer. CA125 is useful for the monitoring of therapy. Elevated CA125 may also be elevated in cervical cancer	213
Development of a multivariate model with three additional biomarkers (apolipoprotein A1, downregulated; a truncated form of transyretin, downregulated; inter-alpha-trypsin inhibitor heavy chain H4, upregulated).	214
Combination of soluble mesothelial-related marker with CA125 for the diagnosis of ovarian cancer	215
CA125 levels of less that 10 U/mL in patients an excellent clinical response to platinum-based chemotherapy are good candidates for paclitaxel maintenance therapy	216

heart failure[218] due to an atrial septal defect where it was suggested that this was due to increased secretion of CA125 from activated peritoneal mesothelium. CA125 is reported to be elevated in benign lesions such as large endometrioma.[219] Krediet[220] has shown that CA125 is a mesothelial cell marker, which can be measured in peritoneal effluent and used as a measure of in vivo biocompatibility of dialysis solutions. DiBaise and Donovan[221] reported elevated CA125 in hepatic cirrhosis. These latter observations have been confirmed and extended by Xiao and Liu[222] and Kim and coworkers.[223]

With respect to the specificity issues, CA125 has been used for the study of ovarian cancer for some time with considerable success.[224–229] There is considerable interest in the use of multiple biomarkers to improve sensitivity and specificity. Current work as shown in Table 3.9 would suggest that sensitivity can be increased by using CA125 in combination with other potential biomarkers identified with proteomic technology. Yurkovetsky and coworkers[245] evaluated a panel of 64 biomarkers using multiplex technology and reported that prolactin was the most powerful marker for early detection of endometrial cancer in a high-risk population. The reader is also directed to some other work on multiple biomarkers.[246–251] The U. S. Food and Drug Administration approved a serological test, OVA1,[152] on September 11, 2009; OVA1 used the levels of five blood proteins for the management of women with pelvic masses. OVA1 is not intended for screening or diagnosis but rather for predictive/theranostic purposes. In this regards, CA125 would appear to be most useful as a predictive biomarker.[253–258]

There is no question that the development of a successful screening protocol for ovarian cancer would be an outstanding accomplishment. In 1994, Droegemueller[259] discussed the various problems associated with screening for ovarian cancer concluding that the effectiveness of periodic screening has been established for even the

TABLE 3.9

Use of Multiple Biomarkers for Ovarian Cancer

Study	Biomarker(s)[a]	References
Review	CA125, tumor-associated trypsin inhibitor	230
Two-dimensional gel electrophoresis	OP-18, PCNA, triosephosphate isomerase, elongation factor-2, GST all upregulated; TM2 and laminin c downregulated	231, 232
Serum proteomic analysis, SELDI	CA125 + four other biomarkers	210
Serum proteomic analysis, two-dimensional gel electrophoresis	52 kDa FK506 binding protein, Rho G-protein dissociation inhibitor (RhoGD1), glyoxalase 1	233
Serum proteomics analysis, SELDI	CA125 + haptoglobin α-subunit	212
Serum proteomics analysis, SELDI[b]	CA125; CA125 + apolipoprotein A1, truncated transthyretin, inter-α-trypsin inhibitor heavy chain H4	214
Serum proteomic analysis	CA125 + soluble mesothelin-related marker	215
Serum protein analysis	Hemoglobin	234
Immunohistocytochemistry	Type IV collagen, CD44v6, P53, Ki-67	235
Review on value of multiple biomarkers for improving specificity for screening	CA125 multiplexed with other serum proteins—Use of multiplex bead technology (Luminex microsphere technology)	236
Glycosylation analysis (electrophoresis; hydrolysis with N-glycosidase F; analysis of fluorescent-labeled glycans by exoglycosidase digestion)	Difference in glycosylation of CA125, haptoglobin, IgG heavy chains in ovarian cancer	237
Multiple biomarkers	Use of artificial neural network (ANN) for improving the accuracy of the diagnosis of ovarian cancer	238
Early detection of ovarian cancer	CA125 combined with transvaginal sonography (TVS)	239
Glycosylation in ovarian cancer	Use of glycoforms of CA125 as biomarker	240
Screening of at-risk postmenopausal women	CA125, mesothelin, and HE4[b]	241
Risk of endothelial ovarian cancer in women with a pelvic mass	CA125 and HE4[b] as biomarkers to triage subjects	242
Validation of multiple biomarkers for detection of early-stage ovarian cancer	CA125 in combination with apolipoprotein A01, transthyretin, and transferrin	243
Differentiation of ovarian cancer from benign gynecological disease	Afamin[c] in combination with CA125	244

[a] Biomarkers are separate when separated by a comma; combined with + sign to give profiling.

[b] HE4, a member of the WFDC domain family or WAP family. See Bouchard, D., Morisset, D., Bourbonnais, Y., and Tremblay, G.M., Proteins with whey-acidic-protein motifs and cancer, *Lancet Oncol.* 7, 167–174, 2006; Hellstrom, I. and Hellstrom, K.E., SMRP and HE4 as biomarkers for ovarian carcinoma when used alone and in combination with CA125 and/or each other, *Adv. Exp. Med. Biol.* 622, 15–21, 2008.

[c] Afamin is a vitamin D-binding glycoprotein is human plasma (Jerkovic, L., Voegele, A.F., Chwatal, S. et al., Afamin is a novel human vitamin E-binding glycoprotein characterization and in vitro expression, *J. Proteome Res.* 4, 889–899, 2005).

highest risk subjects; with a specificity of 99.6, there would be nine false-positives for every positive diagnosis. Look[188] provided an excellent review of the epidemiology, etiology, and screening for ovarian cancer in 2001 noting the problems associated with biomarker specificity in the development of screening. As of this writing, the combination of transvaginal sonography (TVS) and CA125 appears to have the greatest promise.[260] More recently, Jacobs and Menon,[261] while emphasizing the value of proteomic technology in identifying new biomarkers, note that high specificity will be required. However, at present, there is no test that merits the screening of the general population for ovarian cancer.[262] Without being pessimistic, these investigators note that the price of a false positive is a laparotomy or laparoscopy to rule out ovarian cancer. Another group urges considerable caution on the use of CA125 to guide clinical decisions.[263]

Finally, the reader is directed to some recent discussions of the issue of ovarian cancer screening.[248–251] There is interest in other biomarkers as shown in Table 3.10. The reader is also referred to a recent review on biomarker discovery in ovarian cancer.[282]

PANCREATIC CANCER

Similar in magnitude (Table 3.11), pancreatic cancer poses a similar diagnostic problem as ovarian cancer in that early detection is essential for successful therapy and there is a lack of methods for early detection, which results in high mortality.[283–287] The same statistical issues for the value of screening that are discussed above in general and for ovarian cancer are true for pancreatic cancer; in order for screening to be of value, it will be necessary to identify a high-risk population with anticipated higher prevalence.[288–290] Some approaches to the use of biomarkers for screening in pancreatic cancer are listed in Table 3.12. Some prognostic biomarkers for pancreatic cancer are listed in Table 3.13.

PROSTATE CANCER

Prostate cancer has the highest incidence of cancers for men in the United States (Table 3.11). Prostate-specific antigen (PSA) has a long background of use in diagnosis for prostate cancer. While there is some question as to its value in diagnosis,[319–322] there seems to be consensus that PSA measurements, either static or rate (PSA doubling time, PSADT) is useful for predictive and prognostic purposes.[323–329] Biochemical relapse is defined as a change in the concentration of PSA after posttreatment decrease; specifically, a rise in the blood level of PSA in prostate cancer patients after treatment with surgery or radiation. Biochemical relapse may occur in patients who do not have symptoms. It may mean that the cancer has come back; also called biochemical recurrence or PSA failure.[330–332] Other potential biomarkers for prostate cancer are being evaluated (Table 3.14). The reader is directed to several more comprehensive works on prostate cancer.[357,358] Given the considerable importance in the early diagnosis of prostate cancer, there is considerable interest in screening. However, the issues with effective screening for prostate cancer are similar to those for other cancers such as ovarian cancer; a low prevalence. Some 25 years ago,

TABLE 3.10
Recent Studies in Biomarker Development of Ovarian Cancer[a]

Biomarker	Reference
Hypermethylation of genes for secreted frizzled receptor proteins[b] as a prognostic biomarker	264
Expression of survivin, p21, and p53 is prognostic for FIGO[c] stage 1 and 2 endometroid–type endometrial cancer	265
Bcl-2[d] in urine as potential biomarker for ovarian cancer	266
Lysophosphatidic acid as a biomarker for ovarian cancer in blood plasma	267
Serum levels of IL-10 are suggested as a surrogate tumor marker for tumor grading in advanced ovarian cancer	268
Sialylation of kallikrein 6 as a diagnostic biomarker for ovarian cancer; a unique isoform is secreted in ovarian cancer	269
Differences in glycosylation of serum proteins as a biomarker for ovarian cancer. There is a decrease in galactosylation of IgG and an increase in sialyl Lewis X on haptoglobin β-chain, α-acid glycoprotein, and α-antichymotrypsin. There is also sialylation of acute phase proteins	270
Insulin-like growth factor 2 mRNA-binding protein 3 (IGF2BP3) is a prognostic biomarker (immunohistochemistry) for ovarian clear cell carcinoma	271
Lipocalin2[e] is a biomarker ovarian cancer as detected by expression in tumor issue (real-time PCR and immunohistocytochemistry) and level in serum	272
Fascin-1[f], cortactin[g], and survivin[h] are prognostic indicators for ovarian mucinous, serous, and clear cell adenocarcinoma	273
SALL4[i] (detected by immunohistochemistry) is a biomarker for ovarian primitive germ cell tumors, and distinguishes yolk sac tumor from clear cell carcinoma	274
Autoantibody to p53 is increased in type II ovarian cancer but not in type I ovarian cancer patients. Autoantibody to p53 is associated with mutation of TP53[j]	275
Serum CA 19-9[k] levels suggested as a diagnostic marker for ovarian mature cystic teratoma. Elevated CA 19-9 levels appear to correlate with larger tumor diameter and higher rates of ovarian torsion[l]	276
miR-200 microRNAs[m] are a prognostic marker in advanced ovarian cancer	277
Expression of GPR30 (immunohistochemistry) is a prognostic biomarker for ovarian cancer	278
Use serum samples to measure differences in microRNA (miRNA) expression patterns as biomarker for cancer. There is sufficient miRNA in 1 mL of serum to use a high-density microarray without amplification	279
BubR1[n] is a prognostic biomarker for epithelial ovarian cancer	280
Use of metabolomics (LC/MS) as diagnostic biomarker for ovarian cancer	281

[a] The studies shown were selected from a PubMed search for 2009. The reader is directed to some recent reviews for coverage of earlier years; see Diamandis, E.P., Mass spectrometry as a diagnostic and a cancer biomarker discovery tool: Opportunities and potential limitations, *Mol. Cell. Proteomics* 3, 367–378, 2004; Gogoi, R., Srinivasan, S., and Fishman, D.A., Progress in biomarker discovery for diagnostic testing in epithelial ovarian cancer, *Expert Rev. Mol. Diagn.* 6, 627–637, 2006; Mok, S.C., Elias, K.M., Wong, K.K. et al., Biomarker discovery in epithelial ovarian cancer by genomics approaches, *Adv. Cancer Res.* 96, 1–22, 2007; Tung, C.S., Wong, K.K., and Mok, S.C., Biomarker discovery in ovarian cancer, *Women's Health* 4, 27–40, 2008.

[b] See Kikuchi, A., Yamamoto, H., and Kishida, S., Multiplicity of the interactions of Wnt proteins and their receptors, *Cell Signal.* 19, 659–671, 2007; Hendrickx, M. and Leyns, L., Non-conventional frizzled ligands and Wnt receptors, *Dev. Growth Differ.* 50, 229–243, 2008; Bovolenta, P., Esteve, P., Ruiz, J.M. et al., *J. Cell. Sci.* 121, 737–746, 2008.

TABLE 3.10 (continued)
Recent Studies in Biomarker Development of Ovarian Cancer[a]

[c] International Federation of Gynecology and Obstetrics (FIGO). See Goldstein, D.P., Zanten-Przybysz, I.V., Bernstein, M.R., and Berkowitz, R.S., Revised FIGO staging system for gestational trophoblastic tumors. Recommendations regarding therapy, *J. Reprod. Med.* 43, 37–43, 1998.

[d] Bcl-2, an outer membrane protein from mitochondria (Szegezdi, E., Macdonald, D.C., Chonghaile, T. et al., Bcl-2 family on guard at the ER, *Am. J. Physiol. Cell Physiol.* 296, C941–C953, 2009). Bcl-2 is not specific for ovarian cancer but is a biomarker for other tumors (Sekine, I., Shimizu, C., Nishio, K. et al., A literature review of molecular markers predictive of clinical response to cytotoxic chemotherapy in patients with breast cancer, *Int. J. Clin. Oncol.* 14, 112–119, 2009; Roach III M., Waldman, F., and Pollack, A., Predictive models in external beam radiotherapy for clinically localized prostate cancer, *Cancer* 115 (13 Suppl), 3112–3120, 2009).

[e] Lipocalin2 (neutrophil gelatinase-associated lipocalin) is a member of the lipocalin family (Flower, D.R., The lipocalin protein family: Structure and function, *Biochem. J.* 318, 1–14, 1996) which is associated with various disease states (Xu, S. and Venge, P., Lipocalins as biochemical markers of disease, *Biochim. Biophys. Acta* 1482, 298–307, 2000).

[f] Fascin-1 is an actin-bundling protein involved in cell protrusions (see Adams, J.C., Fascin protrusions in cell interactions, *Trends Cardiovasc. Med.* 14, 221–226, 2004).

[g] Cortactin is a signaling protein involved in cell adhesion and migration (Ammer, A.G. and Weed, S.A., Cortactin branches out: Roles in regulating protrusive actin dynamics, *Cell. Motil. Cytoskeleton*, 65, 687–707, 2008) which is associated with tumor invasiveness (Weaver, A.M., Cortactin in tumor invasiveness, *Cancer Lett.* 265, 157–166, 2008). Cortican modulates actin function (Moore, O.L., Kotova, T.I., Moore, A.J., and Schafer, D.A., Dyanmin2 GTPase and cortactin remodel actin filaments, *J. Biol. Chem.* 284, 23995–24005, 2009).

[h] Survivin is a regulatory of mitosis and apoptosis and is a general characteristic of tumors, which is suppressed to inhibit tumor growth (Mita, A.C., Mita, M.M., Nawrocki, S.T., and Giles, F.J., Survivin: Key regulator of mitosis and apoptosis and novel target for cancer therapeutics, *Clin. Cancer Res.* 14, 5000–5005, 2008; Guha, M. and Altieri, D.C., Survivin as a global target of intrinsic tumor suppression networks, *Cell Cycle* 8, 2708–2710, 2009).

[i] SALL4 may be a zinc finger protein involved in regulation of transcription (see Tsubooka, N., Ischisaka, T., Okita, K. et al., Roles of Sall4 in the generation of pluripotent stem cells from blastocysts and fibroblasts, *Genes Cells* 14, 983–694, 2009; Chen. X., Vega, V.B., and Ng, H.H., Transcriptional regulatory networks in embryonic stem cells, *Cold Spring Harb. Symp. Quant. Biol.* 73, 203–209, 2008).

[j] TP32 is the gene-producing p53 tumor suppressor protein (see Olivier, M., Petitjean, A., Marcel, A. et al., Recent advances in p53 research: An interdisciplinary perspective, *Cancer* 16, 1–12, 2009; Whibley, C., Pharoah, P.D., and Hollstein, M., p53 polymorphisms: Cancer implications, *Nat. Rev. Cancer* 9, 95–107, 2009; Hrstka, R., Coates, P.J., and Vojtesek, B., Polymorphisms in p53 and the p53 pathway: Roles in cancer susceptibility and response to treatment, *J. Cell. Mol. Med.* 13, 440–453, 2009).

[k] CA19-9 is a carbohydrate antigen, which is a biomarker for pancreatic cancer and some nonmalignant gastrointestinal diseases (Steinberg, W., The clinical utility of the CA 19-9 tumor-associated antigen, *J. Gastroenterol.* 85, 350–355, 1990; Mathurin, P., Cadranel, J.F., Bouaya, D. et al., Marked increase in serum CA 19-9 level in patients in alcoholic cirrhosis: Report of four cases, *Eur. J. Gastroenterol. Hepatol.* 8, 1129–1131, 1996; Goonetilleke, K.S. and Siriwardena, A.K., Systematic review of carbohydrate antigen (CA 19-9) as a biochemical marker in the diagnosis of pancreatic cancer, *Eur. J. Surg. Oncol.* 33, 266–270, 2007).

[l] Ovarian torsion (Becker, J.H., de Graaff, J., and Vos, C.M., Torsion of the ovary: A known but frequently missed diagnosis, *Eur. J. Emerg. Med.* 16, 124–126, 2009).

(continued)

TABLE 3.10 (continued)
Recent Studies in Biomarker Development of Ovarian Cancer[a]

[m] See Peter, M.E., Let-7 and miR-200 microRNAs: Guardians against pluripotency and cancer progression, *Cell Cycle* 8, 843–852, 2009; Kim, M. and Kao, G.D., Newly identified roles for an old guardian: Profound deficiency of the mitotic spindle checkpoint protein BubR1 leads to early aging and infertility, *Cancer Biol. Ther.* 4, 164–165, 2005.

[n] BubR1 is a spindle control checkpoint protein (see Hoyt, M.A., A new view of the spindle checkpoint, *J. Cell Biol.* 154, 909–911, 2001; Sczaniecka, M.M. and Hardwick, K.G., The spindle checkpoint: How do cells delay anaphase onset?, *SEB Exp. Biol. Ser.* 59, 243–256, 2008).

TABLE 3.11
Cancer Data for the United States[a]

Male (Incidence/Year)	Female (Incidence/Year)
587,641 malignant neoplasms	674,251 malignant neoplasms
Prostate, 79/100,000	Breast, 104/100,000
Lung, trachea, bronchi, 57/100,000[b]	Lung, trachea, bronchi, 33/100,000[b]
Colorectal, 43/100,000	Colorectal, 30/100,000
Melanoma, 33/100,000	Cervix uteri, 16/100,000
Bladder, 30/100,000	Corpus uteri, 15/100,000
Pancreas, 8/100,000	Ovarian, 11/100,000
Esophageal, 7/100,000	Bladder, 6/100,0000

[a] Adapted from World Health Organization (http://www.who.int/whosis/en/) data for 2005.
[b] Lung, bronchi, trachea cancers are the leading causes of cancer deaths.

Podolsky[359] observed that even with the at-risk population of men over the age of 50, the prevalence in 0.05%; thus, even with 95% sensitivity and 95% specificity, there will be 100 false-positive determinations for every true-positive determination.

BREAST CANCER

Breast cancer has the greatest prevalence of occurrence in women in the United States and is the second leading cause of cancer mortality (Table 3.11). Imaging technologies (mammography and MRI) are the primary technologies for screening for breast cancer.[360–362] Molecular diagnostics are extremely valuable in defining breast cancer risk.[363–368] The reader is also directed to larger, more comprehensive works on molecular medicine, molecular diagnostics, and genetic testing.[369–373] Notwithstanding the great progress in the use of molecular diagnostics for breast cancer, there is still interest in the identification and development of new biomarkers.[374] Table 3.15 presents selected studies on the development of biomarkers for breast cancer. Emphasis is placed on the recent work on the development of serological markers; the reader is referred to several selected sources for a consideration

TABLE 3.12
Development of Biomarkers for Screening/Diagnosis[a] in Pancreatic Cancer

Biomarker(s)	Reference
Review of the development of biomarkers for pancreatic cancer	291
Hepatocarcinoma-intestine-pancreas/pancreatic-associated protein I[b] in pancreatic juice as a diagnostic biomarkers for pancreatic adenocarcinoma	292
Serum osteopontin may be a diagnostic biomarker for pancreatic adenocarcinoma. Osteopontin expression (in situ hybridization) was not observed in pancreatic cancer cells or normal cancer cells but is a product of tissue-infiltrating macrophages	293
NAD(P)H:quinone oxidoreductase 1 expression (PCR-RFLP) in pancreatic tissue as a potential biomarker for pancreatic cancer	294
CEACAM1[c] (carcinoembryonic antigen–cell adhesion molecule 1) as serum biomarker for pancreatic cancer; CEACAM1 is expressed in pancreatic adenocarcinoma using RT-PCR	295
NAD(P)H:quinine oxidoreductase (NQO1) expression (immunohistochemical and immunocytochemical) in pancreatic tissue as biomarker for pancreatic cancer	296
MUC1[d] as detected by MAB (immunohistochemistry) is a diagnostic biomarker for pancreatic cancer	297
Hepatic nuclear factor-1β (transcription factor 2 hepatic)[e] as biomarker for clear cell carcinoma of the pancreas	298
Use of solid phase lipophilic extraction[f] followed by MALDI-MS to identify urinary biomarkers differentiating pancreatic cancer from chronic inflammation	299
Type IV collagen[g] in plasma (ELISA assay) is a biomarker for pancreatic cancer. Other studies demonstrate production and secretion of type IV collagen by pancreatic cancer cells	300
Telomerase reverse transcriptase expression (immunohistochemistry) is a biomarker in pancreatic juice for pancreatic ductal adenocarcinoma	301
K-ras mutation in stool samples as detected by mutant-enriched PCR[h] and allele hybridization as biomarker for pancreatic cancer	302

[a] Screening and diagnosis are not separated in this consideration.

[b] Hepatocarcinoma-intestine-pancreas/pancreatic-associated protein I is a lactose-binding C-type lectin involved in cell adhesion, which has antibacterial activity (Christa, L., Pauloin, A., Simon, T. et al., High expression of the human hepatocarcinoma-intestine-pancreas/pancreatic-associated protein (HIP/PAP) gene in the mammary gland of lactating transgenic mice. Secretion into the milk and purification of the HIP/PAP lectin, *Eur. J. Biochem.* 267, 1665–1671, 2000; Dann, S.M. and Eckmann, L., Innate immune defenses in the intestinal tract, *Curr. Opin. Gastroenterol.* 23, 115–120, 2007).

[c] Carcinoembryonic antigen-cell adhesion molecule 1 is involved in cell adhesion (see Gray-Owen, S.D. and Blumberg, R.S., CEACAM1: Contact-dependent control of immunity, *Nat. Rev. Immunol.* 6, 433–446, 2006; Nagaishi, T., Iijima, H., Nakajima, A. et al., Role of CEACAM1 as a regulator of T cells, *Ann. N. Y. Acad. Sci.* 1072, 155–175, 2006).

[d] MUC1, cell-associated mucin, which has been suggested a therapeutic target (Singh, R. and Bandyopadhyay, D., MUC1: A target molecule for cancer therapy, *Cancer Biol. Ther.* 6, 481–486, 2007; Mall, A.S., Analysis of mucins: Role in laboratory diagnosis, *J. Clin. Pathol.* 61, 1018–1024, 2008; Yonezawa, S., Goto, M., Yamada, N. et al., Expression profiles of MUC1, MUC2, and MUC4 mucins in human neoplasms and their relationship with biological behavior, *Proteomics* 8, 3329–3341, 2008).

(*continued*)

TABLE 3.12 (continued)

Development of Biomarkers for Screening/Diagnosis[a] in Pancreatic Cancer

e   Hepatic nuclear factor-1β is also suggested to be involved in diabetes and polycystic kidney disease (Carette, C., Vaury, C., Barthélémy, A. et al., Exonic duplication of the hepatocyte nuclear factor-1β gene (transcription factor 2, hepatic) as a cause of maturity onset diabetes of the young type 5, *J. Clin. Endocrinol. Metab.* 92, 2844–2847, 2007; Gresh, L., Fischer, E., Reismann, A. et al., A transcriptional network in polycystic kidney disease, *EMBO J.* 23, 1657–1668, 2004).

f   Sample captured on a C_{18} 96-well microplate filter and extracted with TFA-acetonitrile.

g   Type IV collagen is implicated a variety of cancers (see Pasco, S., Brassart, B., Ramont, L. et al., *Cancer Detect. Prev.* 29, 260–266, 2005; Jarzembowski, J., Lloyd, R., and McKeever, P., Type IV collagen immunostaining is a simple, reliable diagnostic tool for distinguishing between adenomatous and normal pituitary glands, *Arch. Pathol. Lab. Med.* 131, 931–935, 2007).

h   Based on selective enhancement of mutant genes (Hruban, R.H., van Mansfield, A.D., Offerhaus, G.J. et al., K-ras oncogene activation in adenocarcinoma of the human pancreas. A study of 82 carcinomas using a combination of mutant-enriched polymerase chain reaction analysis and allele-specific oligonucleotide hybridization, *Am. J. Pathol.* 143, 545–554, 1993; Nollau, P., Jung, R., Neumaier, M. et al., Tumour diagnosis by PCR-based detection of tumour cells, *Scand. J. Clin. Lab. Invest. Suppl.* 221, 116–121, 1995).

of earlier work.[390–393] It is the sense of the author that the transition from discovery of biomarker to a validated assay in a clinical laboratory is far easier with a serological biomarker.[394–401] A serological sample is considered to be noninvasive.[402,403] It is likely that saliva would be even more noninvasive.[404] It would be remiss not to mention the work done on expression biomarkers both for prognostic purposes and mechanistic purposes.[405–409] A small selection of such studies is provided in Table 3.16.

Lung cancer has the second highest prevalence of cancer for either male or female but is the leading cause for cancer death in the United States. A variety of modalities are suggested for lung cancer screening.[418–431] Eddy presented an excellent analysis[432] in 1989 of screening technologies (chest roentgenograms and sputum cytology), which showed little impact on mortality. More recent reports[433,434] are no more optimistic about the value of screening; there is continued interest in CT spiral scanning,[435–438] but more time will be required. It should be noted that many studies on the development of biomarkers for screening are negative but this should be taken as lack of value,[439] and it is also understood that a considerable amount of work is required for the development of truly useful biomarkers for lung cancer.[440] This is an area where it is essential to understand the difference between prognostic biomarkers and predictive biomarkers.[441]

A recurring theme in this work is the difficulty in identifying a unique biomarker or pattern of biomarkers for a given disease. The reader again is directed to the comments of Roulston,[154,155] who suggests that tumor cells do not produce qualitatively different materials but rather differ in quantitative expression. Several related studies of interest include the work on the computational prediction of human proteins that can be secreted into the circulation,[442] and work on the cancer secretome.[443]

TABLE 3.13
Prognostic Biomarkers for Pancreatic Cancer[a]

Biomarker	Reference
Valosin-containing protein (p97)[b] expression (immunohistochemistry) is a prognostic biomarker (lymph node metastasis) for pancreatic ductal adenocarcinoma	303
CEACAM6[c] (determined by immunohistocytochemistry with monoclonal antibody using tissue microarray) may be a prognostic biomarker for pancreatic adenocarcinoma	304
Matrix metalloproteinase-7 (MMP-7) in combination with CA 19-9 as predictive biomarker in pancreatic carcinoma in distinguishing from chronic pancreatitis or benign disease	305
Sp1[d] expression (immunohistochemistry) in pancreatic ductal adenocarcinoma tissue as prognostic biomarker in pancreatic cancer	306
Heparanase expression in pancreatic adenocarcinoma tissue (RT-PCR with paraffin-embedded tissue samples) is a prognostic biomarker (lymph node metastasis) for pancreatic cancer	307
Syndecan-1[e] (immunoreactivity) is a biomarker for pancreatic carcinoma. Epithelial syndecan-1 expression predicts better recovery in resectable pancreatic cancer. Lack of stromal expression predicts better recovery independent of disease stage	308
Serum macrophage inhibitory cytokine 1[f] as a prognostic biomarker for pancreatic cancer	309
Cyclin E[g] expression (immunohistochemistry with tissue microarrays) is a prognostic biomarker for pancreatic ductal adenocarcinoma	310
A review of prognostic factors for pancreatic cancer suggesting that the combination of biomarkers might be the most useful for prognosis. The strongest independent biomarkers are p16, MMP7, and VEGF	311
Serum chromogranin A[h] is prognostic biomarker for pancreatic cancer	312
CD133[i] expression (immunohistochemical) suggested as prognostic biomarker (lymphatic metastasis) for pancreatic cancer. CD133 expression was significantly associated with VEGF-C expression	313
Caveolin-1[j] is a prognostic biomarker for pancreatic cancer	314
Rate of change in serum concentration of CA19-9 (CA19-9 velocity) as prognostic biomarker for pancreatic cancer after pancreatectomy	315
Metastin[k] is an independent prognostic biomarker for pancreatic cancer	316
S100A2 protein is a prognostic biomarker for pancreatectomy for pancreatic cancer	317
ALCAM (activated leukocyte cell adhesion molecule, CD166) is an independent prognostic biomarker for pancreatic cancer	318

[a] It is recognized that prognostic biomarkers might also be used for diagnostic purposes.

[b] Valosin-containing protein (p97) is type II AAA ATPase, which acts as a molecular chaperone associated with a variety of physiological functions (Wang, Q., Song, C., and Li, C.C., Molecular perspectives on p97-VCP: Progress in understanding its structure and diverse biological functions, *J. Struct. Biol.* 146, 44–57, 2004; Meyer, H. and Popp, O., Role(s) of Cdc48/p97 in mitosis, *Biochem. Soc. Trans.* 36, 126–130, 2008; Neumann, H., Tolnay, M., and Mackenzie, I.R., The molecular basis of frontotemporal dementia, *Expert Rev. Mol. Med.* 11, e23, 2009).

[c] CEACAM6, carcinoembryonic antigen–related cell adhesion molecule 6, also known as CD66c (Skubitz, K.M., Kuroki, M., Jantscheff, P. et al., CD66c, *J. Biol. Regul. Homeost. Agents* 13, 244–245, 1999; Yasui, W., Oue, N., Ito, R. et al., Search for new biomarkers of gastric cancer through serial analysis of gene expression and its clinical implications, *Cancer Sci.* 95, 385–392, 2004; Barnich, N. and Darfeuille-Michaud, A., Role of bacteria in the etiopathogenesis of inflammatory bowel disease, *World J. Gastroenterol.* 13, 5571–5576, 2007).

(*continued*)

TABLE 3.13 (continued)
Prognostic Biomarkers for Pancreatic Cancer[a]

[d] Spl is a sequence-specific DNA binding zinc finger protein involved in the regulation of transcription (Ooi, G.T., Cohen, F.J., Hsieh, S. et al., Structure and regulation of the ALS gene, *Prog. Growth Factor Res.* 6, 151–157, 1995; Webster, K.A., Prentice, H., and Bishopric, N.H., Oxidation of zinc finger transcription factors: Physiological consequences, *Antioxid. Redox Signal.* 3, 535–548, 2001; Urnov, F.D., A feel for the template: Zinc finger protein transcription factors and chromatin, *Biochem. Cell Biol.* 80, 321–333, 2002).

[e] Syndecan-1 is a transmembrane proteoglycan serving a regulatory function with interaction with receptors for extracellular matrix proteins (Götte, M. and Echtermeyer, F., Syndecan-1 as a regulator of chemokines function, *Scientific World J.* 2, 1327–1331, 2003; Sanderson, R.D. and Yang, Y., Syndecan-1: A dynamic regulator of the myeloma microenvironment, *Clin. Exp. Metastasis* 25, 149–159, 2008; Bass, M.D., Morgan, M.R., and Humphries, M.J., Syndecans shed their reputation as inert molecules, *Sci. Signal.* 2, pe18, 2009).

[f] Macrophage inhibitory cytokine 1 is a member of the TGFβ superfamily implicated in tumor formation and inflammation (Karan, D., Holzbeierlein, J., and Thrasher, J.B., Macrophage inhibitory cytokine-1 (CD138): Possible bridge molecule of inflammation and prostate cancer, *Cancer Res.* 69, 2–5, 2009; Yamashita, T., Yoneta, A., and Hida, T., Macrophage inhibitory cytokine-1: A new player in melanoma development, *J. Invest. Dermatol.* 129, 262–264, 2009).

[g] Cyclin E is a protein involved in cell-cycle control by activated Cdk2 kinase (Möröy, T. and Geisen, C., Cyclin E, *Int. J. Biochem. Cell Biol.* 36, 1424–1439, 2004; Hwang, H.C. and Clurman, B.E., Cyclin E in normal and neoplastic cell cycles, *Oncogene* 24, 2776–2786, 2005; Musgrove, E.A., Cyclins: Roles in mitogenic signaling and oncogenic transformation, *Growth Factors* 24, 13–19, 2006).

[h] Chromogranin A is a glycoprotein, which may yield peptides with regulatory properties (Ferrari, L., Seregni, E., Bajetta, E. et al., The biological characteristics of chromogranin A and its role as a circulating marker in neuroendocrine tumours, *Anticancer Res.* 19, 3415–3427, 1999; O'Connor, D.T., Mahata, S.K, Taupenot, L. et al., Chromogranin A in human disease, *Adv. Exp. Biol. Med.* 482, 377–388, 2000; Helle, K.B., Corti, A., Metz-Boutique, M.H., and Total, B., The endocrine role of chromogranin A: A prohormone for peptides with regulatory properties, *Cell. Mol. Life Sci.* 64, 2863–2886, 2007).

[i] CD133 is a marker of endothelial progenitor cells and is found on trophoblasts (Piechaczek, C., CD133, *J. Biol. Regul. Homeost. Agents* 15, 101–102, 2001; Pötgens, A.J., Bolte, M., Huppertz, B. et al., Human trophoblast contains an intracellular protein reactive with an antibody against CD133—A novel marker for trophoblast, *Placenta* 22, 639–645, 2001; Hristov, M., Erl, W., and Weber, P.C., Endothelial progenitor cells: Isolation and characterization, *Trends Cardiovasc. Med.* 13, 201–206, 2003).

[j] Cavelolin-1 is a specialized membrane scaffold protein, which stabilizes lipid raft domains and has been implicated in transport functions and signal transduction (see Schwanke, C., Braun-Dullaeus, R.C., Wunderlich, C., and Strasser, R.H., Caveolae and caveolin in transmembrane signaling: Implications for human disease, *Cardiovasc. Res.* 70, 42–49, 2006; Quest, A.F., Gutierrez-Pajares, J.L., and Torres, V.A., Caveolin-1: An ambiguous partner in cell signaling and cancer, *J. Cell. Mol. Med.* 12, 1130–1150, 2008; Lajoie, P., Goetz, J.G., Dennis, J.W., and Nabi, I.R., Lattices, rafts, and scaffolds: Domain regulation of receptor signaling at the plasma membrane, *J. Cell Biol.* 185, 381–385, 2009).

[k] Metastin is the product of KiSS-1 gene (a metastasis-suppressing gene), which is ligand for GPR54, a G-protein-coupled receptor (see Harms, J.F., Welch, D.R., and Miele, M.E., KISS1 metastasis suppression and emergent pathways, *Clin. Exp. Metastasis* 20, 11–18, 2003; Seminara, S.B., Metastin and its G protein-coupled receptor, GPR54: Critical pathway modulating GnRH secretion, *Front. Neuroendocrinol.* 26, 131–138, 2005; Colledge, W.H., GPR54 and kisspeptins, *Results Probl. Cell Differ.* 46, 117–143, 2008).

TABLE 3.14

Biomarkers for Prostate Cancer[a]

Biomarker(s)	References
Annexin A3 in urine as biomarker for prostate cancer	333
Sarcosin in urine as biomarker for prostate cancer	334, 335
Hepatocyte growth factor as possible biomarker for prostate cancer	336
Circulating tumor cells as biomarker for prostate cancer mortality	337
Cell-free DNA as prognostic biomarker for prostate cancer. Cell-free DNA correlates with circulating tumor cells	338
Prostate-specific membrane antigen is over-expressed in prostate cancer and is related to poor prognosis, which can be detected with PET imaging using Cu-antibody	339, 340
Matrix metalloproteinases as biomarkers for prostate cancer	341
Serum levels of adipocytokines (ELISA) as biomarkers for prostate cancer	342
Plasma osteopontin (ELISA assay[b]) as a prognostic biomarker for castration-resistant prostate cancer	343
Specific RNA in exosomes (PCA-3 and TMPRSS2:ERG) present in urine as biomarkers for prostate cancer	344
MUC1 (mucin 1) expression (immunohistochemistry) as a possible biomarker for prostate cancer	345
Bcl-2,[c] p53, or high microvessel density (immunohistochemistry) as prognostic biomarker in prostate cancer biopsy sample	346
Peroxisome proliferator-activated receptor (PPAR) expression (immunohistochemistry) as possible prognostic biomarker for prostate cancer. PPAR also correlates with serum PSA levels	347
Staphylococcal nuclease domain–containing protein 1[d] expression as potential biomarker (immunohistochemistry and in situ hybridization) for prostate cancer	348
Magnetic resonance imaging/magnetic resonance spectroscopy imaging (MRI/MRSI) combined with molecular markers (PSA, androgen receptor by immunohistochemistry) to increase concordance index for evaluation of prediction of recurrence	349
COX-2, VEGF, KDR, HIF-1, MIB-1, Ki-67[e] were evaluated (immunohistochemistry) for the effect of celecoxib (a COX-2 inhibitor) on clinically localized prostate cancer. Microvessel density and apoptosis were also evaluated	350
17β-hydroxysteroid dehydrogenase type 11 (Pan1b) expression is a potential biomarker for prostate cancer	351
Serum calcium as prognostic biomarker for prostate cancer	352
Son of sevenless homolog 1 (SOS1) expression (immunohistochemistry in tissue microarray) as prognostic biomarker for prostate cancer in African-American men	353
Apoptopic chromatin condensation inducer in the nucleus (ACINUS) as prognostic biomarker (imaging) for prostate cancer	354
Small heat shock protein HSP-27 expression (immunohistochemistry) as prognostic biomarker for aggressive prostate cancer	355
Ki-67 (immunohistochemistry) is a prognostic biomarker for biochemical recurrence in prostate cancer	356

(continued)

TABLE 3.14 (continued)
Biomarkers for Prostate Cancer[a]

[a] This list is taken from almost 6000 citations obtained from PubMed; the search excluded PSA. The selections were taken from 2009 only.

[b] The strong association of osteopontin with survival is dependent on the use of a specific monoclonal antibody, which recognizes the thrombin cleavage site region (see Klee, G.G., Assay configuration and analytic specificity may have major effects on prediction of clinical outcomes—Implications for reference materials, *Clin. Chem.* 55, 848–849, 2009).

[c] Bcl-2, B-cell lymphoma protein-2 controlling apoptosis (Rong, Y. and Distelhorst, C.W., Bcl-2 protein family members: Versatile regulators of calcium signaling in cell survival and apoptosis, *Annu. Rev. Physiol.* 70, 73–91, 2008; Lessene, G., Czabotar, P.E., and Colman, P.M., BCL-2 family antagonists for cancer therapy, *Nat. Rev. Drug Discov.* 13, 7254–7263, 2008; Szegezdi, E., Macdonald, D.C., Ní Chonghaile, T. et al., Bcl-2 family on guard at the ER, *Am. J. Physiol. Cell Physiol.* 296, C941–C953, 2009).

[d] Staphylococcal nuclease domain-containing protein 1 (SND 1) is involved in the RNA splicing process resulting in RNA interference (see Tong, X., Drapkin, R., Yalamachili, R. et al., The Epstein-Barr virus nuclear protein 2 acidic domain forms a complex with a novel cellular coactivator that can interact with TFIIE, *Mol. Cell. Biol.* 15, 4735–4744, 1995; Yang, J., Välineva, T., Hong, J. et al., Transcriptional co-activator protein p100 interacts with snRNP proteins and facilitates the assembly of the spliceosome, *Nucleic Acids Res.* 35, 4485–4494, 2007).

[e] COX-2, cyclooxygenase-2; VEGF, vascular endothelial growth factor; KDR, kinase insert domain-containing receptor; HIF-1, hypoxia-inducible factor 1; MIB-1, a monoclonal antibody developed for the study of cancer cell proliferation; Ki-67, a antibody for the study of cancer cell proliferation (see Spyratos, F., Ferrero-Poüs, M., Trassard, M. et al., Correlation between MIB-1 and other proliferation markers: Clinical implications of MIB-1 cutoff value, *Cancer* 94, 2151–2159, 2002).

While the current work has focused on the identification of biomarkers in human tissues and fluids, it is clear that in vitro work with cell culture systems can be useful.[444–447] Examples of recent progress on the development of serological markers are listed in Table 3.17 while a limited presentation of some gene-expression biomarkers are shown in Table 3.18. It is recognized that the coverage of the literature for lung cancer biomarkers in Table 3.18 is extremely limited considering the large amount of work in this area (greater than 12,000 citations were obtained for a search combining lung cancer and biomarker; the great majority of these were histochemical and expression biomarkers).

There is reason to think that there are useful biomarkers being developed for prognostic and predictive purposes, there are still issues with biomarkers for diagnosis and screening. There are a number of biomarkers that demonstrate nonspecificity with respect to a specific malignancy but may be unique to oncology as opposed, for example, to neurology or digestive inflammatory disease. An example may be provided by tissue (glandular) kallikrein. Kallikrein is an enzyme activity, which is responsible for the conversion of kininogens to kinins.[473] There is a plasma kallikrein and a glandular kallikrein; plasma kallikrein can convert high-molecular weight kininogen to kinins; only glandular kallikrein can convert low-molecular-weight kininogen to kinins.[474] Kallikrein activity has been suggested to be involved in the carcinoid syndrome for

TABLE 3.15
Serological Biomarkers for Breast Cancer[a,b]

Biomarker(s)	References
CA15-3, XA 27.29, α-fetoprotein, and CEA were evaluated as biomarkers in serum for breast cancer. CA15-3 was the best tumor antigen a diagnostic/predictive biomarker. More recent work from another group has also shown an elevation of CA15-3 in saliva associated with breast cancer. Another group has reported the value of CA15-3 as a predictive biomarker in breast cancer. A PubMed search showed more than 300 citations for the study of CH15-3 in oncology	375–377
HER-2/neu in blood plasma as predictive biomarker for trastuzumab therapy in breast cancer. More recent work has presented some qualification for the use of HER-2 extracellular domain as a biomarker in breast cancer	378, 379
Nectin-4[c] ectodomain in serum as a biomarker for breast cancer	380
Mutant p53 protein in serum is a biomarker for breast cancer. There is a good correlation between p53 mutant protein in serum and that detected by immunohistochemistry	381
Antibodies against p53 are prognostic biomarkers in breast cancer	382
Serum thymidine kinase 1 suggested as a prognostic biomarker for post-surgical breast cancer patients	383
Atopy (allergen-specific IgE) as biomarker for breast cancer risk	384
Multiplex analysis of 55 plasma proteins as predictive biomarkers for breast cancer	385
Metastasin[d] mRNA is serum as a prognostic biomarker for cancer	98
Serum activated leukocyte cell adhesion molecule (ALCAM) as a biomarker for breast cancer. Strength improved by combination with CA15-3	386
α-2-HS glycoprotein and antibodies to α-2-HS glycoprotein may be a useful biomarkers for breast cancer	387
Review of the use of soluble epidermal growth factor receptor (sEGFR) as predictive/theranostic biomarker in breast cancer	388
Identification of diagnostic/prognostic biomarkers for breast cancer in serum using prefractionation with weak cation exchange followed by MALDI-TOF mass spectrometry	389

[a] The reader is directed to an excellent article on the development of biomarkers for breast cancer (see Hinestrosa, M.C., Dickersin, K., Klein, P. et al., Shaping the future of biomarker research in breast cancer to ensure clinical relevance, *Nat. Rev. Cancer* 7, 309–315, 2007).

[b] The studies are selected from a period of January 2003 to September 2009.

[c] Nectin-4 is a 66 kDA adhesion protein, which has been demonstrated to be a histological marker for breast carcinoma (Fabre-Lafay, S., Monville, F., Garrido-Urbani, S. et al., Nectin-4 is a new histological and serological tumor associated marker for breast cancer, *BMC Cancer* 7, 73, 2007). Recent work has shown nectin-4 associated with lung carcinoma (Takano, A., Ishikawa, N., Nishinon, R. et al., Identification of nectin-4 oncoprotein as a diagnostic and therapeutic target for lung cancer, *Cancer Res.* 69, 6694–6703, 2009).

[d] Metastasin is also known as S100A4, placental calcium protein. See Tarabykina, S., Griffiths, T.R., Tulchinsky, E. et al., Metastasis-associated protein S100A4: Spotlight on its role in cell migration, *Curr. Cancer Drug Targets* 7, 217–228, 2007.

TABLE 3.16

Expression Biomarkers for Breast Cancer[a]

Biomarker(s)	References
Expression of Ki-67 (immunohistochemistry) as predictive biomarker for neoadjuvant sequential treatment (exemestane and anastrozole) in hormone-receptor positive breast cancer	410
Expression of phosphatidylinositol 3-kinase (PIK3CA) (immunohistochemistry with tissue microarrays) as prognostic biomarker in breast cancer	411
IMP3[b] is a prognostic biomarker (immunohistochemistry) of triple negative (basal-like) invasive mammary carcinoma	412
Methylation of kallikrein 10 exon 3 (methylation–specific PCR with bisulfite-modified DNA) as prognostic biomarker in early breast cancer	413
Expression of metallothionein (immunohistochemistry) as prognostic biomarker for breast cancer	414
HCCR (human cervical cancer oncogene) oncoprotein expression (immunohistochemistry, Western blotting, FACS, confocal microscopy) as diagnostic biomarker for breast cancer	415
Expression of Rho-GTPase (immunohistochemistry with high-density tissue microarrays) as prognostic biomarker for breast cancer	416, 417

[a] Selected from a PubMed search.
[b] IMP3 protein is a member of the IGF-II mRNA binding protein family (see Nielsen, F.C., Nielsen, J., and Christiansen, J., A family of IGF-II mRNA binding proteins (IMP) involved in RNA trafficking, *Scand. J. Clin. Lab. Invest. Suppl.* 234, 93–99, 2001).

some time.[475] Jenzano and colleagues[476,477] reported elevated levels of glandular kallikrein activity in whole saliva obtained from patients with tumors distant from the oral cavity. Another group reported a decrease in parotid gland–derived kallikrein activity in patients with head and neck cancer.[478] Jenzano and colleagues also studied the effect of age, sex, and race on salivary kallikrein[477] and noted an increase in kallikrein concentration with age; there was a slightly higher concentration in women. While not extensive, this study showed that there was no increase with heart disease, hypertension, inactive cancer, or several other disease states. In unrelated studies, the Jenzano group reported the expression of kallikrein in neutrophils stimulated by thrombin[479] and in human endometrial stromal cells.[480] PSA is a biomarker for prostate cancer[481] with homology to glandular kallikrein.[482] PSA does have enzyme activity[483] and has been found in saliva.[484] Human kallikrein 7, a chymotryptic-like enzyme, is also found in saliva.[485] The interest of glandular kallikrein as a cancer biomarker increased with the discovery of multiple members of the glandular kallikrein family.[486–493] The salivary kallikrein is human kallikrein 1 (KLK1), which is also present in kidney and pancreas. PSA is KLK3 while the other prostate kallikrein is KLK2 (human glandular kallikrein 2). Some studies on the use of glandular kallikrein as a biomarker for cancer are listed in Table 3.19. Other biomarkers common to tumors are the inflammation biomarkers (Table 3.3) the S100 proteins (Tables 3.4 and 3.5), and C-reactive protein (Table 3.6). Table 3.20 contains a listing of some biomarkers that are found in tumors with different tissue origins.

TABLE 3.17
Serological and Other Biomarkers for Lung Cancer[a]

Biomarker(s)	Reference
Raman microspectroscopy for the identification and classification of lung neoplasias[b]	448
Plasma DNA[c] as biomarker for lung cancer. The combination of plasma DNA with spiral computed tomography (spiral CT) did not improve accuracy of screening; increased DNA concentration in plasma at surgery may be a prognostic biomarker	449
Plasma DNA as prognostic biomarker in patients recalled for non-calcified nodule (NCN)	450
Validation of plasma DNA as possible prognostic biomarker in spiral CT-detected lung cancer	451
Plasma DNA as prognostic biomarker in non-small-cell lung cancer	452
Plasma DNA as risk factor for lung cancer	453
Circulating exosomal microRNA (miRNA) as possible screening biomarker for lung cancer	454
Pattern of volatile organic compounds (VOC) exhaled breath as biomarkers for lung cancer	425
VEGF as predictive biomarker for use of chemotherapy in non-small-cell lung cancer	455
Use of FITC-labeled wheat germ agglutinin to identify differentially expressed glycoproteins separated by 1D and 2D electrophoresis. Altered glycoproteins as prognostic biomarkers for lung cancer	456
Soluble B7-H3 (sB7-H3) in serum as prognostic biomarker for non-small-cell lung cancer	457
Serum N-methyltransferase as biomarker for lung cancer; sensitivity is increased in combination with CEA	458
Development of a panel of serum proteins as a biomarker for lymph node metastases in non-small-cell lung cancer	459
Serum cytokeratin 19 fragments[d] as prognostic biomarker for non-small-cell lung cancer	460
Serum ADAM8[e] as a diagnostic biomarker for lung cancer; combination with CEA increased sensitivity	461
Serum thymidine kinase 1 as prognostic biomarker of non-small-cell lung cancer	462

[a] There is a vast literature on biomarkers for lung cancer. This table contains a collection of biomarker literature that exclude expression, hybridization, and immunohistochemical biomarkers, which require a tissue sample. An extremely limited consideration of those biomarkers is given in Table 3.18.

[b] Raman microscopy of normal bronchial tissue (Koljenović, S., Bakker Shut, T.C., van Meerbeeck, J.P. et al., Raman microspectroscopic mapping studies of human bronchial tissue, *J. Biomed. Opt.* 9, 1187–1197, 2004).

[c] There is considerable interest in the use of DNA as a biomarker in lung cancer (Xue, X., Zhu, Y.M., and Woll, P.J., Circulating DNA and lung cancer, *Ann. N. Y. Acad. Sci.* 1075, 154–164, 2006). Only several examples of work are cited in this table.

[d] See also Ekman, S., Eriksson, P., Bergström, S. et al., Clinical value of using serological cytokeratins as therapeutic markers in thoracic malignancies, *Anticancer Res.* 27, 3545–3553, 2007; Brattström, D., Wagenius, G., Sandström, P. et al., Newly developed assay measuring cytokeratins 8, 18 and 19 in serum is correlated to survival and tumor volume in patients with esophageal carcinoma, *Dis. Esophagus* 18, 298–303, 2005; Jerome Marson, V., Mazieres, J., Groussard, O. et al., Expression of TTF-1 and cytokeratins in primary and secondary epithelial lung tumours: Correlation with histological type and grade, *Histopathology* 45, 125–134, 2004.

[e] ADAM, a disintegrin and metalloproteinase (Wolfsberg, T.G., Primakoff, P., Myles, D.G. et al., ADAM, a novel family of membrane proteins containing A disintegrin and metalloproteinase domain: Multipotential functions in cell-cell and cell-matrix interactins, *J. Cell Biol.* 131, 275–278, 1995; Mochizuki, S. and Okada. Y., ADAMs in cancer cell proliferation and progression, *Cancer Sci.* 98, 621–628, 2007).

TABLE 3.18

Expression Biomarkers for Lung Cancer

Biomarker(s)	Reference
Epidermal growth factor receptor (EGFR) and HER2 (oncogene c-ERB-2) expression (increased gene number by FISH) as prognostic biomarker for non-small-cell lung cancer and predictive for EGFT inhibitors	463
Review of the use of DNA microarrays for the identification of gene expression as prognostic, predictive, and/or diagnostics biomarkers	426
MicroRNAs (miRNAs) as biomarkers as detected in tissue by microarrays or RT-PCR	464
Microarray technique for high-throughput screening for gene-expression biomarkers for lung cancer	465
BRCA1(breast cancer susceptibility gene 1) expression as biomarker for non-small-cell lung cancer	466
Expression of urokinase plasminogen activator receptor (UPAR) and C4.4A[a] as biomarkers for lung cancer	467
Discussion of the value of microarray analysis of gene expression for diagnosis and prediction in non-small-cell lung cancer	468
Microarray analysis of gene expression in lung cancer as an approach to a personalized medicine	469
Use of EGFR and HER2 gene expression as predictive biomarkers for EGFR inhibitors in the treatment of non-small-cell lung cancer	470
Quantitative real-time polymerase chain reaction (qRT-PCR) for gene expression as prognostic and predictive biomarker	471
A review of gene expression as biomarker in non-small-cell lung cancer	472

[a] C4.4A, see Esselens, C.W., Malapeira, J., Colomé, N. et al., Metastasis-associated C4.4A, a GPI-anchored protein cleaved by ADAM10 and ADAM17, *Biol. Chem.* 389, 1075–1084, 2008.

TABLE 3.19
Use of Tissue Kallikrein as a Biomarker for Cancer[a]

Study	Reference
Human glandular kallikrein 5 (KLK5) as a prognostic biomarker for breast and ovarian cancer	494
Human glandular kallikrein 13 (KLK13) as a prognostic biomarker for cancer (tumor invasion and metastasis)	495
Development of a prodrug for prostate cancer, which could be activated by human glandular kallikrein 2	496
Human glandular kallikrein 9 (KLK9) was obtained by recombinant DNA technology and used to prepare a polyclonal antisera for use in an immunoassay. There is differential expression of KLK09 in ovarian and breast cancer as prognostic biomarker for ovarian and breast cancer	497
Development of an immunofluorometric assay for human kallikrein 15 (HLK15); suggested as a biomarker for cancer	498
Human glandular kallikrein 2 as an independent prognostic biomarkers for prostate cancer[b]	499
Kallikrein-related peptidase 8 as a biomarker for advanced ovarian cancer. Levels of kallikrein-related peptide 8 were measured by immunofluorescence in tissue microarrays. Processing was aided by automated in situ protein analysis (AQUA)	500
Human glandular kallikrein 5 (KLK5) expression (qRT-PCR) as a prognostic biomarker in prostate needle biopsy	501
Expression (RT-PCR) of human glandular kallikrein 7 (KLK7) as a prognostic biomarker in colon cancer	502
Expression of human glandular kallikrein 6 (mRNA) as a prognostic biomarker in non-small-cell lung cancer	503
Human glandular kallikrein 6 (KLK6) and human glandular kallikrein 13 (KLK13) as prognostic biomarkers in ovarian cancer	504

[a] The reader is directed to Lundwall, A. and Brattsand, M., Kallikrein-related peptidases, *Cell Mol. Life Sci.* 65, 2019–2038; Debela, M., Beaufort, N., Magdolen, V. et al., Structures and specificity of the human kallikrein-related peptidases KLK 4, 5, 6, and 7, *Biol. Chem.* 389, 623–632, 2008; Clements, J.A., Reflections on the tissue kallikrein and kallikrein-related peptidase family—From mice to men—What have we learnt in the last two decades?, *Biol. Chem.* 389, 1447–1454, 2008; Emami, N. and Diamandis, E.P., Utility of kallikrein-related peptidases (KLKs) as cancer biomarkers, *Clin. Chem.* 54, 1600–1607, 2008.

[b] Eliminated assay interference by enzymatic digestion of capture antibodies to produce $F(ab')_2$, which eliminates assay interference by plasma components which bind to the Fc domain such as rheumatoid factor (see Towbin, H., Schmitz, A., van Oostrum, J. et al., Monoclonal antibody based enzyme-linked and chemiluminescent assays for the human interleukin-1 receptor antagonist. Application to measure hIL-1ra levels in monocyte cultures and synovial fluids, *J. Immunol. Methods* 170, 125–135, 1994).

TABLE 3.20
Degeneracy of Biomarkers in Oncology[a]

	Breast	Prostate	Pancreas	Lung	Colorectal	Ovarian	Head/Neck	References
M2 pyruvate kinase	X	?	X	X	X	X	X	505
MUC[b]	X	X	X	X	X	X	X	506, 507
MicroRNA	X	X	X	X	X	X	X	508, 509
Circulating DNA	X	X	X	X	X	X	X	510
Hsp70[c]	X	X	X	X	X	X	X	511, 512
COX-2[d]	X	X	X	X	X	X	X	513, 514
Circulating tumor cells	X	X	-	X	X	X	X	515, 516
CA 19-9[e]	X	X[b]	X	X	X	X	X	214, 517
CEA[f]	X	X	X	X	X	X	X	

[a] This information is only to indicate that an observation suggesting the presence of a biomarker in a specific cancer type has been reported. This is not to suggest that the observation has been validated or accepted as a diagnostic/prognostic/predictive/ theranostic biomarker for the indicated disease state.

[b] MUC, mucin (Van Klinken, B.J., Einerhand, A.W., Büller, H.A. et al., Strategic biochemical analysis of mucins, *Anal. Biochem.* 265, 103–116, 1998; Dekker, J., Rossen, J.W., Büller, H.A., and Einerhand, A.W., The MUC family: An obituary, *Trends Biochem. Sci.* 27, 126–131, 2002; Senapati, S., Sharma, P., Bafna, S. et al., The MUC gene family: Their role in the diagnosis and prognosis of gastric cancer, *Histol. Histopathol.* 23, 1541–1552, 2008).

[c] Hsp70, heat shock protein 70.

[d] COX-2, cyclooxygenase-2.

[e] The greatest interest for CA19-9 is in pancreatic cancer as a prognostic marker after initial diagnosis (Grote, T. and Logsdon, C.D., Progress on molecular markers of pancreatic cancer, *Curr. Opin. Gastroenterol.* 23, 508–514, 2007).

[f] CEA, carcinoembryonic antigen is one of the oldest serological biomarkers for cancer (Thompson, D.M., Krupey, J. Freedman, S.O., and Gold, P., The radioimmunoassay of circulating carcinoembryonic antigen of the human digestive system, *Proc. Natl. Acad. Sci. USA* 64, 161–167, 1969) which is expressed in a broad range of tumors (Karyampudi, L., Krco, C.J., Kalli, K.R. et al., Identification of a broad coverage HLA-DR degenerate epitope pool derived from carcinoembryonic antigen, *Cancer Immunol. Immunother.* 59, 161–171, 2010). There is current effort to use CEA as a therapeutic target (Sarobe, P., Huarte, E., Lasarte, J.J., and Borrás-Cuesta, F., Carcinoembryonic antigen as a target to induce anti-tumor immune responses, *Curr. Cancer Drug Targets* 4, 443–454, 2004; Wang, D., Rayani, S., and Marshall, J.L., Carcinoembryonic antigen as a vaccine target, *Expert Rev. Vaccines* 7, 987–993, 2008).

REFERENCES

1. Lambert, A., The importance of the early diagnosis in the surgical treatment of carcinoma of the lung, *N. Y. State J. Med.* 47, 2688, 1947.
2. Scott, W.G., Significance of rectal bleeding and the importance of diagnosing early cancer of the colon, *Illinois Med. J.* 96, 252–254, 1949.
3. Myers, H.C., Importance of early diagnosis and treatment of gastric carcinoma, *Clin. Med.* 8, 1083–1086, 1961.
4. Lashgari, A.R. and Friedlander, S.F., The importance of early diagnosis in multiple endocrine neoplasia III: Report of a case with thyroid C-cell hyperplasia, *J. Am. Acad. Dermatol.* 36, 296–300, 1997.
5. Turhani, D., Krapfenbauer, K., Thrunher, D. et al., Identification of differentially expressed, tumor-associated proteins in oral squamous cell carcinoma by proteomic analysis, *Electrophoresis* 27, 1417–1423, 2006.
6. Suri, A., Cancer testis antigens—Their importance in immunotherapy and in the early detection of cancer, *Expert Opin. Biol. Ther.* 6, 379–389, 2006.
7. Ebert, M.P.A. and Roecken, C., Molecular screening of gastric cancer by proteome analysis, *Eur. J. Gastroenterol. Hepatol.* 18, 847–853, 2006.
8. Hermann, K., Walch, A., Balluff, B. et al., Proteomic and metabolic prediction of response to therapy in gastrointestinal cancers, *Nat. Clin. Pract. Gastroenterol. Hepatol.* 6, 170–183, 2009.
9. Gast, M.-C., Schellens, J.H., and Beijnen, J.H., Clinical proteomics in breast cancer: A review, *Breast Cancer Res. Treat.* 116, 17–29, 2009.
10. Jones, H.B., On a new substance occurring in the urine of a patient with mollities ossium, *Philos. Trans. R. Soc. Lond.* 138, 55–62, 1848.
11. Sell, S. (ed.), *Serological Cancer Markers*, Humana Press, Totowa, NJ, 1992.
12. Raulston, J.E. and Leonard, R.C.F., *Serological Tumor Markers: An Introduction*, Churchill Livingstone, Edinburgh, U.K., 1993.
13. Cotter, R.E. (ed.), *Molecular Diagnosis of Cancer*, Humana Press, Totowa, NJ, 1996.
14. Wu, J.T. and Nakamura, R.M., *Human Circulatory Tumor Markers. Current Concepts and Clinical Applications*, American Society for Clinical Pathology, Chicago, IL, 1997.
15. Gospodarowicz, M., O'Sullivan, B., and Sobin, L.H. (eds.), *Prognostic Factors in Cancer*, 3rd edn., John Wiley & Sons, Hoboken, NJ, 2006.
16. Hamdan, M.H., *Cancer Biomarkers Analytical Techniques for Discovery*, Wiley-Interscience, Hoboken, NJ, 2007.
17. Chan, D.W., Booth, R.A., and Diamandis, E.P., Tumor markers, in *Tietz Textbook of Clinical Chemistry and Molecular Diagnostics*, C.A. Burtis, E.R. Ashwood, and D.E. Burns (eds.), Elsevier Saunders, St. Louis, MO, Chapter 23, pp. 745–795, 2006.
18. Airley, R., Loncaster, J., Davidson, S. et al., Glucose transporter Glut-1 expression correlates with tumor hypoxia and predicts metastasis-free survival in advanced carcinoma of the cervix, *Clin. Cancer Res.* 7, 928–934, 2001.
19. Eissa, S., Kassim, S., and El-Ahmady, O., Detection of bladder tumors: Role of cytology, morphology-based assays, biochemical and molecular markers, *Curr. Opin. Obstet. Gynecol.* 16, 395–403, 2003.
20. Amiel, G.E., Shu, T., and Lerner, S.P., Alternatives to cytology in the management of non-muscle invasive bladder cancer, *Curr. Treat. Options Oncol.* 5, 377–389, 2004.
21. Jankowski, J.A. and Odze, R.D., Biomarkers in gastroenterology: Between hope and hype comes histopathology, *Am. J. Gastroenterol.* 104, 1093–1096, 2009.
22. Thames, H., Kuban, D., Levy, L. et al., Comparison of alternative biochemical failure definitions based on clinical outcome in 4839 prostate cancer patients treated by external beam radiotherapy between 1986–1995, *Int. J. Radiat. Oncol. Biol. Phys.* 57, 929–943, 2003.

23. Michiels, S., Le Maltre, A., Buyse, M. et al., Surrogate endpoints for overall survival in locally advanced head and neck cancer: Meta-analyses of individual patients data, *Lancet Oncol.* 10, 341–350, 2009.

24. Nass, S.J. and Moses, H.L. (eds.), *Cancer Biomarkers. The Promises and Challenges of Improving Detection and Treatment*, National Academy of Science, Washington, DC, 2007.

25. Hayat, M.A. (ed.), *Cancer Imaging*, Elsevier, Amsterdam, the Netherlands, 2008.

26. Adami, H.O., Hunter, D., and Trichopoulos, D. (eds.), *Textbook of Cancer Epidemiology*, 2nd edn., Oxford University Press, Oxford, U.K., 2008.

27. Savage, L., Forensic bioinformatician aims to solve mysteries of biomarker studies, *J. Natl. Cancer Inst.* 100, 983–987, 2008.

28. Kaiser, J., Research, two universities sued over validity of prostate cancer test, *Science* 325, 1484, 2009.

29. Waxman, A.G., Cervical cancer screening in the early post vaccine era, *Obstet. Gynecol. Clin. North Am.* 35, 537–548, 2008.

30. Waxman, A.G. and Zsemlye, M.M., Preventing cervical cancer: The Pap test and the HPV vaccine, *Med. Clin. North Am.* 92, 1059–1082, 2008.

31. Lipkin, M. and Newmark, H., Application of intermediate biomarkers and the prevention of cancer of the large intestine, *Prog. Clin. Biol. Res.* 279, 135–150, 1988.

32. van der Kwast, T.H., Intermediate biomarkers for chemoprevention of prostate cancer, *IARC Sci. Publ.* 154, 199–205, 2001.

33. Kosmeder II, J.W. and Pezzuto, J.M., Intermediate biomarkers, *Cancer Treat. Res.* 106, 31–61, 2001.

34. Rundle, A. and Schwartz, S., Issues in the epidemiological analysis and interpretation of intermediate biomarkers, *Cancer Epidemiol. Biomarkers Prev.* 12, 491–496, 2003.

35. Boffetta, P. and Trichopoulos, D., Biomarkers in cancer epidemiology, in *Textbook of Cancer Epidemiology*, 2nd edn., H.-O. Adami, D. Hunter, and D. Trichopoulos (eds.), Oxford University Press, Oxford, U.K., Chapter 5, pp. 109–126, 2008.

36. Scalmati, A. and Lipkin, M., Intermediate biomarkers of increased risk for colorectal cancer: Comparison of different methods of analysis and modifications by chemopreventive interventions, *J. Cell. Biochem. Suppl.* 16G, 65–71, 1992.

37. Lipkin, M., Early development of cancer chemoprevention clinical trials: Studies of dietary calcium as a chemopreventive agent for human subjects, *Eur. J. Cancer Prev.* 11(Suppl 2), S65–S70, 2002.

38. Lippman, S.M., Hittelman, W.N., Lotan, R. et al., Recent advances in cancer chemoprevention, *Cancer Cells* 3, 59–65, 1991.

39. Boffetta, P., Molecular epidemiology, *J. Intern. Med.* 248, 447–454, 2000.

40. Grant, S.G., Molecular epidemiology of human cancer: Biomarkers of genotoxiexposure and susceptibility, *J. Environ. Pathol. Toxicol. Oncol.* 20, 245–261, 2001.

41. Rundle, A., Molecular epidemiology of physical activity and cancer, *Cancer Epidemiol. Biomarkers Prev.* 14, 227–236, 2005.

42. Potter, J.D., Epidemiology informing clinical practice: From bills of mortality to population laboratories, *Nat. Clin. Pract. Oncol.* 2, 625–634, 2005.

43. Merlo, D.F., Sormani, M.P., and Bruzzi, P., Molecular epidemiology: New rules for new tools?, *Mutat. Res.* 600, 3–11, 2006.

44. Biasco, G., Paganelli, G.M., Brandi, G. et al., Rectal cell proliferation as an intermediate biomarker of risk of colorectal cancer, *Eur. J. Cancer Prev.* 2(Suppl 2), 89–93, 1993.

45. Jones, D.R., Davidson, A.G., Summers, C.L. et al., Potential application of p53 as an intermediate biomarker in Barrett's esophagus, *Ann. Thorac. Surg.* 57, 598–603, 1994.

46. Indulski, J.A. and Lutz, W., The biomarkers detecting early changes in the human organism exposed to occupational carcinogens, *Cent. Eur. J. Public Health* 7, 221–224, 1999.

47. Cravo, M., Fidalgo, P., Pereira, A.D. et al., DNA methylation as an intermediate biomarker in colorectal cancer: Modulation by folic acid supplementation, *Eur. J. Cancer Prev.* 3, 473–479, 1994.

48. Donat, T.L., Sakr, W., Lehr, J.E., and Pienta, K.J., Unique nuclear matrix protein alternations in head and neck squamous cell carcinomas: Intermediate biomarker candidates, *Otolaryngol. Head Neck Surg.* 114, 387–393, 1996.

49. Zeng, Q., Smith, D.C., Suscovich, T.J. et al., Determination of intermediate biomarker expression levels by quantitative reverse transcription-polymerase chain reaction in oral mucosa of cancer patients treated with liarozole, *Clin. Cancer Res.* 6, 2245–2251, 2000.

50. Einspahr, J.G., Alberts, D.S., Gapstur, S.M. et al., Surrogate end-point biomarkers as measures of colon cancer risk and the use in cancer chemoprevention trials, *Cancer Epidemiol. Biomarkers Prev.* 6, 37–48, 1997.

51. Georgakoudi, I., Jacobson, B.C., Müller, M.G. et al., NAD(P)H and collagen as in vivo quantitative fluorescent biomarkers of epithelial precancerous changes, *Cancer Res.* 62, 682–687, 2002.

52. Hirsch, F.R., Merrick, D.T., and Franklin, W.A., Role of biomarkers for early detection of lung cancer and chemoprevention, *Eur. Respir. J.* 19, 1151–1158, 2002.

53. Jonsson, S., Varella-Garcia, M., Miller, Y.E. et al., Chromosomal aneusomy in bronchial high-grade lesions is associated with invasive lung cancer, *Am. J. Respir. Crit. Care Med.* 177, 342–347, 2008.

54. Massion, P.P., Zou, Y., Uner, H. et al., Recurrent genomic gains in preinvasive lesions as a biomarker of risk for lung cancer, *PLoS One* 4, e5611, 2009.

55. Einspahr, J.G., Xu, M.J., Warneke, J. et al., Reproducibility and expression of skin biomarkers in sun-damaged skin and actinic keratoses, *Cancer Epidemiol. Biomarkers Prev.* 15, 1841–1848, 2006.

56. Khan, A.M. and Singer, A., Biomarkers in cervical precancer management: The new frontiers, *Future Oncol.* 4, 515–524, 2008.

57. Bronner, M.P., O'Sullivan, J.N., Rabinovitch, P.S. et al., Genomic biomarkers to improved ulcerative colitis neoplasia surveillance, *Am. J. Pathol.* 173, 1853–1860, 2008.

58. Santella, R.M., DNA damage as an intermediate biomarker in intervention studies, *Proc. Soc. Exp. Biol. Med.* 216, 166–171, 1997.

59. Matter, B., Malejka-Giganti, D., Csallany, A.S. et al., Quantitative analysis of the oxidative DNA lesion, 2,2-diamino-4-[(2-deoxy-β-D-erythro-pentofuranosyl)]amino-5(2H)-oxazolone (oxazolone), in vitro and in vivo by isotope dilution-capillary HPLC-ESI-MS/MS, *Nucleic Acids Res.* 34, 5449–5460, 2006.

60. Greenwald, P. and Dunn, B.K., Do we make optimal use of the potential of cancer prevention?, *Recent Results Cancer Res.* 181, 3–17, 2009.

61. Keith, R.L., Chemoprevention of lung cancer, *Proc. Am. Thorac. Soc.* 6, 187–193, 2009.

62. *Cancer and Inflammation*, Novartis Foundation Symposium, John Wiley & Sons, Chichester, U.K., 2004.

63. Smedby, K.E., Askling, J., Mariette, X., and Baecklund, E., Autoimmune and inflammatory disorder and risk of malignant lymphomas—An update, *J. Intern. Med.* 264, 514–527, 2008.

64. Ahmadi, A., Polyak, S., and Draganov, P.V., Colorectal cancer surveillance in inflammatory bowel disease: The search continues, *World J. Gastroenterol.* 15, 61–66, 2009.

65. Gerner, E.W. and Meyskens Jr. F.L., Combination chemoprevention for colon cancer targeting polyamine synthesis and inflammation, *Clin. Cancer Res.* 15, 758–761, 2009.

66. Berasain, C., Castillo, J., Perugorria, M.J. et al., Inflammation and liver cancer: New molecular links, *Ann. N. Y. Acad. Sci.* 1155, 206–221, 2009.

67. Feagins, L.A., Souza, R.F., and Spechler, S.J., Carcinogenesis in IBD: Potential targets for the prevention of colorectal cancer, *Nat. Rev. Gastroenterol. Hepatol.* 6, 297–305, 2009.

68. Colotta, F., Allavena, P., Sica, A. et al., Cancer-related inflammation, the seventh hallmark of cancer: Links to genetic instability, *Carcinogenesis* 30, 1073–1081, 2009.

69. Balkwill, F. and Mantovani, A., Inflammation and cancer: Back to Virchow, *Lancet* 357, 539, 2001.

70. Coussens, L.M. and Werb, Z., Inflammatory cells and cancer: Think different!, *J. Exp. Med.* 193, F23, 2001.

71. Blankenstein, T., The role of inflammation in tumour growth and tumour suppression, in *Cancer and Inflammation*, Novartis Foundation Symposium, John Wiley & Sons, Chichester, U.K., pp. 205–214, 2004.

72. Brown, D.J., Milroy, R., Preston, T., and McMillan, D.C., The relationship between an inflammation-based prognostic score (Glasgow Prognostic Score) and changes in serum biochemical variables in patients with advanced lung and gastrointestinal cancer, *J. Clin. Pathol.* 60, 705–708, 2007.

73. Kawanishi, S. and Hiraku, Y., Oxidative and nitrative DNA damage as biomarker for carcinogenesis with special reference to inflammation, *Antioxid. Redox Signal.* 8, 1047–1058, 2006.

74. Hoki, Y., Murata, M., Hiraku, Y. et al., 8-Nitroguanine as a potential biomarker for progression of malignant fibrous histiocytoma, a model of inflammation-related cancer, *Oncol. Rep.* 18, 1165–1169, 2007.

75. Son, J., Pang, B., McFaline, J.L. et al., Surveying the damage: The challenges of developing nucleic acid biomarkers of inflammation, *Mol. Biosyst.* 4, 902–908, 2008.

76. Hiraku, Y. and Kawanishi, S., Immunohistochemical analysis of 8-nitroguanine, a nitrative DNA lesion, in relation to inflammation-associated carcinogenesis, *Methods Mol. Biol.* 512, 3–13, 2009.

77. Mills, P.J., Ancoli-Israel, S., Parker, B. et al., Predictors of inflammation in response to anthracycline-based chemotherapy for breast cancer, *Brain Behav. Immun.* 22, 98–104, 2008.

78. Loeb, S., Gashti, S.N., and Catalona, W.J., Exclusion of inflammation in the differential of an elevated prostate-specific antigen (PSA), *Urol. Oncol.* 27, 64–66, 2009.

79. van Waarde, A. and Elsinga, P.H., Proliferation markers for the differential diagnosis of tumor and inflammation, *Curr. Pharm. Des.* 14, 3326–3339, 2008.

80. Roslind, A. and Johansen, J.S., YKL-40: A novel marker shared by chronic inflammation and oncogenic transformation, *Methods Mol. Biol.* 511, 159–184, 2009.

81. Pacova, H., Astl, J., and Martinek, J., The pathogenesis of chronic inflammation and malignant transformation in the human upper airways: The role of β-defensins, eNOS, cell proliferation and apoptosis, *Histol. Histopathol.* 24, 815–820, 2009.

82. Moore, B.W., A soluble protein characteristic of the nervous system, *Biochem. Biophys. Res. Commun.* 19, 739–744, 1965.

83. Marenholz, I., Heizmann, C.W., and Fritz, G., S100 proteins in mouse and man from evolution to function and pathology (including an update of the nomenclature), *Biochem. Biophys. Res. Commun.* 322, 1111–1122, 2004.

84. Marenholz, I., Lovering, R.C., and Heizmann, C.W., An update of the S100 nomenclature, *Biochim. Biophys. Acta* 1763, 1282–1283, 2006.

85. Kligman, D. and Hilt, D.C., The S100 protein family, *Trends Biochem. Sci.* 13, 437–443, 1988.

86. Santamaria-Kisiel, L., Rintala-Dempsey, A.C., and Shaw, G.S., Calcium-dependent and -independent interactions of the S100 protein family, *Biochem. J.* 396, 201–214, 2006.

87. Steinbakk, M., Naess-Andresen, C.F., Lingaas, E. et al., Antimicrobial actions of calcium binding leukocyte L1 protein, calprotectin, *Lancet* 336, 763–765, 1990.

88. Brandtzaeg, P., Dale, I., and Gabrielsen, T.O., The leucocyte L1 protein (calprotectin): Usefulness as an immunohistochemical marker antigen and putative biological function, *Histopathology* 21, 191–196, 1992.

89. Brandtzaeg, P., Gabrielsen, T.O., Dale, I. et al., The leucocyte protein L1 (calprotectin): A putative nonspecific defence factor at epithelial surfaces, *Adv. Exp. Med. Biol.* 371A, 201–206, 1995.

90. Andersson, K.B., Sletten, K., Berntzen, H.B. et al., The leucocyte L1 protein: Identity with the cystic fibrosis antigen and the calcium-binding MRP-8 and MRP-14 macrophage components, *Scand. J. Immunol.* 27, 241–245, 1988.

91. Longbottom, D., Sallenave, J.M., and van Heyninger, V., Subunit structure of calgranulins A and B obtained from sputum, plasma, granulocytes and cultured epithelial cells, *Biochim. Biophys. Acta* 1120, 215–222, 1992.

92. Roth, J., Vogl, T., Sorg, C., and Sunderkötten, C., Phagocyte-specific S100 proteins: A novel group of proinflammatory molecules, *Trends Immunol.* 24, 155–158, 2003.

93. Ott, H.W., Lindner, H., Sorg, B. et al., Calgranulins in cystic fluid and serum from patients with ovarian carcinomas, *Cancer Res.* 63, 7507, 2003.

94. Luu, H.H., Zhou, L., Haydon, R.C. et al., Increased expression of S100A6 is associated with decreased metastasis and inhibition of cell migration and anchorage independent growth in human osteosarcoma, *Cancer Lett.* 229, 135–148, 2005.

95. Lee, O.J., Hong, S.M., Razvi, M.H. et al., Expression of calcium-binding proteins S100A2 and S100A4 in Barrett's adenocarcinoma, *Neoplasia* 8, 843–850, 2006.

96. Melo-Junior, M.R., Araujo, J.L.S., Cavalcanti, C.L.B. et al., Detection of S100 protein from prostatic cancer patients using anti-S100 protein antibody immobilized on POS-PVA discs, *Biotechnol. Bioeng.* 97, 182–187, 2007.

97. Yao, R., Davidson, D.D., Lopez-Beltran, A. et al., The S100 proteins for screening and prognostic grading of bladder cancer, *Histol. Histopathol.* 22, 1025–1032, 2007.

98. El-Abd, E., El-Tahan, R., Fahmy, L. et al., Serum mestastasin mRNA is an important survival predictor in breast cancer, *Br. J. Biomed. Sci.* 65, 90–94, 2008.

99. S100A8 and S100A9 overexpression is associated with poor pathological parameters in invasive ductal carcinoma of the breast, *Curr. Cancer Drug Targets* 8, 243–252, 2008.

100. Xie, R., Schlumbrecht, M.P., Shipley, G.L. et al., S100A4 mediates endometrial cancer invasion and is a target of TGF-β1 signaling, *Lab. Invest.* 89, 937–947, 2009.

101. Guo, H.B., Stoffel-Wagner, B., Bierwirth, T. et al., Clinical significance of serum S100 in metastatic malignant melanoma, *Eur. J. Cancer* 31A, 1989–1902, 1995.

102. Abraha, H.D., Fuller, L.C., Du Vivier, A.W. et al., Serum S-100 protein: A potentially useful prognostic marker in cutaneous melanoma, *Br. J. Dermatol.* 137, 381–385, 1997.

103. Hansson, L.O., von Schloultz, E., Djureen, E. et al., Prognostic value of serum analyses of S-100 protein beta in malignant melanoma, *Anticancer Res.* 17, 3071–3073, 1997.

104. Tofani, A., Cioffi, R.P., Sciuto, R. et al., S-100 and NSE as serum makers in melanoma, *Acta Oncol.* 36, 761–764, 1997.

105. Schultz, E.S., Diepgen, T.L., and Von Den Driesch, P., Clinical and prognostic relevance of serum S-100 beta protein in malignant melanoma, *Br. J. Dermatol.* 138, 426–430, 1998.

106. Seregni, E., Massaron, S., Martinetti, A. et al., S100 protein serum levels in cutaneous malignant melanoma, *Oncol. Rep.* 5, 601–604, 1998.

107. Hauschild, A., Engel, G., Brenner, W. et al., S100B protein detection in serum is a significant prognostic factor in metastatic melanoma, *Oncology* 56, 338–344, 1999.

108. Jury, C.S., McAllister, E.J., and McKie, R.M., Rising levels of serum S100 protein precede other evidence of disease progression in patients with malignant melanoma, *Br. J. Dermatol.* 143, 269–274, 2000.

109. Schalgenhauff, B., Schittek, B., Ellwanger, U. et al., Significance of serum protein S100 levels in screening for melanoma metastasis: Does protein S100 enable early detection of melanoma recurrence?, *Melanoma Res.* 10, 451–459, 2000.

110. Mårtenson, E.D., Hansson, L.O., Nilsson, B. et al., Serum S100b protein as a prognostic marker in malignant melanoma, *J. Clin. Oncol.* 19, 824–831, 2001.

111. Mohammed, M.Q., Abraha, H.D., Sherwood, R.A. et al., Serum S100β protein as a marker of disease activity in patients with malignant melanoma, *Med. Oncol.* 18, 109–120, 2001.

112. Ghanem, G., Loir, B., Morandini, R. et al., On the release and half-life of S100B protein in the peripheral blood of melanoma patients, *Int. J. Cancer* 94, 586–590, 2001.

113. Harpio, R. and Einarsson, R., S100 proteins as cancer biomarkers with focus on S100B in malignant melanoma, *Clin. Biochem.* 37, 512–518, 2004.

114. Smit, L.H., Korse, C.M., and Bonfrer, J.M., Comparison of four different assays for determination of serum S-100B, *Int. J. Biol. Markers* 20, 34–42, 2005.

115. Beyeler, M., Waldispuhl, S., Strobel, K. et al., Detection of melanoma relapse: First comparative analysis on imaging techniques versus S100 protein, *Dermatology* 213, 187–191, 2006.

116. Domingo-Domènech, J., Castel, T., Auge, J.M. et al., Prognostic implications of protein S-100β serum levels in the clinical outcome of high-risk melanoma patients, *Tumour Biol.* 28, 264–272, 2007.

117. Schiltz, P.M., Dillman, R.O., Korse, C.M. et al., Lack of elevation of serum 100B in patients with metastatic melanoma as a predictor of outcome after induction with an autologous vaccine of proliferating tumor cells and dendritic cells, *Cancer Biother. Radiopharm.* 23, 214–221, 2008.

118. Egberts, F., Pollex, A., Egberts, J.H. et al., Long-term survival analysis in metastatic melanoma: Serum S100B is an independent prognostic marker and superior to LDH, *Onkologie* 31, 380–384, 2008.

119. Smit, L.H., Nieweg, O.E., Mool, W.J. et al., Value of serum S100B for prediction of distant relapse and survival in stage III B/C melanoma, *Anticancer Res.* 28, 2297–2302, 2008.

120. Mocellini, S., Zavagno, G., and Nitti, D., The prognostic value of serum S100B in patients with cutaneous melanoma: A meta-analysis, *Int. J. Cancer* 123, 2370–2376, 2008.

121. Oberholzer, P.A., Urosevic, M., Steinert, H.C., and Dummer, R., Baseline staging of melanoma with unknown primary site: The value of serum s100 protein and positron emission tomography, *Dermatology* 217, 351–355, 2008.

122. Egberts, R., Hitschler, W.N., Weichenthal, M., and Hauschild, A., Prospective monitoring of adjuvant treatment in high-risk melanoma patients: Lactate dehydrogenase and protein S-100B as indicators of relapse, *Melanoma Res.* 19, 31–35, 2009.

123. Perez, D., Demartines, N., Meier, K. et al., Protein S100 as prognostic marker for gastrointestinal stromal tumors: A clinicopathological risk factor analysis, *J. Invest. Surg.* 20, 181–186, 2007.

124. Ohuchida, K., Mizumoto, K., Miyasaka, Y. et al., Over-expression of S100A2 in pancreatic cancer correlates with progression and poor prognosis, *J. Pathol.* 213, 275–282, 2007.

125. Fullen, D.R., Garrisi, A.J., Sanders, D., and Thomas, D., Expression of S100A6 protein in a broad spectrum of cutaneous tumors using tissue microarrays, *J. Cutan. Pathol.* 35(Suppl 2), 28–34, 2008.

126. Min, H.S., Choe, G., Kim, S.W. et al., S100A4 expression is associated with lymph node metastasis in papillary microcarcinoma of the thyroid, *Mod. Pathol.* 21, 748–755, 2008.

127. Sapkota, D., Bruland, O., Bøe, O.E. et al., Expression profile of the S100 gene family members in oral squamous cell carcinomas, *J. Oral Pathol.* 37, 607–615, 2008.

128. Blankin, A.V., Kench, J.G., Colvin, A.K. et al., Expression of S100A2 calcium-binding protein predicts response to pancreatectomy for pancreatic cancer, *Gastroenterology* 137, 558–568, 2009.

129. Kim, J.H., Kim, C.N., Kim, S.Y. et al., Enhanced S100A4 protein expression is clinicopathologically significant to metastatic potential and p53 dysfunction in colorectal cancer, *Oncol. Rep.* 22, 41–47, 2009.
130. Starkey, B.J., Screening for colorectal cancer, *Ann. Clin. Biochem.* 39, 351–365, 2002.
131. Sumerston, C.B., Longlands, M.G., Wiener, K. et al., Faecal calprotectin: A marker of inflammation throughout the intestinal tract, *Eur. J. Gastroenterol. Hepatol.* 14, 8, 841–845, 2002.
132. von Roon, A.C., Karamoutzos, L., Purkayastha, S. et al., Diagnostic precision of fecal calprotectin for inflammatory bowel disease and colorectal malignancy, *Am. J. Gastroenterol.* 102, 803–813, 2007.
133. Karl, J., Wild, N., Tacke, M. et al., Improved diagnosis of colorectal cancer using a combination of fecal occult blood and novel fecal protein markers, *Clin. Gastroenterol. Hepatol.* 6, 1122–1128, 2008.
134. Kim, H.J., Kang, H.J., Lee, H. et al., Identification of S100A8 and S100A9 as serological markers for colorectal cancer, *J. Proteome Res.* 8, 1368–1379, 2009.
135. Canani, A. and Troncone, R., Faecal calprotectin as reliable non-invasive marker to assess the severity of mucosal inflammation in children with inflammatory bowel disease, *Dig. Liver Dis.* 40, 547–553, 2008.
136. Otten, C.M., Kok, L., Witteman, B.J. et al., Diagnostic performance of rapid tests for detection of fecal calprotectin and lactoferrin and their ability to discriminate inflammatory from irritable bowel syndrome, *Clin. Chem. Lab. Med.* 46, 1275–1280, 2008.
137. Foell, D., Wittkowski, H., and Roth, J., Monitoring disease activity by stool analyses: From occult blood to molecular markers of intestinal inflammation and damage, *Gut* 58, 859–868, 2009.
138. Tursi, A., Brandimarte, G., Elisei, W. et al., Faecal calprotectin in colonic diverticular disease: A case-control study, *Int. J. Colorectal Dis.* 24, 49–55, 2009.
139. Pine, J.K., Fusai, K.G., Young, R. et al., Serum C-reactive protein concentration and the prognosis of ductal adenocarcinoma of the head of the pancreas, *Eur. J. Surg. Oncol.* 35, 605–610, 2009.
140. Koch, A., Fohlin, H., and Sörenson, S., Prognostic significance of C-reactive protein and smoking in patients with advanced non-small cell lung cancer treated with first-line palliative chemotherapy, *J. Thorac. Oncol.* 4, 326–332, 2009.
141. Wang, C.Y., Hsieh, M.J., Chiu, Y.C. et al., Higher serum C-reactive protein concentration and hypoalbuminemia are poor prognostic indicators in patients with esophageal cancer undergoing radiotherapy, *Radiother. Oncol.* 92, 270–275, 2009.
142. Willaims, D.K. and Muddiman, D.C., Absolute quantification of C-reactive protein in human plasma derived from patients with epithelial ovarian cancer utilizing protein cleavage isotope dilution mass spectrometry, *J. Proteome Res.* 8, 1085–1090, 2009.
143. Allin, K.H., Bojesen, S.E., and Nordestgaard, B.G., Baseline C-reactive protein is associated with incident cancer and survival in patients with cancer, *J. Clin. Oncol.* 27, 2217–2224, 2009.
144. Lundin, E., Dossus, L., Clendenen, T. et al., C-reactive protein and ovarian cancer: A prospective study nested in three cohorts (Sweden, USA, Italy), *Cancer Causes Control* 20, 1151–1159, 2009.
145. Ki, Y., Kim, W., Nam, J. et al., C-reactive protein levels and radiation-induced mucositis in patients with head-and-neck cancer, *Int. J. Radiat. Oncol. Biol. Phys.* 75, 393–398, 2009.
146. Koukourakis, M.I., Kambouromiti, G., Pitsiava, D. et al., Serum C-reactive protein (CRP) levels in cancer patients are linked with tumor burden and are reduced by antihypertensive medication, *Inflammation* 32, 169–175, 2009.
147. Otake, T., Uezono, K., Takahashi, R. et al., C-reactive protein and colorectal adenomas: Self Defense Forces Health Study, *Cancer Sci.* 100, 709–714, 2009.

148. Khandavilli, S.D., Ceallaigh, P.O., Lloyd, C.J., and Whitaker, R., Serum C-reactive protein as a prognostic indicator in patients with oral squamous cell carcinoma, *Oral Oncol.* 45, 912–914, 2009.

149. Hefler-Frischmuth, K., Hefler, L.A., Heitze, G. et al., Serum C-reactive protein in the differential diagnosis of ovarian masses, *Eur. J. Obstet. Gynecol. Reprod. Biol.* 147, 65–68, 2009.

150. Ishizuka, M., Nagata, H., Takagi, K., and Kubota, K., Influence of inflammation-based prognostic score on mortality of patients undergoing chemotherapy for far advanced or recurrent unresectable colorectal cancer, *Ann. Surg.* 250, 268–272, 2009.

151. Trichopoulos, D., Psaltopoulou, T., Orfanos, P. et al., Plasma C-reactive protein and risk of cancer: A prospective study from Greece, *Cancer Epidemiol. Biomarkers Prev.* 15, 381–384, 2006.

152. Heikkilä, K., Ebrahim, S., and Lawlor, D.A., A systematic review of the association between circulating concentrations of C reactive protein and cancer, *J. Epidemiol. Community Health* 61, 824–833, 2007.

153. Briggs, C.D., Neal, C.P., Mann, C.D. et al., Prognostic molecular markers in cholangiocarcinoma: A systematic review, *Eur. J. Cancer* 45, 33–47, 2009.

154. Roulston, J.E. and Leonard, R.C.F., *Serological Tumour Markers: An Introduction,* Chapter 1, Churchill Livingstone, Edinburgh, U.K., 1993.

155. Roulston, J.E., Screening with tumor markers, *Mol. Biotechnol.* 20, 153–162, 2002.

156. Gospodarowicz, M.K., O'Sullivan, B., and Koh, E.-W., Prognostic factors: Principles and applications, in *Prognostic Factors in Cancer,* 3rd edn., M. Gospodarowicz, B. O'Sullivan, and L.H. Sobin (eds.), John Wiley & Sons, Hoboken, NJ, Chapter 2, pp. 23–38, 2006.

157. Dowsett, M. and Dunbier, A.K., Emerging biomarkers and new understanding of traditional markers in personalized therapy for breast cancer, *Clin. Cancer Res.* 14, 8019–8026, 2008.

158. Duffy, M.J. and Crown, J., A personalized approach to cancer treatment: How biomarkers can help, *Clin. Chem.* 54, 1770–1779, 2008.

159. Santos, E.S., Perez, C.A., Raez, L.E. et al., How is gene-expression profiling going to challenge the future management of lung cancer?, *Future Oncol.* 5, 827–835, 2009.

160. Mandrekar, S.J. and Sargent, D.J., Clinical trial designs for predictive biomarker validation: One size does not fit all, *J. Biopharm. Stat.* 19, 530–542, 2009.

161. Kubicek, G.J., Wang, F., Reddy, E. et al., Importance of treatment institution in head and neck cancer radiotherapy, *Otolaryngol. Head Neck Surg.* 141, 172–176, 2009.

162. http://www.cancer.gov/cancertopics/factsheet/Sites-Types/head-and-neck; National Cancer Institute Head and Neck Cancer Fact Sheet.

163. Rhee, J.C., Khuri, F.R., and Shin, D.M., Emerging drugs for head and neck cancer, *Expert Opin. Emerg. Drugs* 9, 91–104, 2004.

164. Almadori, G., Bussu, F., and Paludetti, G., Should there be more molecular staging of head and neck cancer to improve the choice of treatments and thereby improve survival?, *Curr. Opin. Otolaryngol. Head Neck Surg.* 16, 117–126, 2008.

165. Akervall, J., Genomic screening of head and neck cancer and its implications for therapy planning, *Eur. Arch. Otorhinolaryngol.* 263, 297–304, 2006.

166. Bossi, P., Locati, L.D., and Licitra, L., Biological agents in head and neck cancer, *Expert Rev. Anticancer Ther.* 7, 1643–1650, 2007.

167. Hunter, K.D., Parkinson, E.K., and Harrison, P.R., Profiling early head and neck cancer, *Nat. Rev. Cancer* 5, 127–135, 2005.

168. Chai, R.L. and Grandis, J.R., Advances in molecular diagnostics and therapeutics in head and neck cancer, *Curr. Treat. Options Oncol.* 7, 3–11, 2006.

169. Radhakrishnan, R., Solomon, M., Satyamoorthy, K. et al., Tissue microarray—A high-throughput molecular analysis in head and neck cancer, *J. Oral Pathol. Med.* 37, 166–176, 2008.

170. Chang, S.S. and Califano, J., Current status of biomarkers in head and neck cancer, *J. Surg. Oncol.* 97, 640–643, 2008.

171. Nagaraj, N.S., Evolving 'omics' technologies for diagnostics of head and neck cancer, *Brief Funct. Genomic Proteomic* 97, 640–643, 2008.
172. Pai, S.I. and Westra, W.H., Molecular pathology of head and neck cancer: Implications for diagnosis, prognosis, and treatment, *Annu. Rev. Pathol.* 4, 49–70, 2009.
173. Yoo, G.H., Piechocki, M.P., Ensley, J.F. et al., Docetaxel induced gene expression patterns in head and neck squamous cell carcinoma using cDNA microarray and PowerBlot, *Clin. Cancer Res.* 8, 3910–3921, 2002.
174. Cho, W.C., Yip, T.T., Yip, C. et al., Identification of serum amyloid a protein as a potentially useful biomarker to monitor relapse of nasopharyngeal cancer by serum proteomic profiling, *Clin. Cancer Res.* 10, 43–52, 2004.
175. Lee, B.J., Wang, S.G., and Choi, J.S., The prognostic value of telomerase expression in peripheral blood mononuclear cells of head and neck cancer patients, *Am. J. Clin. Oncol.* 29, 163–167, 2006.
176. Righini, C.A., de Fraipont, F., Timsit, J.F. et al., Tumor-specific methylation in saliva: A promising biomarker for early detection of head and neck cancer recurrence, *Clin. Cancer Res.* 13, 1179–1185, 2007.
177. Franzmann, E.J., Reategui, E.P., Carraway, K.L. et al., Salivary soluble CD44: A potential molecular marker for head and neck cancer, *Cancer Epidemiol. Biomarkers Prev.* 14, 735–739, 2005.
178. Franzmann, E.J., Reategui, E.P., Pedroso, F. et al., Soluble CD44 is a potential marker for the early detection of head and neck cancer, *Cancer Epidemiol. Biomarkers Prev.* 16, 1348–1355, 2007.
179. Büntzel, J., Bruns, F., Glatzel, M. et al., Zinc concentrations in serum during head and neck cancer progression, *Anticancer Res.* 27, 1941–1943, 2007.
180. El Houda Agueznay, N., Badoual, C., Hans, S. et al., Soluble interleukin-2 receptor and metalloproteinase-9 expression in head and neck cancer: Prognostic value and analysis their relationships, *Clin. Exp. Immunol.* 150, 114–123, 2007.
181. Lin, H.S., Talwar, H.S., Tarca, A.L. et al., Autoantibody approach for serum-based detection of head and neck cancer, *Cancer Epidemiol. Biomarkers Prev.* 16, 2396–2405, 2007.
182. Matta, A., DeSouza, L.V., Skukla, N.K. et al., Prognostic significance of head-and-neck cancer biomarkers previously discovered and identified using iTRAQ-labeling and multidimensional liquid chromatography-tandem mass spectrometry, *J. Proteome Res.* 7, 2078–2087, 2008.
183. Brunner, M., Thurnher, D., Heiduschka, G. et al., Elevated levels of circulating endothelial progenitor cells in head and neck cancer patients, *J. Surg. Oncol.* 98, 545–550, 2008.
184. Kappler, M., Taubert, H., Holzhausen, H.J. et al., Immunohistochemical detection of HIG-1α an CAIX in advanced head-and-neck cancer. Prognostic role and correlation with tumor markers and tumor oxygenation parameters, *Strahlenther. Onkol.* 184, 393–399, 2008.
185. Wineland, A.M. and Stack Jr. B.C., Modern methods to predict costs for the treatment and management of head and neck cancer patients: Examples of methods used in the current literature, *Curr. Opin. Otolaryngol. Head Neck Surg.* 16, 113–116, 2008.
186. Chu, C.S. and Rubin, S.C., Epidemiology, staging, and clinical characteristics, in *Surgery for Ovarian Cancer. Principles and Practice*, R.F. Bristow and B.Y. Karlan (ed.), Taylor & Francis, London, U.K., Chapter 1, 2006.
187. Elmasry, K. and Gaythen, S.A., Epidemiology of ovarian cancer, in *Cancer of the Ovary*, R.H. Reznek (ed.), Cambridge University Press, Cambridge, U.K., Chapter 1, pp. 1–19, 2007.
188. Look, K.Y., Epidemiology, etiology, and screening of ovarian cancer, in *Ovarian Cancer*, 2nd edn., S.C. Rubin and G.P. Sutton (eds.), Lippincott Williams & Wilkins, Philadelphia, PA, Chapter 8, pp. 167–180, 2001.
189. World Health Organization; http://www.who.int/selection_medicines/committees/expert/17/application/ifosfamide_inclusion.pdf

190. Sankaranarayanan, R. and Ferlay, J., Worldwise burden of gynecological cancer: The size of the problem, *Best Pract. Res. Clin. Obstet. Gynaecol.* 20, 207–225, 2006.
191. Klein, G., Tumor antigens, *Annu. Rev. Microbiol.* 20, 223–252, 1966.
192. Gold, P. and Freedman, S.L., Demonstration of tumor-specific antigens in human colonic carcinoma by immunological tolerance and absorption techniques, *J. Exp. Med.* 121, 439–462, 1965.
193. Thompson, D.M.P., Kruper, J., Freedman, S.O., and Gold, P., The radioimmunoassay of circulating carcinoembryonic antigen of the human digestive system, *Proc. Natl. Acad. Sci. USA* 64, 161–167, 1968.
194. Rosai, J., Immunohistochemical markers of thyroid tumors: Significance and diagnostic applications, *Tumori* 89, 517–519, 2003.
195. Tarro, G., Perna, A., and Esposito, C., Early diagnosis of lung cancer by detection of tumor liberated protein, *J. Cell. Physiol.* 203, 1–5, 2005.
196. Duffy, M.J., Serum tumor markers in breast cancer: Are they of clinical value?, *Clin. Chem.* 52, 345–352, 2006.
197. King, J.E., Thatcher, N., Pickering, C.A., and Hasleton, P.S., Sensitivity and specific of immunohistochemical markers used in the diagnosis of epithelioid mesothelioma: A detailed systematic analysis using published data, *Histopathology* 48, 223–232, 2006.
198. Sung, H.J. and Cho, J.Y., Biomarkers for the lung cancer diagnosis and their advances in proteomics, *BMB Rep.* 41, 615–625, 2008.
199. Grigoriu, B.D., Grigoriu, C., Chahine, B. et al., Clinical utility of diagnostic markers for malignant pleural mesothelioma, *Monaldi Arch. Chest Dis.* 71, 31–38, 2009.
200. Bhattacharya, M. and Barlow, J.J., Tumor-specific antigens associated with human ovarian cystadenocarcinoma, in *Handbook of Cancer Immunology*, H. Waters (ed.), Garland, STPM Press, New York, Chapter 10, pp. 277–295, 1968.
201. Gall, S.A., Walling, J., and Pearl, J., Demonstration of tumor-associated antigens in human gynecological malignancies, *Am. J. Obstet. Gynecol.* 115, 387–393, 1973.
202. Order, S.E., Thurston, J., and Knapp, R., Ovarian tumor antigens: A new potential for therapy, *Natl. Cancer Inst. Monogr.* 42 (*Symposium on Ovarian Carcinoma*), 33–43, 1975.
203. Lu, K.H., Patterson, A.P., Wang, L. et al., Selection of potential markers for epithelial ovarian cancer with gene expression arrays and recursive descent partition analysis, *Clin. Cancer Res.* 10, 3291–3300, 2004.
204. Singh, A.P., Senapati, S., Ponnusamy, M.P. et al., Clinical potential of mucins in diagnosis, prognosis, and therapy of ovarian cancer, *Lancet Oncol.* 9, 1076–1085, 2008.
205. Bast Jr. R.C., Feeney, M., Lazarus, H. et al., Reactivity of a monoclonal antibody with human ovarian cancer, *J. Clin. Invest.* 68, 1331–1337, 1981.
206. Bast Jr. R.C., Klug, T.L., St. John, E. et al., A radioimmunoassay using a monoclonal antibody to monitor the course of epithelial ovary cancer, *New Engl. J. Med.* 309, 883–887, 1983.
207. Nilhoff, J.M., Klug, T.L., Schaetzl, E. et al., Elevation of serum CA125 in carcinomas of the fallopian tube, endometrium, and endocervix, *Am. J. Obstet. Gynecol.* 148, 1057–1058, 1984.
208. Jones, M.B., Krutzsch, H., Shu, H. et al., Proteomic analysis and identification of new biomarkers and therapeutic targets for invasive ovarian cancer, *Proteomics* 2, 76–84, 2002.
209. Ardekami, A.M., Liotta, L.A., and Petricoin III E.F., Clinical potential of proteomics in the diagnosis of ovarian cancer, *Exp. Rev. Mol. Diagnost.* 2, 313–320, 2002.
210. Rai, A.J., Zhang, Z., Rosenzweig, J. et al., Proteomic approaches to tumor marker discovery, *Arch. Pathol. Lab. Med.* 126, 1518–1526, 2002.
211. Steinert, R., van Hogen, P., Fels, L.M., Gunther, K., Lippert, H., and Reymond, M.A., Proteomic prediction of disease outcome in cancer: Clinical framework and current status, *Am. J. Pharmacogenomics* 3, 107–115, 2003.

212. Ye, B., Cramer, D.W., Skates, S.J. et al., Haptoglobin-alpha subunit as potential serum biomarker in ovarian cancer: Identification and characterization using proteomic profiling and mass spectrometry, *Clin. Cancer Res.* 9, 2904–2911, 2003.
213. Gadducci, A., Cosier, S., Carpi, A., Nicolini, A., and Genazzani, A.R., Serum tumor markers in the management of ovarian endometrial and cervical cancer, *Biomed. Pharmacol.* 58, 24–38, 2004.
214. Zhang, Z., Bast Jr. R.C., Yu, Y. et al., Three biomarkers identified from serum proteomic analysis for the detection of early stage ovarian cancer, *Cancer Res.* 64, 5882–5890, 2004.
215. McIntosh, M.W., Drescher, C., Karlan, B. et al., Combining CA 125 and SMR serum markers for diagnosis and early detection of ovarian cancer, *Gynecol. Oncol.* 95, 9–15, 2004.
216. Micha, J.P., Goldstein, B.H., Rettenmaier, M.A. et al., Clinical utility of CA-125 for maintenance therapy in the treatment of advanced stage ovarian carcinoma, *Int. J. Gynecol. Cancer* 19, 239–241, 2009.
217. Canney, P.A., Moore, M., Wilkinson, P.M., and Janes, R.D., Ovarian cancer antigen CA125: A prospective clinical assessment of its role as a tumor marker, *Br. J. Cancer* 50, 765–769, 1984.
218. Matthew, B., Bhatia, V., Mahy, I.R., Ahmed, I., and Francis, L., Elevation of the tumor marker CA125 in right heart failure, *South. Med. J.* 97, 1013, 2004.
219. Phopong, V., Chen, O., and Ultchaswadi, P., High level of CA125 due to large endometrioma, *J. Med. Assoc. Thai* 87, 1108, 2004.
220. Krediet, R.T., Dialyzate cancer antigen 125 concentration as marker of periotoneal membrane status in patients treated with chronic peritoneal dialysis, *Periotoneal Dialysis Int.* 21, 560, 2001.
221. DiBaise, J.K. and Donovan, J.P., Markedly elevated CA125 in hepatic cirrhosis: Two case illustrations and review of the literature, *J. Clin. Gastroenterol.* 28, 159, 1999.
222. Xiao, W.B. and Liu, Y.L., Elevation of serum and ascites cancer antigen 125 levels in patients with liver cirrhosis, *J. Gastroenterol. Hepatol.* 18, 1315, 2003.
223. Kim, Y.S., Kim, D.Y., and Ryu, K.H., Clinical significance of serum CA 125 in patients with chronic liver diseases, *Korean J. Gastroenterol.* 42, 409, 2003.
224. Cannistra, S.A., Cancer of the ovary, *New Engl. J. Med.* 351, 2519, 2004.
225. Gara, S., Boussen, H., Ghanem, A., and Guemira, F., Use of common seric tumor markers in patients with solid cancers, *Tunis Med.* 86, 579–583, 2008.
226. Medeiros, L.R., Rosa, D.D., Bozzetti, M.C. et al., Laparoscopy versus laparotomy for benign ovarian tumour, *Cochrane Database Syst. Rev.* Issue 2:CD004751, 2009.
227. Tian, C., Markman, M., Zaino, R. et al., CA-125 change after chemotherapy in prediction of treatment outcome among advanced mucinous and clear cell epithelial ovarian cancers: A Gynecologic Oncology Group study, *Cancer* 115, 1395–1403, 2009.
228. Gadducci, A. and Cosio, S., Surveillance of patients after initial treatment of ovarian cancer, *Crit. Rev. Oncol. Hematol.* 71, 43–52, 2009.
229. Medieros, L.R., Rosa, D.D., da Rosa, M.I. et al., Accuracy of CA 125 in the diagnosis of ovarian tumors: A quantitative systematic review, *Eur. J. Obstet. Gynecol. Reprod. Biol.* 142, 99–105, 2009.
230. Saksela, E., Prognostic markers in epithelial ovarian cancer, *Int. J. Gynecol. Pathol.* 12, 156–161, 1993.
231. Franzén, B., Okuzawa, K., Linder, S., Katz, H., and Aver, G., Non-enzymatic extraction of cells from clinical tumor material for analysis of gene expression by two-dimensional gel electrophoresis, *Electrophoresis* 14, 382–390, 1993.
232. Alaiya, A.A., Franzén, B., Fujioka, K. et al., Phenotypic analysis of ovarian carcinoma: Polypeptide expression in benign, borderline, and malignant tumors, *Int. J. Cancer* 73, 678–683, 1997.

233. Janes, M.B., Krutzsch, H., Shu, H. et al., Proteomic analysis and identification of new biomarkers and therapeutic targets for invasive ovarian cancer, *Proteomics* 2, 76–84, 2002.
234. Ferrero, A., Zola, P., Mazzola, S. et al., Pretreatment serum hemoglobin level and a preliminary investigation of intratumoral microvessel density in advanced ovarian cancer, *Gynecol. Oncol.* 95, 323–329, 2004.
235. Bar, J.K., Grelewski, P., Popiela, A., Naga, L., and Rabxzyski, J., Type IV collagen and CD44v6 expression in benign, malignant primary and metastatic ovarian tumors: Correlation with Ki-67 and p53 immunoreactivity, *Gynecol. Oncol.* 95, 23–31, 2004.
236. Yurkovetsky, Z.R., Linkov, F.Y., Malehorn, D., and Lokshin, A.E., Multiple biomarker panels for early detection of ovarian cancer, *Future Oncol.* 2, 733–741, 2006.
237. Saldova, R., Royle, L., Radcliffe, C.M. et al., Ovarian cancer is associated with changes in glycosylation in both acute-phase proteins and IgG, *Glycobiology* 17, 1344–1356, 2007.
238. Zhang, Z., Yu, Y., Xu, F. et al., Combining multiple serum tumor markers improves detection of stage 1 epithelial ovarian cancer, *Gynecol. Oncol.* 107, 526–531, 2007.
239. Badgwell, D. and Bast Jr. R.C., Early detection of ovarian cancer, *Dis. Markers* 23, 397–410, 2007.
240. Jankovic, M.M. and Milutinovic, B.S., Glycoforms of CA125 antigen as a possible cancer marker, *Cancer Biomark.* 4, 35–42, 2008.
241. Lowe, K.A., Shah, C., Wallace, E. et al., Effects of personal characteristics on serum CA125, mesothelin, and HE4 levels in healthy postmenopausal women at high-risk for ovarian cancer, *Cancer Epidemiol. Biomarkers Prev.* 17, 2480–2487, 2008.
242. Moore, R.G., McMeekin, D.S., Brown, A.K. et al., A novel multiple marker bioassay utilizing HE4 and CA125 for the prediction of ovarian cancer in patients with a pelvic mass, *Gynecol. Oncol.* 112, 40–46, 2009.
243. Nosov, V., Su, F., Amneus, M. et al., Validation of serum biomarkers for detection of early-stage ovarian cancer, *Am. J. Obstet. Gynecol.* 200, 639, e1–5, 2009.
244. Dieplinger, H., Ankerst, D.P., Burges, A. et al., Afamin and apolipoprotein A-IV: Novel protein markers for ovarian cancer, *Cancer Epidemiol. Biomarkers Prev.* 18, 1127–1133, 2009.
245. Yurkovetsky, Z.R., Ta'asan, S., Skates, S. et al., Development of multimarker panel for early detection of endometrial cancer. High diagnostic power of prolactin, *Gynecol. Oncol.* 107, 58–65, 2007.
246. Tsukishiro, S., Suzumori, N., Nishikawa, H. et al., Use of serum secretory leukocyte protease inhibitor levels in patients to improve specificity of ovarian cancer diagnosis, *Gynecol. Oncol.* 96, 516, 2005.
247. Havrilesky, L.J., Whitehead, C.M., Rubatt, J.M. et al., Evaluation of biomarker panels for early stage ovarian cancer detection and monitoring for disease recurrence, *Gynecol. Oncol.* 110, 374–382, 2008.
248. Cho, C.S. and Rubin, S.C., Epidemiology, staging, and clinical characteristics, in *Surgery for Ovarian Cancer Principles and Practice*, R.E. Bristow and B.Y. Karlam (eds.), Taylor & Francis, London, U.K., Chapter 1, 2006.
249. Jacobs, I., van Nagell Jr. J.R., and Depriest, P.D., Screening for epithelial cancer, in *Ovarian Cancer. Controversies in Management*, D.M. Gershenson and W.P. McGuire (eds.), Churchill Livingstone, New York, Chapter 1, 1998.
250. Rosenthal, A., Menon, U., and Jacobs, I., Ovarian cancer screening, in *Cancer of the Ovary*, R.H. Reznek (ed.), Cambridge University Press, Cambridge, U.K., Chapter 3, 2007.
251. Skates, S.J., Menon, U., MacDonald, N. et al., Calculation of risk of ovarian cancer from serial CA-125 values for preclinical detection in postmenopausal women, *J. Clin. Oncol.* 21(Suppl 10), 206–210, 2003.
252. Vemillion, Inc., http://www.vermillion.com/

253. Gronlund, B., Hogdall, C., Hilden, J., Engelholm, S.A., Høgdall, E.V.S., and Hansen, H.H., Should CA-125 response criteria be preferred to response criteria in solid tumors (RECIST) for prognostication during second-line chemotherapy of ovarian carcinoma, *J. Clin. Oncol.* 22, 4051, 2004.

254. Senapad, S., Neungton, S., Thirapakawong, C. et al., Predictive value of the combined serum CA125 and TPS during chemotherapy and before second-look laparotomy in epithelial ovarian cancer, *Anticancer Res.* 20, 1297–1300, 2000.

255. Andreopoulou, E., Andreopoulos, D., Adamidis, K. et al., Tumor volumetry as predictive and prognostic factor in the management of ovarian cancer, *Anticancer Res.* 22, 1903–1908, 2002.

256. Gadducci, A., Cosio, S., Fanucchi, A. et al., The predictive and prognostic value of serum CA 125 half-life during paclitaxel/platinum-based chemotherapy in patients with advanced ovarian carcinoma, *Gynecol. Oncol.* 93, 131–136, 2004.

257. Martínez-Said, H., Rincón, D.G., Montes de Oca, M.M. et al., Predictive factors for irresectability in advanced ovarian cancer, *Int. J. Gynecol. Cancer* 14, 423–430, 2004.

258. Lenhard, M.S., Nehring, S., Nagel, D. et al., Predictive value of CA 125 and CA 72-4 in ovarian borderline tumors, *Clin. Chem. Lab. Med.* 47, 537–542, 2009.

259. Droegemueller, W., Screening for ovarian carcinoma: Hopeful and wistful thinking, *Am. J. Obstet. Gynecol.* 170, 1095, 1994.

260. Partridge, E., Kreimer, A.R., Greenlee, R.T. et al., Results for four rounds of ovarian cancer screening in a randomized trial, *Obstet. Gynecol.* 113, 775–782, 2009.

261. Jacobs, I.J. and Menon, U., Progress and challenges in screening for early detection of ovarian cancer, *Mol. Cell. Proteomics* 3, 355, 2004.

262. Hensley, M.L., Alektior, K.M., and Chi, D.S., Ovarian and fallopian-tube cancer, in *Gynecologic Oncology*, 2nd edn., R.P. Borakat, M.W. Bevers, D.M. Gershenson, and W.J. Hoskin (eds.), Martin Dunitz Ltd., London, U.K., Chapter 14, pp. 249–269, 2000.

263. Markman, M., Webster, K., Zanotti, K. et al., Examples of the marked variability in the relationship between the serum CA-125 antigen level and cancer-related symptoms in ovarian cancer, *Gynecol. Oncol.* 93, 715–717, 2004.

264. Su, H.Y., Lai, H.C, Lin, Y.W. et al., An epigenetic marker panel for screening and prognostic prediction of ovarian cancer, *Int. J. Cancer* 124, 387–393, 2009.

265. Steinbakk, A., Skaland, I., Gudlaugsson, E. et al., The prognostic value of molecular biomarkers in tissue removed by curettage from FIGO stage 1 and 2 endometroid type endometrial cancer, *Am. J. Obstet. Gynecol.* 200, 78, e1–8, 2009.

266. Anderson, N.S., Bermudez, Y., Badgwell, D. et al., Urinary levels of Bcl-2 are elevated in ovarian cancer patients, *Gynecol. Oncol.* 112, 60–67, 2009.

267. Sedláková, I., Várová, J., Tosner, J., and Hanousek, L., Lysophosphatidic acid: An ovarian cancer marker, *Eur. J. Gynecol. Oncol.* 29, 511–514, 2008.

268. Mustea, A., Braicu, E.I., Koensgen, D. et al., Monitoring of IL-10 in the serum of patients with advanced ovarian cancer: Results from a prospective pilot-study, *Cytokine* 45, 8–11, 2009.

269. Kuzmanov, U., Jiang, N., Smith, C.R. et al., Differential N-glycosylation of kallikrein 6 derived from ovarian cancer cells or the central nervous system, *Mol. Cell. Proteomics* 8, 791–798, 2009.

270. Saldova, R., Wormald, M.R., Dwek, R.A., and Rudd, P.M., Glycosylation changes on serum glycoproteins in ovarian cancer may contribute to disease pathogenesis, *Dis. Markers* 25, 219–232, 2008.

271. Köbel, M., Xu, H., Bourne, P.A. et al., IGF2BP3 (IMP3) expression is a marker of unfavorable prognosis in ovarian carcinoma of clear cell subtype, *Mod. Pathol.* 22, 469–475, 2009.

272. Cho, H. and Kim, J.H., Lipocalin2 expressions correlate significantly with tumor differentiation in epithelial ovarian cancer, *J. Histochem. Cytochem.* 57, 513–521, 2009.

273. Lin, C.K., Chao, T.K., Yu, C.P. et al., The expression of six biomarker in the four most common ovarian cancers: Correlation with clinicopathological parameters, *APMIS* 117, 162–175, 2009.

274. Cao, D., Guo, S., Allan, R.W. et al., SALL4 is a novel sensitive and specific marker of ovarian primitive germ cell tumors and is particularly useful in distinguishing yolk sac tumors from clear cell carcinoma, *Am. J. Surg. Pathol.* 33, 894–904, 2009.

275. Tsai-Turton, M., Santillan, A., Lu, D. et al., p53 autoantibodies, cytokine levels and ovarian carcinogenesis, *Gynecol. Oncol.* 114, 12–17, 2009.

276. Kyung, M.S., Choi, J.S., Hong, S.H., and Kim, H.S., Elevated CA 19-9 levels in mature cystic teratoma of the ovary, *Int. J. Biol. Markers* 24, 52–56, 2009.

277. Hu, X., Macdonald, D.M., Huettner, P.C. et al., A miR-200 microRNA cluster as prognostic marker in advanced ovarian cancer, *Gynecol. Oncol.* 114, 457–464, 2009.

278. Smith, H.O., Arias-Pulido, H., Kuo, D.Y. et al., GPR30 predicts poor survival for ovarian cancer, *Gynecol. Oncol.* 114, 465–471, 2009.

279. Lodes, M.J., Caraballo, M., Suciu, D. et al., Detection of cancer with serum miRNAs on an oligonucleotide microarray, *PLoS One* 4, e6229, 2009.

280. Lee, Y.K., Choi, E., Kim, M.A. et al., BubR1 as a prognostic marker for recurrence-free survival rates in epithelial ovarian cancer, *Br. J. Cancer* 101, 504–510, 2009.

281. Guan, W., Zhou, M., Hampton, C.Y. et al., Ovarian cancer detection from metabolomic liquid chromatography/mass spectrometry data by support vector machines, *BMC Bioinformatics* 10, 259, 2009.

282. Tung, C.S., Wong, K.K., and Mok, S.C., Biomarker discovery in ovarian cancer, *Womens Health* 4, 27–40, 2008.

283. von Wichert, G., Seufferlein, T., and Adler, G., Palliative treatment of pancreatic cancer, *J. Dig. Dis.* 9, 1–7, 2008.

284. Dunphy, E.P., Pancreatic cancer: A review and update, *Clin. J. Oncol. Nurs.* 12, 735–741, 2008.

285. Von Hoff, D.D., Evans, D.B., and Hruban, R.H. (eds.), *Pancreatic Cancer*, Jones and Bartlett Publishers, Sudbury, MA, 2005.

286. Pliarchopoulou, K. and Pectasides, D., Pancreatic cancer: Current and future treatment strategies, *Cancer Treat. Rev.* 35, 431–436, 2009.

287. Fryer, R.A., Galustian, C., and Dalgelish, A.G., Recent advances and developments in treatment strategies against pancreatic cancer, *Curr. Clin. Pharmacol.* 4, 102–112, 2009.

288. Lowenfels, A.B. and Maisonneuve, P., Pancreatico-biliary malignancy: Prevalence and risk factors, *Ann. Oncol.* 10(Suppl 4), 1–3, 1999.

289. Micheli, A., Yancik, R., Krogh, V. et al., Contrasts in cancer prevalence in Connecticut, Iowa, and Utah, *Cancer* 95, 430–439, 2002.

290. Kim, D.H., Crawford, B., Ziegler, J., and Beattie, M.S., Prevalence and characteristics of pancreatic cancer in families with BRCA1 and BRCA2 mutations, *Fam. Cancer* 8, 153–158, 2009.

291. Jimeno, A. and Hidalgo, M., Molecular biomarkers: Their increasing role in the diagnosis, characterization, and therapy guidance in pancreatic cancer, *Mol. Cancer Ther.* 5, 787–796, 2006.

292. Rosty, C., Christa, L., Kuzdzal, S. et al., Identification of hepatocarcinoma-intestine-pancreas/pancreatitis-associated protein I as a biomarker for pancreatic ductal adenocarcinoma by protein biochip technology, *Cancer Res.* 62, 1868–1875, 2002.

293. Koopmann, J., Fedarko, N.S., Jain, A. et al., Evaluation of osteopontin as biomarker for pancreatic adenocarcinoma, *Cancer Epidemiol. Biomarkers Prev.* 13, 487–491, 2004.

294. Lyn-Cook, B.D., Yan-Sanders, Y., Moore, S. et al., Increased levels of NAD(P)H:Quinine oxidoreductase 1 (NQO1) in pancreatic tissues from smokers and pancreatic adenocarcinomas: A potential biomarker of early damage in the pancreas, *Cell Biol. Toxicol.* 22, 73–80, 2006.

295. Simeone, D.M., Ji, B., Banerjee, M. et al., CEACAM1, a novel serum biomarker for pancreatic cancer, *Pancreas* 34, 436–443, 2007.
296. Awadallah, N.S., Dehn, D., Shah, R.J. et al., NQO1 expression in pancreatic cancer and its potential use as a biomarker, *Appl. Immunohistochem. Mol. Morphol.* 16, 24–31, 2008.
297. Gold, D.V., Karanjawala, Z., Modrak, D.E. et al., PAM4-reactive MUC1 is a biomarker for early pancreatic adenocarcinoma, *Clin. Cancer Res.* 13, 7380–7387, 2007.
298. Kim, L., Liao, J., Zhang, M. et al., Clear cell carcinoma of the pancreas: Histopathologic features and a unique biomarker: Hepatocyte nuclear factor-1β, *Mod. Pathol.* 21, 1075–1083, 2008.
299. Kojima, K., Asmellash, S., Klug, C.A. et al., Applying proteomic-based biomarker tools for the accurate diagnosis of pancreatic cancer, *J. Gastrointest. Surg.* 12, 1683–1690, 2008.
300. Ohlund, D., Lundin, C., Ardnor, B. et al., Type IV collagen is a tumor stroma-derived biomarker for pancreas cancer, *Br. J. Cancer* 101, 91–97, 2009.
301. Nakashima, A., Murakami, Y., Uemura, K. et al., Usefulness of human telomerase reverse transcriptase in pancreatic juice as a biomarker of pancreatic malignancy, *Pancreas* 38, 527–533, 2009.
302. Haug, U., Hillebrand, T., Bendzko, P. et al., Mutant-enriched PCR and allele-specific hybridization reaction to detect K-ras mutations in stool DNA: High prevalence in a large sample of older adults, *Clin. Chem.* 53, 787–790, 2007.
303. Yamamoto, S., Tomita, Y., Hoshida, Y. et al., Increased expression of valosin-containing protein (p97) is associated with lymph node metastasis and prognosis of pancreatic ductal adenocarcinoma, *Ann. Surg. Oncol.* 11, 165–172, 2004.
304. Duxbury, M.S., Matros, E., Clancy, T. et al., CEACAM6 is a novel biomarker in pancreatic adenocarcinoma and PanIN lesions, *Ann. Surg.* 241, 491–496, 2005.
305. Kuhlmann, K.F., van Till, J.W., Boermeester, M.A. et al., Evaluation of matrix metalloproteinase 7 in plasma and pancreatic juice as a biomarker for pancreatic cancer, *Cancer Epidemiol. Biomarkers Prev.* 16, 886–891, 2007.
306. Jiang, N.Y., Woda, B.A., Banner, B.F. et al., Spl, a new biomarker that identifies a subset of aggressive pancreatic ductal adenocarcinoma, *Cancer Epidemiol. Biomarkers Prev.* 17, 1648–1652, 2008.
307. Hoffmann, A.-C., Nori, R., Vallböhmer, D. et al., High expression of heparanase is significantly associated with differentiation and lymph node metastasis in patients with pancreatic ductal adenocarcinomas and correlated to PDGFA and via HIF1a and bFGF, *J. Gastrointest. Surg.* 12, 1674–1682, 2008.
308. Juuti, A., Nordling, S., Lundin, J. et al., Syndecan-1 expression—A novel prognostic marker in pancreatic cancer, *Oncology* 68, 97–106, 2005.
309. Koopmann, J., Rosenzweig, C.N., Zhang, Z. et al., Serum markers in patients with resectable pancreatic adenocarcinoma: Macrophage inhibitory cytokine 1 versus CA19-9, *Clin. Cancer Res.* 12, 442–445, 2006.
310. Skalicky, D.A., Kench, J.G., Segara, D. et al., Cyclin E expression and outcome in pancreatic ductal adenocarcinoma, *Cancer Epidemiol. Biomarkers Prev.* 15, 1941–1947, 2006.
311. Tonini, G., Pantano, F., Vincenzi, B. et al., Molecular prognostic factors in patients with pancreatic cancer, *Expert Opin. Ther. Targets* 11, 1553–1569, 2007.
312. Malaguarnera, M., Cristaldi, E., Cammalleri, L. et al., Elevated chromogranin A (CgA) serum levels in the patients with advanced pancreatic cancer, *Arch. Gerontol. Geriatr.* 48, 213–217, 2009.
313. Maeda, S., Shinchi, H., Kurahara, H. et al., CD133 expression is correlated with lymph node metastasis and vascular endothelial growth factor-C expression in pancreatic cancer, *Br. J. Cancer* 98, 1389–1397, 2008.

314. Tanase, C.P., Caveolin-1: A marker for pancreatic cancer diagnosis, *Expert Rev. Mol. Diagn.* 8, 395–404, 2008.

315. Hernandez, J.M., Cowgill, S.M., Al-Saadi, S. et al., CA 19-9 velocity predicts disease-free survival and overall survival after pancreatectomy of curative intent, *J. Gastrointest. Surg.* 13, 349–353, 2009.

316. Nagai, K., Doi, R., Katagiri, E. et al., Diagnostic value of metastin expression in human pancreatic cancer, *J. Exp. Clin. Cancer Res.* 28, 9, 2009.

317. Biankin, A.V., Kench, J.G., Colvin, E.K. et al., Expression of S100A2 calcium-binding protein predicts response to pancreatectomy for pancreatic cancer, *Gastroenterology* 137, 558–568, 2009.

318. Kahlert, C., Weber, H. Mogler, C. et al., Increased expression of ALCAM/CD166 in pancreatic cancer is an independent prognostic marker for poor survival and early tumour relapse, *Br. J. Cancer* 101, 457–464, 2009.

319. Fradet, Y., Biomarkers in prostate cancer diagnosis and prognosis: Beyond prostate-specific antigen, *Curr. Opin. Urol.* 19, 243–246, 2009.

320. Ito, K., Prostate-specific antigen-based screening for prostate cancer: Evidence, controversies and future perspectives, *Int. J. Urol.* 16, 458–464, 2009.

321. Pienta, K.J., Critical appraisal of prostate-specific antigen in prostate cancer screening, *Urology* 73(5 Suppl), S11–S20, 2009.

322. Lin, D.W., Beyond PSA: Utility of novel tumor markers in the setting of elevated PSA, *Urol. Oncol.* 27, 315–321, 2009.

323. Schüller, P., Schäfer, U., Micke, O. et al., PSA course after definitive high-dose radiotherapy of localized prostate cancer, *Anticancer Res.* 25, 1555–1557, 2005.

324. Taylor III J.A., Koff, S.G., Dauser, D.A., and McLeod, D.G., The relationship of ultrasensitive measurements of prostate-specific antigen levels to prostate cancer recurrence after radical prostectomy, *BJU Int.* 98, 540–543, 2006.

325. Meier, R. and Brawer, M.K., Selecting treatment for high-risk, localized prostate cancer, *Rev. Urol.* 4, 141–146, 2002.

326. Jacinto, A.A., Fede, A.B., Fagundes, L.A. et al., Salvage radiotherapy for biochemical relapse after complex PSA response following radical prostatectomy: Outcome and prognostic factors for patients who have never received hormonal therapy, *Radiat. Oncol.* 2, 8, 2007.

327. Duffy, M.J., Role of tumor markers in patients with solid cancers: A critical review, *Eur. J. Intern. Med.* 18, 175–184, 2007.

328. Jereczek-Fossa, B.A. and Orecchia, R., Evidence-based radiation oncology: Definitive, adjuvant and salvage radiotherapy for non-metastatic prostate cancer, *Radiother. Oncol.* 84, 197–215, 2007.

329. Teeter, A.E., Presti, J.C., Aronson, W.J. et al., Does early prostate-specific antigen doubling time (ePSADT) after radical prostatectomy, calculated using PSA values for the first detectable until the first recurrence value, correlate with standard PSADT? A report from the Shared Equal Access Regional Cancer Hospital Database Group, *BJU Int.* 104, 1604–1609, 2009.

330. National Cancer Institute, Definition of Cancer Terms; http://www.cancer.gov/Templates/db_alpha.aspx?CdrID=542440

331. Kupelian, P.A., Reddy, C.A., Carlson, T.P. et al., Preliminary observations on biochemical relapse-free survival rates after short-course intensity-modulated radiotherapy (70 Gy at 2.5 Gy/fraction) for localized prostate cancer, *Int. J. Radiat. Oncol.* 53, 904–912, 2002.

332. Nguyen, C.T., Reuther, A.M., Stephenson, A.J. et al., The specific definition of high risk prostate cancer has minimal impact on biochemical relapse-free survival, *J. Urol.* 181, 75–80, 2008.

333. Schostak, M., Schwall, G.P., Poznanovic, S. et al., Annexin A3 in urine: A highly specific noninvasive marker for prostate cancer early detection, *J. Urol.* 18, 343–353, 2009.

334. Sreekumar, A., Poisson, L.M., Rajendiran, T.M. et al., Metabolomic profiles delineate potential role for sarcosine in prostate cancer progression, *Nature* 457, 910–914, 2009.
335. Kuehn, B.M., Promising marker found for deadly prostate cancer, *JAMA* 301, 1008, 2009.
336. Yasuda, K., Nagakawa, O., Akashi, T. et al., Serum active hepatocyte growth factor (AHGF) in benign prostatic disease and prostate cancer, *Prostate* 69, 346–351, 2009.
337. Panteleakou, Z., Lembessis, P., Sourla, A. et al., Detection of circulating tumor cells in prostate cancer patients: Methodological pitfalls and clinical relevance, *Mol. Med.* 15, 101–114, 2009.
338. Schwarzenbach, H., Alix-Panabières, C., Müller, I. et al., Cell-free tumor DNA in blood plasma as a marker for circulating tumor cells in prostate cancer, *Clin. Cancer Res.* 15, 1032–1038, 2009.
339. Colombatti, M., Grasso, S., Porzia, A. et al., The prostate specific membrane antigen regulates the expression of IL-6 and CCL5 in prostate tumour cells by activating the MAPK pathways, *PLoS One* 4, e4608, 2009.
340. Elsässe-Beile, U., Reishcl, G., Wiehr, S. et al., PET imaging of prostate cancer xenografts with a highly specific antibody against the prostate-specific membrane antigen, *J. Nucl. Med.* 50, 958, 2009.
341. Dos Reis, S.T., Pontes Jr. J., Villanova, F.E. et al., Genetic polymorphisms of matrix metalloproteinases: Susceptibility and prognostic implications for prostate cancer, *J. Urol.* 181, 2320–2325, 2009.
342. Arisan, E.D., Arisan, S., Atis, G. et al., Serum adipocytokine levels in prostate cancer patients, *Urol. Int.* 82, 203–208, 2009.
343. Anborgh, P.H., Wilson, S.M., Tuck, A.B. et al., New dual monoclonal ELISA for measuring plasma osteopontin as a biomarker associated with survival in prostate cancer: Clinical validation and comparison of multiple ELISAs, *Clin. Chem.* 55, 895–903, 2009.
344. Nilsson, J., Skog, J., Nordstrand, A. et al., Prostate cancer-derived urine exosomes: A novel approach to biomarkers for prostate cancer, *Br. J. Cancer* 100, 1603–1607, 2009.
345. Rabiau, N., Dechelotte, P., Guy, L. et al., Immunohistochemical staining of mucin 1 in prostate tissue, *In Vivo* 23, 203–207, 2009.
346. Concato, J., Jain, D., Uchio, E. et al., Molecular markers and death from prostate cancer, *Ann. Intern. Med.* 150, 595–603, 2009.
347. Nakamura, Y., Susuki, T., Sugawara, A. et al., Peroxisome proliferator-activated receptor gamma in human prostate cancer, *Pathol. Int.* 59, 288–293, 2009.
348. Kuruma, H., Kamata, Y., Takahashi, H. et al., Staphylococcal nuclease domain-containing protein 1 as a potential tissue marker for prostate cancer, *Am. J. Pathol.* 174, 2044–2050, 2009.
349. Shukla-Dave, A., Hricak, H., Ishill, N. et al., Prediction of prostate cancer recurrence using magnetic resonance imaging and molecular profiles, *Clin. Cancer Res.* 15, 3842–3849, 2009.
350. Sooriakumaran, P., Coley, H.M., Fox, S.B. et al., A randomized controlled trial investigating the effects of celecoxib in patients with localized prostate cancer, *Anticancer Res.* 29, 1483–1488, 2009.
351. Nakamura, Y., Suzuki, T., Arai, Y. et al., 17β-Hydroxysteroid dehydrogenase type 11 (Pan1b) expression in human prostate cancer, *Neoplasma* 56, 317–320, 2009.
352. Schwartz, G.G., Is serum calcium a biomarker of fatal prostate cancer, *Future Oncol.* 5, 577–580, 2009.
353. Timofeeva, O.A., Zhang, X., Ressom, R.W. et al., Enhanced expression of SOS1 is detected in prostate cancer epithelial cells from African-American men, *Int. J. Oncol.* 35, 751–760, 2009.
354. Singh, S.S., Mehedint, D.C., Ford III O.H. et al., Comparison of ACINUS, caspase-3, and TUNEL as apoptotic markers in determination of tumor growth rates of clinically localized prostate cancer using image analysis, *Prostate* 69, 1603–1610, 2009.

355. Foster, C.S., Dodson, A.P., Ambroisine, L. et al., Hsp-27 expression at diagnosis predicts poor clinical outcome in prostate cancer independent of ETS-gene rearrangement, *Br. J. Cancer* 101, 1137–1144, 2009.

356. Parker, A.S., Heckman, M.G., Wu, K.J. et al., Evaluation of Ki-67 staining levels as an independent biomarker of biochemical recurrence after salvage radiation therapy for prostate cancer, *Int. J. Radiat. Oncol. Biol. Phys.* 75, 1364–1370, 2009.

357. Kirby, R.S., Partin, A.W., Feneley, M., and Parsons, J.K. (eds.), *Prostate Cancer Principles and Practice*, Taylor & Francis, London, U.K., 2006.

358. Waldman, A.R., Screening and early detection in contemporary issues, in *Prostate Cancer*, J. Held-Warmkessel (ed.), Jones and Bartlett Publishers, Sudbury, MA, Chapter 2, pp. 44–59, 2006.

359. Podolsky, D.K., Serological markers in the diagnosis and management of pancreatic carcinoma, *World J. Surg.* 8, 822–830, 1984.

360. Tice, J.A. and Kerlikowske, K., Screening and prevention of breast cancer in primary care, *Primary Care* 36, 533–558, 2009.

361. Buttimer, A., Customer focus in breast cancer screening services, *Int. J. Health Care Qual. Assur.* 22, 514–524, 2009.

362. Lee, J.M., Halpern, E.F., Rafferty, E.A., and Gazelle, G.S., Evaluating the correlation between film mammography and MRI for screening women with increased breast cancer risk (1), *Acad. Radiol.* 16, 1323–1328, 2009.

363. Frank, T.S., Laboratory determination of hereditary susceptibility to breast and ovarian cancer, *Arch. Pathol. Lab. Med.* 123, 1023–1026, 1999.

364. Pusztai, L., Cristofanilli, M., and Paik, S., New generation of molecular prognostic and predictive tests for breast cancer, *Semin. Oncol.* 34 (2 Suppl 3), S10–S16, 2007.

365. Lea, P. and Ling, M., New molecular assays for cancer diagnosis and targeted therapy, *Curr. Opin. Mol. Ther.* 10, 251–259, 2008.

366. Kolesar, J.M., Assessing therapeutically developed assays, *Manag. Care* 17 (7 Suppl 7), 9–12, 2008.

367. Pupa, S.M., Tagliabue, E., Ménard, S., and Anichini, A., HER-2: A biomarker at the crossroad of breast cancer immunotherapy and molecular medicine, *J. Cell Physiol.* 205, 10–18, 2005.

368. James, C.R., Quinn, J.E., Mullan, P.B. et al., BRCA1, a potential predictive biomarker in the treatment of breast cancer, *Oncologist* 12, 142–150, 2007.

369. Gasparini, G. and Hayes, D., *Biomarkers in Breast Cancer: Molecular Diagnostics for Predicting and Monitoring Therapeutic Effect*, Humana Press, Totowa, NJ, 2006.

370. Serre, J.L. and Heath, I. (eds.), *Diagnostic Techniques in Genetics*, Wiley, Chichester, U.K., 2006.

371. Chhieng, D.C. and Siegal, G.P. (eds.), *Updates in Diagnostic Pathology*, Springer, New York, 2005.

372. DeVita, V.T. and Hellman, S. (eds.), *Cancer, Principles & Practice of Oncology*, Lippincott Williams & Wilkins, Philadelphia, PA, 2005.

373. Runge, M.S. and Patterson, C. (eds.), *Principles of Molecular Medicine*, Humana Press, Totowa, NJ, 2006.

374. Bertucci, F. and Goncalves, A., Clinical proteomics and breast cancer: Strategies for diagnostic and therapeutic biomarker discovery, *Future Oncol.* 4, 271–287, 2008.

375. Clinton, S.R., Beason, K.L., Bryant, S. et al., A comparative study of four serological tumor markers for the detection of breast cancer, *Biomed. Sci. Instrum.* 39, 408–414, 2003.

376. Agha-Hosseini, F., Mirzaii-Diezgah, I., and Rahimi, A., Correlation of serum and salivary CA15-3 levels in patients with breast cancer, *Med. Oral Pathol. Oral. Cir. Bucal.* 14, e521–e524, 2009.

377. Kim, H.S., Park, Y.H., Park, M.J. et al., Clinical significance of a serum CA15-3 surge and the usefulness of CA15-3 kinetics in monitoring chemotherapy response in patients with metastatic breast cancer, *Breast Cancer Res. Treat.* 118, 89–97, 2009.

378. Hoopmann, M., Neumann, R., Tanasale, T., and Schöndorf, T., HER-2/neu determination in blood plasma of patients with HER-2/neu overexpressing metastasized breast cancer: A longitudinal study, *Anticancer Res.* 23, 1031–1034, 2003.

379. Leary, A.F., Hanna, W.M., van de Vijver, M.J. et al., Value and limitations of measuring HER-2 extracellular domain in the serum of breast cancer patients, *J. Clin. Oncol.* 27, 1694–1705, 2009.

380. Fabre-Lafay, S., Garrido-Urbani, S., Reymond, N. et al., Nectin-4, a new serological breast cancer marker, is a substrate for tumor necrosis factor-α-converting enzyme (TACE/ADAM-17), *J. Biol. Chem.* 280, 19543–19550, 2005.

381. Balogh, G.A., Mailo, D.A., Corte, M.M. et al., Mutant p53 protein in serum could be used as a molecular marker in human breast cancer, *Int. J. Oncol.* 28, 995–1002, 2006.

382. Müller, M., Meyer, M., Schilling, T. et al., Testing for anti-p53 antibodies increases the diagnostic sensitivity of conventional tumor markers, *Int. J. Oncol.* 29, 973–980, 2006.

383. He, Q., Fornander, T., Johansson, H. et al., Thymidine kinase 1 in serum predicts increased risk of distant of loco-regional recurrence following surgery in patients with early breast cancer, *Anticancer Res.* 26, 4753–4759, 2006.

384. Petridou, E.T., Chavelas, C., Dikalioti, S.K. et al., Breast cancer risk in relation to most prevalent IgE specific antibodies: A case control study in Greece, *Anticancer Res.* 27, 1709–1713, 2007.

385. Nolen, B.M., Marks, J.R., Ta'san, S. et al., Serum biomarker profiles and response to neoadjuvant chemotherapy for locally advanced breast cancer, *Breast Cancer Res.* 10, R45, 2008.

386. Kulasingam, V., Zheng, Y., Soosaipillai, A. et al., Activated leukocyte cell adhesion molecule: A novel biomarker for breast cancer, *Int. J. Cancer* 125, 9–14, 2009.

387. Yi, J.K., Chang, J.W., Han, W. et al., Autoantibody to tumor antigen, alpha-2-HS glycoprotein: A novel biomarker of breast cancer screening and diagnosis, *Cancer Epidemiol. Biomarkers Prev.* 18, 1357–1364, 2009.

388. Baron, A.T., Wilken, J.A., Haggstrom, D.E. et al., Clinical implementation of soluble EGFR (sEGFR) as a theragnostic serum biomarker of breast, lung and ovarian cancer, *IDrugs* 12, 302–308, 2009.

389. Schaub, N.P., Jones, K.J., Nyalwidhe, J.O. et al., Serum proteomic biomarker discovery reflective of stage and obesity in breast cancer patients, *J. Am. Coll. Surg.* 208, 970–978, 2009.

390. Zusman, I. and Ben-Hur, H., Serological markers for detection of cancer, *Int. J. Mol. Med.* 7, 547–556, 2001.

391. Sölétormos, G., Serological tumor markers for monitoring breast cancer, *Dan. Med. Bull.* 48, 229–255, 2001.

392. Rui, Z., Jian-Guo, J., Yuan-Peng, T. et al., Use of serological proteomic methods to find biomarkers associated with breast cancer, *Proteomics* 3, 433–439, 2003.

393. Hardouin, J., Lasserre, J.P., Sylvius, L. et al., Cancer immunomics: From serological proteome analysis to multiple affinity protein profiling, *Ann. N. Y. Acad. Sci.* 1107, 223–230, 2007.

394. Reymaux, D., Sendid, B., Poulain, D. et al., Serological markers in inflammatory bowel diseases, *Best Pract. Res. Clin. Gastroenterol.* 17, 19–35, 2003.

395. Yuen, M.F. and Lai, C.L., Serological markers of liver cancer, *Best Pract. Res. Clin. Gastroenterol.* 205, 19, 91–99, 2005.

396. Papp, M., Norman, G.L., Altorjay, I., and Lakatos, P.L., Utility of serological markers in inflammatory bowel diseases: Gadget or magic?, *World J. Gastroenterol.* 13, 2028–2036, 2007.

397. Invernizzi, P., Lleo, A., and Podda, M., Interpreting serological tests in diagnosing autoimmune liver disease, *Semin. Liver Dis.* 27, 161–172, 2007.

398. di Mario, F. and Cavallaro, L.G., Non-invasive tests in gastric diseases, *Dig. Liver Dis.* 40, 523–530, 2008.

399. Jungbluth, A.A., Serological reagents for the immunohistochemical analysis of melanoma metastases in sentinel lymph nodes, *Semin. Diagn. Pathol.* 24, 120–125, 2008.

400. Li, X., Conklin, L., and Alex, P., New serological biomarkers of inflammatory bowel disease, *World J. Gastroenterol.* 14, 5115–5124, 2008.

401. Sun, S., Day, P.J., Lee, N.P. et al., Biomarkers for early detection of liver cancer: Focus on clinical evaluation, *Protein Pept. Lett.* 16, 473–478, 2009.

402. Fournier, A., Oprisiu, R., Said, S. et al., Invasive versus non-invasive diagnosis of renal bone disease, *Curr. Opin. Nephrol. Hypertens.* 6, 333–348, 1997.

403. Bischoff, F.Z., Sinacori, M.K., Dang, D.D. et al., Cell-free fetal DNA and intact fetal cells in maternal blood circulation: Implications for first and second trimester non-invasive prenatal diagnosis, *Hum. Reprod. Update* 8, 493–500, 2002.

404. Bramer, S.L. and Kallungal, B.A., Clinical considerations in study designs that use cotinine as a biomarker, *Biomarkers* 8, 187–203, 2003.

405. Strand, K.J., Khalak, H., Strovel, J.W. et al., Expression biomarkers for clinical efficacy and outcome prediction in cancer, *Pharmacogenomics* 7, 105–115, 2006.

406. Pusztai, L., Current status of prognostic profiling in breast cancer, *Oncologist* 13, 350–360, 2008.

407. Lucas, J., Carvalho, C., and West, M., A Bayesian analysis strategy for cross-study translation of gene expression biomarkers, *Stat. Appl. Genet. Mol. Biol.* 8, Article 11, 2009.

408. Sotiriou, C. and Pusztai, L., Gene-expression signatures in breast cancer, *New Engl. J. Med.* 360, 790–800, 2009.

409. Ross, J.S., Multigene classifiers, prognostic factors, and predictors of breast cancer clinical outcome, *Adv. Anat. Pathol.* 16, 204–215, 2009.

410. Freedman, O.C., Amir, E., Hanna, W. et al., A randomized trial exploring the biomarker effects of neoadjuvant sequential treatment with exemestane and anastrozole in postmenopausal women with hormone receptor-positive breast cancer, *Breast Cancer Res. Treat.* 119, 155–161, 2010.

411. Aleskandarany, M.A., Rakha, E.A., Ahmed, M.A. et al., PIK3CA expression in invasive breast cancer: A biomarker of poor prognosis, *Breast Cancer Res. Treat.*, 2009. doi:10.1007/s10549-009-0508-9.

412. Walter, O., Prasad, M., Lu, S. et al., IMP3 is a novel biomarker for triple negative invasive mammary carcinoma associated with a more aggressive phenotype, *Hum. Pathol.* 40, 1528–1533, 2009.

413. Kioulafa, M., Kaklamanis, L., Stahopoulos, E. et al., Kallikrein 10 (KLK10) methylation as a novel prognostic biomarker in early breast cancer, *Ann. Oncol.* 20, 1020–1025, 2009.

414. Bay, B.H., Jin, R., Huang, J., and Tan, P.H., Metallothionein as a prognostic biomarker in breast cancer, *Exp. Biol. Med.* (Maywood) 231, 1516–1521, 2006.

415. Jung, S.S., Park, H.S., Lee, I.J. et al., The HCCR oncoprotein as a biomarker for human breast cancer, *Clin. Cancer Res.* 11, 7700–7708, 2005.

416. Kleer, C.G., Griffith, K.A., Sabel, M.S. et al., RhoC-GTPase is a novel tissue biomarker associated with biologically aggressive carcinomas of the breast, *Breast Cancer Res. Treat.* 93, 101–110, 2005.

417. Houchens, N.W. and Merajver, S.D., Molecular determinants of the inflammatory cancer phenotype, *Oncology* (Williston Park) 22, 1556–1561, 2008.

418. Smith, R.A., von Eschenbach, A.C., Wender, R. et al., American Cancer Society guidelines for the early detection of cancer: Update of early detection guidelines for prostate, colorectal, and endometrial cancers. Also: Update 2001—Testing for early lung cancer detection, *CA Cancer J. Clin.* 51, 38–75, 2001.

419. Haberkorn, U. and Schoenberg, S.O., Imaging of lung cancer with CT, MRT and PET, *Lung Cancer* 34(Suppl 3), S13–S23, 2001.

420. Argiris, A. and Murren, J.R., Staging and clinical prognostic factors for small-cell lung cancer, *Cancer J.* 7, 437–447, 2001.
421. Rami-Porta, R., Crowley, J.J., and Goldstraw, P., The revised TNM staging system for lung cancer, *Ann. Thorac. Cardiovasc. Surg.* 15, 4–9, 2009.
422. Gordon, I.O., Sitterding, S., Mackinnon, A.C. et al., Update in neoplastic lung diseases and mesothelioma, *Arch. Pathol. Lab. Med.* 133, 1106–1115, 2009.
423. Detterbeck, F.D., Boffa, D.J., and Tanoue, L.T., The new lung cancer staging system, *Chest* 136, 260–271, 2009.
424. Semenov, S., Microwave tomography: Review of the progress towards clinical applications, *Philos. Trans. A Math. Phys. Eng. Sci.* 367, 3021–3042, 2009.
425. Horváth, I., Lázár, Z., Gyulai, N. et al., Exhaled biomarkers in lung cancer, *Eur. Respir. J.* 34, 261–275. 2009.
426. Santos, E.S., Blaya, M., and Raez, L.E., Gene expression profiling and non-small-cell lung cancer: Where are we now?, *Clin. Lung Cancer* 10, 168–173, 2009.
427. Tanoue, L.T. and Detterbeck, F.C., New TNM classification for non-small-cell lung cancer, *Expert Rev. Anticancer Ther.* 9, 413–423, 2009.
428. Nomori, H., Ohba, Y., Yoshimomot, K. et al., Positron emission tomography in lung cancer, *Gen. Thorac. Cardiovasc. Surg.* 57, 184–191, 2009.
429. Gomez, M. and Silvestri, G.A., Endobronchial ultrasound for the diagnosis and staging of lung cancer, *Proc. Am. Thorac. Soc.* 15, 180–186, 2009.
430. Cho, J.Y. and Sung, H.J., Proteomic approaches in lung cancer biomarker development, *Expert Rev. Proteomics* 6, 27–42, 2009.
431. Smith, R.A., Cokkinides, V., and Brawley, O.W., Cancer screening in the United States, 2009: A review of current American Cancer Society guidelines and issues in cancer screening, *CA Cander J. Clin.* 59, 27–41, 2009.
432. Eddy, D.M., Screening for lung cancer, *Ann. Int. Med.* 111, 232–237, 1989.
433. Bach, P.B., Silvestri, G.A., Hanger, M., and Jett, J.R., Screening for lung cancer. ACCP evidence-based clinical practice guidelines (2nd edition), *Chest* 132, 132, 69S–77S, 2007.
434. Minna, J.D. and Schiller, J.H., Neoplasms of the lung, in *Harrison's Principles of Internal Medicine*, 17th edn., A.S. Fauci, E. Braunwald, D.L. Kasper et al. (eds.), McGraw-Hill Medical, New York, Chapter 85, 2008.
435. Bepler, G., Are we coming full circle for lung cancer screening a second time?, *Am. J. Respir. Crit. Care Med.* 180, 384–385, 2009.
436. Infante, M., Cavuto, S., Lutman, F.R. et al., DANTE study group. A randomized study of lung cancer screening with spiral computed tomography: Three-year results from the DANTE trial, *Am. J. Respir. Crit. Care Med.* 180, 445–453, 2009.
437. Castleberry, A.W., Smith, D., Anderson, C. et al., Cost of a 5-year lung cancer survivor: Symptomatic tumour identification vs. proactive computed tomography screening, *Br. J. Cancer* 101, 882–896, 2009.
438. Reich, J.M., Cost-effectiveness of computed tomography lung cancer screening, *Br. J. Cancer* 101, 879–880, 2009.
439. Hirschowitz, E.A., Biomarkers for lung cancer screening. Interpretation and implications of an early negative advanced validation study, *Am. J. Respir. Crit. Care Med.* 179, 1–3, 2009.
440. Ghosal, R., Kloer, P., and Lewis, K.E., A review of novel biological tools used in screening for the early detection of lung cancer, *Postgrad. Med. J.* 85, 358–363, 2009.
441. Pérez-Soler, R., Individualized therapy in non-small-cell lung cancer: Future versus current clinical practice, *Oncogene* 28(Suppl 1), S38–S45, 2009.
442. Cui, J., Lu., Q., Puett, D., and Xu, Y., Computational prediction of human proteins that can be secreted into the bloodstream, *Bioinformatics* 24, 2370–2375, 2008.
443. Xue, H., Lu, B., and Lai, M., The cancer secretome: A reservoir of biomarkers, *J. Transl. Med.* 6, 52, 2008.

444. Kim, J.E., Koo, K.H., Kim, Y.H. et al., Identification of potential lung cancer biomarkers using an *in vitro* carcinogenesis model, *Exp. Mol. Med.* 40, 709–720, 2008.
445. Ucar, D., Cogle, C.R., Zucali, J.R. et al., Aldehyde dehydrogenase activity as a functional marker for lung cancer, *Chem. Biol. Interact.* 178, 48–55, 2009.
446. Kim, S., Takahashi, H., Lin, W.W. et al., Carcinoma-produced factors activate myeloid cells through TLR2 to stimulate metastasis, *Nature* 457, 102–106, 2009.
447. Dumitriu, I.E., Dunbar, D.R., Howie, S.E. et al., Human dendritic cells produce TGF-β1 under the influence of lung carcinoma cells and prime the differentiation of CD4+CD25+Foxp3+ regulatory T cells, *J. Immunol.* 182, 2795–2807, 2009.
448. Jess, P.R.T., Mazilu, M., Dholakia, K. et al., Optical detection and grading of lung neoplasia by Raman microspectroscopy, *Int. J. Cancer* 124, 376–380, 2009.
449. Sozzi, G., Roz, L., Conte, D. et al., Plasma DNA quantification in lung cancer computed tomography screening, *Am. J. Respir. Crit. Care Med.* 179, 69–74, 2009.
450. Carozzi, F.M., Bisanzi, S., Falini, P. et al., Molecular profile in body fluids in subjects enrolled in a randomized trial for lung cancer screening: Perspectives of integrated strategies for early diagnosis, *Lung Cancer* 68, 216–221, 2010.
451. Roz, L., Verri, C., Conte, D. et al., Plasma DNA levels in spiral CT-detected and clinically detected lung cancer patients: A validation study, *Lung Cancer* 66, 270–271, 2009.
452. van der Drift, M.A., Hol, B.E., Klaassen, C.H. et al., Circulating DNA is a noninvasive prognostic factor for survival in non-small cell lung cancer, *Lung Cancer* 68, 283–287, 2010.
453. Yoon, K.A., Park, S., Lee, S.H. et al., Comparison of circulating plasma DNA levels between lung cancer patients and healthy control, *J. Mol. Diagn.* 11, 182–185, 2009.
454. Rabinowits, G., Gercel-Taylor, C., Day, J.M. et al., Exosomal microRNA: A diagnostic marker for lung cancer, *Clin. Lung Cancer* 10, 42–46, 2009.
455. Hanrahan, E.O., Ryan, A.J., Mann, H. et al., Baseline vascular endothelial growth factor concentration as a potential predictive marker of benefit from vandetanib in non-small cell lung cancer, *Clin. Cancer Res.* 15, 3600–3609, 2009.
456. Hongsachart, P., Huang-Liu, R., Sinchaikul, S. et al., Glycoprotein analysis of WGA-bound glycoprotein biomarkers in sera from patients with lung adenocarcinoma, *Electrophoresis* 30, 1206–1220, 2009.
457. Zhang, G., Xu, Y., Lu., X. et al., Diagnosis value of serum B7-H3 expression in non-small cell lung cancer, *Lung Cancer* 66, 245–259, 2009.
458. Tomida, M., Mikami, I., Takeuchi, S. et al., Serum levels of nicotinamide *N*-methyltransferase in patients with lung cancer, *J. Cancer Res. Clin. Oncol.* 135, 1223–1229, 2009.
459. Borgia, J.A., Basu, S., Faber, L.P. et al., Establishment of a multi-analyte serum biomarker panel to identify lymph node metastases in non-small cell lung cancer, *J. Thorac. Oncol.* 4, 338–347, 2009.
460. Nisman, B., Biran, H., Heching, N. et al., Prognostic role of serum cytokeratin 19 fragments in advanced non-small-cell lung cancer: Association of marker changes after two chemotherapy cycles with different measures of clinical responses and survival, *Br. J. Cancer* 98, 77–79, 2008.
461. Ishikawa, N., Daigo, Y., Yasui, W. et al., ADAM8 as a novel serological and histochemical marker for lung cancer, *Clin. Cancer Res.* 10, 8363–8370, 2004.
462. Li, H.X., Lei, D.S., and Wang, X.Q., Serum thymidine kinase 1 is a prognostic and monitoring factor in patients with non-small cell lung cancer, *Oncol. Rep.* 13, 145–149, 2005.
463. Hirsch, F.R., Varella-Garcia, M., and Cappuzzo, F., Predictive value of EGFR and HER2 overexpression in advanced non-small-cell lung cancer, *Oncogene* 28(Suppl 1), S32–S37, 2009.
464. Bartels, C.L. and Tsongalis, G.J., MicroRNAs: Novel biomarkers for human cancer, *Clin. Chem.* 55, 623–631, 2009.

465. Beane, J., Spira, A., and Lenburg, M.E., Clinical impact of high-throughput gene expression studies in lung cancer, *J. Thorac. Oncol.* 4, 109–118, 2009.

466. Reguart, N., Cardona, A.F., Carrasco, E. et al., BRCA1: A new genomic marker for non-small-cell lung cancer, *Clin. Lung Cancer* 9, 331–339, 2008.

467. Jacobsen, B. and Ploug, M., The urokinase receptor and its structural homologue C4.4A in human cancer: Expression, prognosis and pharmacological inhibition, *Curr. Med. Chem.* 15, 2559–2573, 2008.

468. Newnham, G.M., Thomas, D.M., NcLachlan, S.A. et al., Molecular profiling of non-small cell lung cancer: Of what value in clinical practice?, *Heart Lung Circ.* 17, 451–462, 2008.

469. Singhal, S., Miller, D., Ramalingam, S., and Sun, S.Y., Gene expression profiling of non-small cell lung cancer, *Lung Cancer* 60, 313–324, 2008.

470. Eberhard, D.A., Giaccone, G., Johnson, B.E. et al., Biomarkers of response to epidermal growth factor receptor inhibitors in non-small-cell lung cancer working group: Standardization for use in the clinical trial setting, *J. Clin. Oncol.* 26, 983–994, 2008.

471. Skrzpski, M., Quantitative reverse transcriptase real-time polymerase chain reaction (qRT-PCR) in translational oncology: Lung cancer perspective, *Lung Cancer* 59, 147–152, 2008.

472. Lacroix, L., Commo, F., and Soria, J.C., Gene expression profiling on non-small cell lung cancer, *Expert Rev. Mol. Diagn.* 8, 167–178, 2008.

473. Clements, J.A., Reflections on the tissue kallikrein and kallikrein-related peptidase family—From mice to men—What have we learnt in the last two decades?, *Biol. Chem.* 389, 1447–1454, 2008.

474. Komiya, M., Kato, H., and Suzuki III T., Structural comparison of high molecular weight kininogen and low molecular weight kininogens, *J. Biochem.* 76, 833–845, 1974.

475. Grahame-Smith, D.G., The carcinoid syndrome, *Am. J. Cardiol.* 21, 376–387, 1968.

476. Jenzano, J.W., Courts, N.F., Timko, D.A., and Lundblad, R.L., Levels of glandular kallikrein in whole saliva obtained from patients with solid tumors distant from the oral cavity, *J. Dent. Res.* 65, 67–70, 1986.

477. Jenzano, J.W., Hogan, S.L., and Lundblad, R.L., The influence age, sex, and race on salivary kallikrein levels in human mixed saliva, *Agents Actions* 35, 29–33, 1992.

478. Czokalo, M., Silko, J., and Topczewska, E., The effect of local radiotherapy on kallikrein activity in saliva secreted by parotid gland in patients with head and neck cancers, *Rocz. Akad. Med. Bialymst.* 41, 441–451, 1996.

479. Cohen, W.M., Wu., H.-F., Featherstone, G.L. et al., Linkage between blood coagulation and inflammation: Stimulation of neutrophil tissue kallikrein by thrombin, *Biochem. Biophys. Res. Commun.* 178, 315–320, 1991.

480. Wu., H.-F., Xu, L.-H., Jenzano, J.W. et al., Expression of tissue kallikrein in normal and transfected human endometrial stromal cells, *Pathobiology* 61, 123–127, 1993.

481. Raynor, R.H., Hazra, T.A., Moncure, C.W., and Mohanakumar, T., Characterization of a monoclonal antibody, KP-P8, that detects a new prostate-specific marker, *J. Natl. Cancer Inst.* 73, 617–625, 1984.

482. Lundwall, A., Characterization of the gene for prostate-specific antigen, a human glandular kallikrein, *Biochem. Biophys. Res. Commun.* 161, 1151–1159, 1989.

483. Watt, K.W.K., Lee, P.-J., M'Timkulu, T. et al., Human prostate-specific antigen: Structural and functional similarity with serine proteases, *Proc. Natl. Acad. Sci. USA* 83, 3166–3170, 1986.

484. Turan, T., Demir, S., Aybek, H. et al., Free and total prostate-specific antigen levels in saliva and the comparison with serum levels in man, *Eur. Urol.* 38, 550–554, 2000.

485. Kishi, T., Soosaipillai, A., Grass, L. et al., Development of an immunofluorometric assay and quantification of human kallikrein 7 in tissue extracts and biological fluids, *Clin. Chem.* 50, 709–716, 2004.

486. Shine, J., Mason, A.J., Evans, B.A., and Richards, R.I., The kallikrein multigene family: Specific processing of biologically active peptides, *Cold Spring Harb. Symp. Quant.* 48, 419–426, 1983.

487. Sutherland, G.R., Baker, E., Hyland, V.J. et al., Human prostate-specific antigen (APS) is a member of the glandular kallikrein gene family at 19q13, *Cytogenet. Cell Genet.* 48, 205–207, 1988.

488. Clements, J.A., The glandular kallikrein family of enzymes: Tissue-specific expression and hormonal regulation, *Endocr. Rev.* 10, 393–419, 1989.

489. Berg, T., Bradshaw, R.A., Carretero, O.A. et al., A common nomenclature for members of the tissue (glandular) kallikrein gene families, *Agents Action Suppl.* 38, 19–25, 1992.

490. Clements, J.A., The human kallikrein gene family: A diversity of expression and function, *Mol. Cell. Endocrinol.* 99, C1–C6, 1994.

491. Clements, J.A., Current perspectives on the molecular biology of the renal tissue kallikrein gene and the related tissue kallikrein gene family, *Biol. Res.* 31, 151–159, 1998.

492. Diamandis, E.P., Yousef, G.M., Luo, L.Y. et al., The new human kallikrein gene family: Implications in carcinogenesis, *Trends Endocrinol. Metab.* 11, 54–60, 2000.

493. Yousef, G.M. and Diamandis, E.P., Expanded human tissue kallikrein family—A novel panel of cancer biomarkers, *Tumor Biol.* 23, 185–192, 2002.

494. Diamandis, E.P., Borgoño, C.A., Scorilas, A. et al., Immunofluorometric quantification of human kallikrein 5 expression in ovarian cancer cytosols and its association with unfavorable patient prognosis, *Tumour Biol.* 24, 299–309, 2003.

495. Kapadia, C., Ghosh, M.C., Grass, L., and Diamandis, E.P., Human kallikrein 13 involvement in extracellular matrix degradation, *Biochem. Biophys. Res. Commun.* 323, 1084–1090, 2004.

496. Janssen, S., Jakobsen, C.M., Rosen, D.M. et al., Screening a combinatorial peptide library to develop a human glandular kallikrein 2-activated prodrug as targeted therapy for prostate cancer, *Mol. Cancer Ther.* 3, 1439–1450, 2004.

497. Memari, N., Grass, L., Nakamura, T. et al., Human tissue kallikrein 9: Production of recombinant proteins and specific antibodies, *Biol. Chem.* 387, 733–740, 2006.

498. Shaw, J.L., Grass, L., Sotriopouou, G., and Diamandis, E.P., Development of an immunofluorometric assay for human kallikrein 15 (KLK15) and identification of KLK15 in tissues and biological fluids, *Clin. Biochem.* 40, 104–110, 2007.

499. Väisänen, V., Peltola, M.T., Lilja, H. et al., Intact free prostate-specific antigen and free and total human glandular kallikrein 2. Elimination of assay interference by enzymatic digestion of antibodies to F(ab')$_2$ fragments, *Anal. Chem.* 78, 7809–7815, 2006.

500. Kountourakis, P., Psyrri, A., Scorilas, A. et al., Expression and prognostic significance of kallikrein-related peptidase 8 protein levels in advanced ovarian cancer by using automated quantitative analysis, *Thromb. Haemost.* 101, 541–546, 2009.

501. Korbakis, D., Gregorakis, A.K., and Scorilas, A., Quantitative analysis of human kallikrein 5 (KLK5) expression in prostate need biopsies: An independent cancer biomarker, *Clin. Chem.* 55, 904–913, 2009.

502. Talieri, M., Mathioudaki, K., Prezas, P. et al., Clinical significance of kallikrein-related peptidase 7 (KLK7) in colorectal cancer, *Thromb. Haemost.* 101, 741–747, 2009.

503. Nathalie, H.V., Chris, P., Serge, G. et al., High kallikrein-related peptidase 6 in non-small cell lung cancer cells: An indicator of tumor proliferation and poor prognosis, *J. Cell. Mol. Med.* 13, 4014–4022, 2009.

504. White, N.M., Mathews, M., Yousef, G.M. et al., KLK6 and KLK13 predict tumor recurrence in epithelial ovarian carcinoma, *Br. J. Cancer* 101, 1107–1113, 2009.

505. Hathurusinghe, H.R., Goonetilleke, K.S., and Siriwardena, A.K., Current status of tumor M2 pyruvate kinase (tumor M2-PK) as biomarker of gastrointestinal malignancy, *Ann. Surg. Oncol.* 14, 2714–2720, 2007.

506. Mall, A.S., Analysis of mucins: Role in laboratory diagnosis, *J. Clin. Pathol.* 61, 1018–1024, 2006.
507. Maher, J. and Wilkie, S., CAR mechanics: Driving T cells into the MUC of cancer, *Cancer Res.* 69, 4559–4562, 2009.
508. Ledes, M.J., Carabello, M., Suciu, D. et al., Detection of cancer miRNAs on an oligonucleotide microarray, *PLoS One* 4, e6229, 2009.
509. Ward, T.H., Cummings, J., Dean, E. et al., Biomarkers of apoptosis, *Br. J. Cancer* 99, 841–846, 2008.
510. van der Vaart, M. and Pretarius, P.J., Is the role of circulating DNA as a biomarker of cancer being prematurely overrated?, *Clin. Biochem.* 43, 26–36, 2009.
511. Dakappagari, N., Neely, L., Tangri, S. et al., An investigation into the potential use of serum Hsp70 as a novel tumour biomarker for Hsp90 inhibitors, *Biomarkers* 15, 31–38, 2010.
512. Witkin, S.S., Heat shock protein expression and immunity: Relevance to gynecologic oncology, *Eur. J. Gynaecol. Oncol.* 22, 249–256, 2001.
513. Steffensen, K.D., Waldstrøm, M., Jeppesen, U. et al., The prognostic importance of cyclooxygenase 2 and HER2 expression in epithelial ovarian cancer, *Int. J. Gynecol. Cancer* 17, 798–807, 2007.
514. Harris, R.E., Cyclooxygenase-2 (cox-2) and the inflammation of cancer, *Subcell Biochem.* 42, 93–126, 2007.
515. Denlinger, C.S. and Cohen, S.J., Progress in the development of prognostic and predictive markers for gastrointestinal malignancies, *Curr. Treat. Options Oncol.* 8, 339–351, 2007.
516. Budd, G.T., Let me do more than count the ways: What circulating tumor cells can tell us about the biology of cancer, *Mol. Pharm.* 6, 1307–1310, 2009.
517. Boeck, S., Stieber, P., Holdenrieder, S. et al., Prognostic and therapeutic significance of carbohydrate antigen 19-9 as tumor marker in patients with pancreatic cancer, *Oncology* 70, 255–264, 2006.

4 The Use of Proteomics to Discover Biomarkers

This chapter focuses on (1) various tissue sources for the development and characterization of biomarkers and (2) methods for the discovery of biomarkers. The intent is to provide basic information to the development of a biomarker discovery process, which leads to facile clinical laboratory application of a biomarker (see Chapter 1 for definition). In this sense, it is critical to understand the provenance of the sample and all possible downstream applications of the biomarker. This approach is analogous to Efraim Racker's admonition to his colleagues; "Don't waste clean thinking on dirty enzymes."[1] All of the sophistication in analysis and data processing will not rectify errors in sample preparation and processing. As noted in several other discussions in this book, a biomarker is intended to serve a diagnostic use and is therefore considered in the category of devices by the FDA and other regulator agencies. As a device, it is necessary to adhere to the principles of design control.[2–4] All that is required is that you know where you are going in the development process and can document the process; if it is not recorded, it never happened. Chapter 8 discusses the process of biomarker development in greater detail. Finally, it is useful to remember that the discovery and application of biomarkers is analytical chemistry and it is recommended that the serious investigators review basic texts in this area.

The analytical process may be divided into two parts: sample preparation and the process of analysis, which can be considered with analytical quality management.[5] Here, Burgess separates the analytical process into the sample source, sampling process to obtain sample, the analytical measurement(s), and the analytical results. The objective of this process is to obtain analytical results that meet the requirements of the end user. The control framework for this process is quality management, which includes both quality assurance and quality control. Sample preparation is at least as important as the actual analytical process, and control of the overall process from the sample material (blood, urine, tissue, etc.) to the analytical output (e.g., concentration determined by ELISA assay) is critical. It is suggested that issues in sample preparation can add as much variance as biological variation.[6–10] Sample collection and storage has always been of concern for classical clinical chemistry[11–14] but only recently has there has been concern in biomarker discovery and development.[15–17]

Given the hypothesis that it is possible to obtain a valid biomarker from an organism, what is the specimen that will provide a valid sample of the biomarker for analysis? The answer can be a bit more complicated as one considers the various issues that can influence composition of a sample. Biomarkers can be obtained from biological fluids such as blood, urine, and saliva, or from tissue samples (Table 4.1). The attributes of each of these sources is considered below.

TABLE 4.1

Relative Value of Tissue Sources for Biomarker Discovery and Development[a]

Query	Number of PubMed Citations
Blood and biomarker	220,254
Feces and biomarker	1,110
Lymph and biomarker	16,638
Saliva and biomarker	1,154
Sweat and biomarker	539
Tissue[b] and biomarker	118,205
Urine and biomarker	13,160
Vitreous fluid and biomarker	62

[a] This data was derived from a PubMed search in November 2009.
[b] Tissue is taken to mean solid tissues.

BLOOD

Blood is an extremely popular source for biological samples.[18] It is reasonably easy to obtain samples; the samples are technically and psychologically easy to process (e.g., feces, while easy to obtain, generally is considered difficult to process because of cultural reasons), and they are mostly considered homogeneous when compared to saliva or urine, both of which are somewhat compositionally dependent on fluid flow rates.

The initial blood sample is removed from the circulatory system via venipuncture.[19,20] It is of interest that the quality of the sample is influenced by the site of venipuncture[21,22] and by the length of time of venous occlusion prior to sampling.[23,24] The whole blood may serve directly as the sample but more commonly the blood is fractionated into blood plasma or allowed to form serum.

My personal experience suggests that the time between venipuncture and freezing, process/storage containers, centrifugation speed, and the temperature of storage are the most critical variables for plasma. The material nature of the process and storage containers also influences sample quality.[25–28] There are multiple anticoagulants available for the collection of blood including citrate and heparin*; the choice of anticoagulant and temperature can influence the storage stability of the analyte.[29–34] Anticoagulant can influence the quality of the plasma and the effect varies depending on the analyte; thus, there is no default as each has advantages and disadvantages. For example, both citrate and EDTA allow the collection of blood plasma by the chelation of calcium

---

* Vacuum (evacuated) tubes are used to obtain samples for clinical analysis. These have stoppers that are color-coded to indicate the anticoagulant or lack thereof (lavender, EDTA; green, heparin; red, serum; blue, citrate). Serum separator tubes (red/black) contain an inert polymer gel substance that forms a barrier between the serum and separated cells/fibrin after centrifugation (Brown, L.F. and Fraser, C.G., Assay validation and biological variation of serum receptor for advanced glycation end-products, *Ann. Clin. Biochem.* 45, 518–519, 2008; Mather, G., Zwart, S.R., and Smith, S.M., Stability of blood analytes after storage in BD SST tubes for 12 mo., *Clin. Biochem.*, in press, 2009).

ions required for the process of blood coagulation. EDTA causes platelet activation.[35] Blood coagulation factor VIII is unstable in the presence of citrate or EDTA but antigenic activity is retained.[36,37] Factor VIII activity is more stable in the presence of heparin.[37,38] Heparin does bind to plasma proteins,[39–42] which can influence separation characteristics. Heparin can inhibit the polymerase chain reaction (PCR), which can complicate the use of DNA as a biomarker in blood and tissue samples.[43–51] The inhibition of the PCR reaction in blood samples by heparin can be removed by treatment with the enzyme heparinase.[43–46] Heparinase was also used by Nohara and coworkers[47] to eliminate the inhibition of the PCR reaction by heparin proteoglycan in studies on gene expression in mast cells. One group[48] showed that dilution of a heparinized blood sample with albumin could overcome the heparin inhibition of PCR amplification; this group also observed that PCR amplification was greater with citrated blood than with EDTA blood. Another group[49] used treatment of heparinized plasma with Chelex® 100 resin to reverse the inhibition of the PCR reaction.

Some of the more useful studies on the effect of anticoagulant on biomarkers in blood are shown in Table 4.2. A consideration of the various data suggests that there is no true default anticoagulant; it is the subjective sense of the author that citrate is the least likely to cause significant problems. The removal of calcium ions rather than chelation as discussed below is an option, which should receive more consideration.

Serum is obtained when the sample is obtained in the absence of anticoagulant and the blood clots. A short period of time is required to form a fibrin clot, which can be removed by centrifugation to obtain serum. The blood can also be allowed to stand for longer periods of time and clot retraction occurs,[64,65] where the serum is expressed from the clot. The temperature for clot formation and the time of clotting before the removal of serum from the fibrin and cells are all preanalytical variables.[66–68] It must be emphasized that plasma and serum are not interchangeable terms. There are gross analytical differences between plasma and serum,[12] and some major specific differences such as the loss of fibrinogen resulting from conversion to fibrin[69,70] and matrix-metalloproteinases, some of which are constitutive to plasma and other forms derived from sources during the formation of serum.[60] One study[63] suggests that 40% of the peptides in serum are not present in plasma. There are a number of other changes[71] including the formation of protease–serpin complex and protein fragments such as D-dimer and prothrombin fragment 1. Contrary to at least one statement in the literature,[72] many of the coagulation factors such as factor IX, factor X, and factor XIa are retained in serum with some of the original names such as serum prothrombin conversion accelerator (now known as factor VIIa); as noted below, there are substantial changes in the transition from whole blood to serum that render qualitative and quantitative differences between blood plasma and blood serum, a generalized loss of coagulation factors is not one of such changes. As a result of the clotting of fibrinogen, the protein concentration of serum is less than that of plasma.[73,74] In the process of whole blood coagulation, the cellular elements (erythrocytes, leukocytes, platelets) can secrete components. In particular, platelets contribute a variety of components to blood serum.[75–77] Vascular endothelial growth factor (VEGF) is an excellent example.[78,79] In one study,[77] normal individuals had a serum concentration of 250 pg/mL with a plasma concentration of 30 pg/mL; breast cancer patients with thrombocytosis had a median VEGF concentration of 833 pg/mL compared to 249 pg/mL in other patients.

TABLE 4.2
Anticoagulant Effect on Biomarkers in Human Blood Plasma

Anticoagulant	Study	Reference
EDTA	Measurement of angiogenic cytokines (e.g., VEGF) in blood plasma. EDTA causes variable platelet activation with release of cytokines; cytokines were not released when citrate-theophylline-adenosine-dipyridamole (CTAD) was used as an anticoagulant	52
EDTA; citrate	The level of hyaluronidase activity in EDTA plasma is comparable to the level of hyaluronidase activity in serum; the level of hyaluronidase activity in citrated plasma is lower than the level in serum	53
EDTA; citrate	VEGF concentration was similar in EDTA-PGE1-theophylline plasma and CTAD plasma; VEGF was significantly higher in citrated plasma	54
Citrate	Soluble CD40 ligand (sCD40L) concentration was evaluated in citrated plasma and CTAD plasma. Sample processed in the cold had low concentrations and there was no difference between the two plasmas in either normal subjects or acute coronary syndrome (ACS) patients; for samples processed at room temperature, sCD40L was elevated in ACS patients	55
EDTA; citrate	The concentration of soluble CD40 ligand (sCD40L) was higher in EDTA plasma than citrated plasma	56
EDTA; citrate	Soluble thrombomodulin was unstable in EDTA but was stable in acidified citrated plasma	57
EDTA; citrate; heparin	MALDI/TOF/MS of low-molecular-weight proteins (prefractionation with C_8 or Cu-IMAC). Difference in proteomes depends on anticoagulant with largest difference between EDTA and heparin or citrate	58
Heparin; citrate; EDTA	No difference in IL-6 levels as a function of type of anticoagulant	59
Heparin; citrate; EDTA	A review on the effect of various anticoagulants on the activity of circulating matrix metalloproteinases in plasma	60
EDTA; heparin	A small difference in the immunoreactivity of β-C-telopeptide between potassium EDTA plasma and lithium heparin plasma	61
EDTA, heparin, citrate	Difference in low molecular weight peptides from plasma as determined by MALDI/TOF/MS. Peptides obtained from plasma by filtration through 3 kDa nominal mw cutoff filter. Addition of protease inhibitors prevented degradation	62
Heparin plasma, citrated plasma, EDTA plasma, serum	Emphasis on the preparation of samples for the analysis of peptides. Analysis by RP-HPLC/MS.[a] Pattern depends on anticoagulant choice. The use of EDTA or citrated plasma recommended for study of low-molecular-weight proteome. Heparin plasma pattern was markedly different from either EDTA or citrated plasma	63

[a] Samples prepared by ultrafiltration of plasma or serum after denatured in guanidine (50 kDa cutoff filter) were separated on RP-HPLC (C_5) and effluent fractions analyzed by MALDI-MS.

Both studies suggest that platelets can contribute to VEGF levels in both plasma and serum but more markedly so in serum. It is suggested that the immediate separation of plasma or serum from the cellular elements provide optimal analyte stability.[72,80] Again, critical process variables for serum are process/storage containers, time of clot retraction/removal of the fibrin clot with associated platelets and other cellular elements, centrifugation speed, and temperature of storage. Some studies on the use of serum for the study of biomarkers are shown in Table 4.3. This table also lists some studies that compare serum and plasma for the study of biomarkers.

It is also possible to withdraw blood through a resin-containing device, which depletes the blood of calcium precluding coagulation.[83] Heppinstall[84] reported the use of Chelex 100* to obtain platelets free of calcium and magnesium ions without having to go through the process of separation either by centrifugation or gel filtration. An examination of the literature suggests the majority of Chelex 100 use has been for DNA extraction and water analysis with only a few applications to blood other than for forensic chemistry. Chelex 100 has been used for the analysis of copper in human plasma.[85] Bierau and coworkers[86] showed that the treatment of blood with Chelex 100 decreased inosine triphosphate pyrophosphohydrolase activity. Chelex 100 has also been used to define divalent cation effects in serum and culture media.[87–89] Considering differential effects of various anticoagulants and the differences between plasma and serum, it is a bit surprising that the removal of divalent cations has not been considered more during the biomarker discovery process. It is granted that it might be difficult to incorporate this type of step in the normal blood-drawing process.

URINE

Urine is a source of new biomarkers as it has a considerable history of use in clinical chemistry. The collection of urine does not present the technical challenges that are provided by blood in that venipuncture is not required; however, blood is likely a more reproducible source as issues such as flow rate need not be considered. A PubMed search yielded more than 13,000 citations for urine biomarker. Table 4.4 lists some selected studies on biomarkers in urine with an emphasis on technique. In addition, the reader is directed to the work by Papale and coworkers[90] who reported on the profiling of urine using surface-enhanced laser desorption/ionization-time-of-flight/mass spectrometry (SELDI-TOF/MS). These investigators discuss the importance of sample collection and processing prior to analysis. A progressive degradation of protein was observed on storage at room temperature,† which was only prevented for a period of 2 h by the addition of a protease inhibitor

---

* Chelex 100 is a resin (National Forensic Science Training Center: http://www.nfstc.org/pdi/Subject03/pdi_s03_m03_01.htm) which is used to chelate divalent cations. Chelex 100 is a product of BioRad Laboratories (http://www.bio-rad.com/webmaster/pdfs/9184_Chelex.PDF) and available from Sigma-Aldrich (http://www.sigmaaldrich.com). There is considerable use of Chelex 100 resin for DNA analysis in biological fluids (Willard, J.M., Lee, D.A., and Hollland, M.M., Recovery of DNA for PCR amplification from blood and forensic samples using a chelating resin, *Methods Mol. Biol.* 98, 9–18, 1998).

† The use of the term room temperature is to be discouraged; reference to the actual temperature is preferred. It is frequently assumed that room temperature is 23°C.

TABLE 4.3

Differences in Biomarkers in Serum or Plasma

Anticoagulant	Study	Reference
EDTA, citrate, heparin, serum	No difference in IL-6 levels in serum and plasma[a]	59
EDTA, citrate, serum	The level of hyaluronidase activity in EDTA plasma is comparable to the level of hyaluronidase activity in serum; the level of hyaluronidase activity in citrated plasma is lower than the level in serum	53
EDTA, citrate, serum	The concentration of soluble CD40 ligand (sCD40L) was higher in EDTA plasma than citrated plasma. The concentration in either EDTA plasma or citrated plasma was lower than that in serum	56
Plasma[a], serum	MALDI/TOF/MS analysis of prefractionated plasma or serum (C_8 or Cu-IMAC) demonstrated major differences between plasma and serum	58
Heparin plasma, serum[b]	Evaluation of thyroxine, parathyroid hormone, follicle-stimulating hormone determinations as a function of sample collection conditions	26
Plasma, serum	Differences in the levels of activity of matrix metalloproteinases in serum and plasma	60
Plasma, serum	Difference between low molecular plasma proteome as determined by MALDI-TOF-MS	62
Heparin plasma, serum[c]	Cardiac troponin I not affected by collector tube type; myoglobin and CK-MB did show statistically significant differences with respect to collection tube type.[c]	81
Heparin plasma, serum	Cardiac troponin T(cTnT) was lower in heparinized plasma than serum; addition of heparin to serum decreased cTnT immunoreactivity. Also evaluation of a number of common laboratory analytes	82
Heparin plasma, citrated plasma, EDTA plasma, serum	Major difference between plasma and serum by MS analysis.[d] As would be expected, there are peptides in serum, which are not found in plasma. Statistical analysis of the data suggested that 40% of the peptides in serum are not found in plasma	63

[a] Heparin, EDTA, or citrated plasmas evaluated.
[b] Serum, serum separator tubes compared with lithium heparin anticoagulants.
[c] Heparin plasma and serum, both with and without gel separator.
[d] Samples prepared by ultrafiltration of plasma or serum after denatured in guanidine (50 kDa cutoff filter) were separated on RP-HPLC (C_5) and effluent fractions analyzed by MALDI-MS.

cocktail.[91] These investigators also demonstrated that the sample could be subjected up to four freeze/thaw cycles. It is common to express the concentration of biomarkers in urine relative to creatinine concentration[92–94] to correct for urine dilution. Creatinine, formed from creatine, is a function of muscle mass within an individual and is considered to be constant within that individual.[93] Arndt[95] has reported on

TABLE 4.4
Some Selected Studies on Biomarkers in Urine

Biomarker Study	Reference
Use of urine sample for measurement of collagen crosslinks (N-terminal telopeptide, pyridinium, deoxypyridinoline) as biomarker for bone resorption. Variability could be reduced by collecting urine for longer periods of time. Normalization with creatinine was not useful; 24 h collection recommended	109
Urinary albumin (albumin/creatinine ratio) is a biomarker, which predicts increasing blood pressure in normotensive individuals; urinary albumin was not correlated with baseline ambulatory blood pressure in normotensive individuals. Individuals with hypertension or diabetes were excluded from this study	110
Urine albumin (albuminuria)[a] is used as a biomarker for cardiovascular and renal disease. Albumin in urine shows a circadian variation; it is argued that a specific time-point (spot morning urine sample) should be used. It is suggested that freezing urine sample results in increased variability[b]	111
The ratio of IL-6 to IL-10 in urine is a prognostic biomarker for the recurrence of intermediate risk superficial bladder carcinoma. The assays were solid phase immunoassays	112
Urinary beta-defensin-1 as biomarker for exposure to arsenic. Analysis by SELDI-TOF/MS confirmed by immunoassay. This study is an example of careful sample collection and processing prior to analysis	113
Creatinine correction of biomarkers in urine from a pediatric population is useful; daytime spot sample and overnight urine samples are not comparable	114
Evaluation of S100 in urine as a biomarker for head injury in children; there was no difference in urine levels of S100 in children with head trauma and children with extracranial injury	115
VEGF as a biomarker in urine for cancer. Study on the effect of specimen collection and processing on the assay of VEGF with chemiluminescent immunoassay. Variables evaluated included time to freezing after sample collection, number of freeze/thaw cycles, and type of polypropylene tubes. The sediment from centrifugation of urine from healthy subjects had higher activity than the supernatant fraction	28
Urine nuclear matrix protein 22 as a diagnostic biomarker for transitional cell carcinoma of the bladder	116
MALDI-MS used to identify urinary biomarkers for chronic allograft rejection	117
The pH of urine does not influence separation on 2D gels	118
Urinary NT-proBNP (N-terminal pro-brain natriuretic peptide) immunoreactivity is a product of various degradation products of blood NT-proBNP; results of assay in urine may be unique to individual assay requiring validation on an assay-by-assay basis	119
Development of multidimensional LC system for discovery of biomarker peptides in urine	120
Study on the use of urinary neutrophil gelatinase-associated lipocalin as prognostic biomarker for acute kidney injury in an ICU population	121

[a] Albuminuria is an accepted biomarker for cardiovascular risk (Bakris, G.L. and Kuritzky, L., Monitoring and managing urinary albumin excretion: Practical advice for primary care clinicians, *Postgrad. Med.* 121, 51–60, 2009).

[b] The reader is directed to a commentary on this work (Dyer, A.R., Invited commentary: Evaluation of measures of urinary albumin excretion in epidemiological studies, *Am. J. Epidemiol.* 164, 728–732, 2006).

creatinine concentration in a cohort of 45,000 subjects providing a strong basis for reference levels of creatinine in urine.

The use of urine for the study of exposure to environmental agents is not mentioned in this table but is of considerable interest.[96–104] The use of metabolomics to identify biomarkers in urine is of increasing interest[105–108] but is not within the scope of Table 4.4. The reader is referred to several recent reviews[122–125] on the use of urinary proteomics for biomarker discovery. Nucleic acid biomarkers[126–129] are present in urine but not discussed in Table 4.4.

SALIVA

The author has worked on saliva and salivary proteins and does appreciate some of the difficulties associated with the use of this material for chemical and biological assays; on the other hand, saliva and urine are less invasive than blood, which in turn is less invasive than obtaining a tissue sample. Despite the low number of citations, there are some useful biomarkers in saliva that should be considered. The reader is also directed to the discussion on glandular/tissue kallikrein as a biomarker for cancer in Chapter 3. First, for some definitions: whole saliva is that which would be obtained from the oral cavity by expectoration; whole saliva may be unstimulated, that is, without the action of chewing or any pharmacological intervention, or stimulated with chewing or citric acid application.[130–136] Protein concentration can be lower in stimulated saliva than unstimulated saliva but does not increase with the time of stimulation.[135,137] A normal adult will secrete more than a liter of saliva every 24 h, which is derived primarily from the parotid glands, the submaxillary glands, and the sublingual glands with minor contributions from the gingival crevicular fluid. While glandular saliva is unique in the gland in question, the gingival crevicular fluid is derived from plasma ultrafiltered through the gingival crevice. The composition of the gingival crevicular fluid depends on the periodontal health of the subject.[138,139] Gingival crevicular fluid can be a source of biomarkers for oral health.[140,141] However, the collection of gingival crevicular fluid is a tedious process requiring considerable skill and it is more useful to measure gingival crevicular fluid constituents in whole saliva.[142–145]

It is possible to measure biomarkers in isolated glandular saliva. Parotid saliva and submaxillary saliva can be obtained with devices.[146–152] However, as with gingival crevicular fluid, constituents unique to a glandular source can be measured in whole saliva. Also, glandular saliva is more invasive than the collection of whole saliva and does require no small amount of technical skill for the process. Whole saliva would appear to be the most useful source of biomarkers.[153] A selected list of salivary biomarkers is presented in Table 4.5.

It is useful to consider that care must be taken in the measurement of protein concentration of biological fluids. The protein composition of plasma or serum can be determined with most of the commonly available colorimetric methods; for one thing, the protein concentration of serum or plasma is dominated by the presence of two proteins, albumin and immunoglobulin. Thus, the determination of protein concentration of plasma or serum is relatively insensitive to technique.[171] However, albumin and immunoglobulin do perform differently with various colorimetric assays

TABLE 4.5
Some Selected Studies on Biomarkers in Saliva

Biomarker/Study	References
Salivary kallikrein as diagnostic biomarker for tumors distant from the oral cavity	154
Cortisol in saliva as a biomarker for stress	155–157
Cortisol in saliva after ACTH stimulation as biomarker for adrenal insufficiency in end-stage renal disease	158
Testosterone in saliva as biomarker for androgen function	159–161
Testosterone in saliva as biomarker for male androgen deficiency	162
Toluene in saliva as biomarker for exposure to toluene	162
Lysozyme in saliva[a] as biomarker for hypertension	164
Procalcitonin is a biomarker in saliva for periodontitis and hyperglycemia in type 2 diabetes	165
Urea as a biomarker in saliva for chronic renal failure; there is a good correlation between serum urea and salivary urea	166
Salivary α-amylase as biomarker for stress	167
Salivary leptin as a biomarker for salivary gland tumors	168
CEA and CA-50 as prognostic biomarkers in saliva for oral and salivary malignant tumors	169
Chromogranin A as a biomarker for stress in saliva	170

[a] Salivary lysozyme is derived from several sources including leukocytes, which enter the oral cavity by the gingival crevicular fluid (Raeste, A.M., Lysozyme (muramidase) activity of leukocytes and exfoliated epithelial cells in the oral cavity, *Scand. J. Dent. Res.* 80, 422–427, 1972; Suomalainen, K., Saxén, L., Vilja, P. et al., Peroxidases, lactoferrin and lysozyme in peripheral blood neutrophils, gingival crevicular fluid and whole saliva of patients with localized juvenile periodontitis, *Oral Dis.* 2, 129–134, 1996); lysozyme is also derived from submaxillary gland/sublingual gland (Noble, R.E., Salivary α-amylase and lysozyme levels: A non-invasive technique for measuring parotid vs. submandibular/sublingual gland activity, *J. Oral. Sci.* 42, 83–86, 2000).

required the availability of different commercial standards for these two materials. The determination of protein concentration in urine or saliva by colorimetric techniques can be more problematic.[172–174] The concentration of a biomarker in urine can be (imperfectly) corrected by using the ratio to creatinine.[175] There is no similar correction for salivary biomarker concentration. While there are difficulties, there are some studies on the development of biomarkers in saliva, which are shown in Table 4.5. In addition to the information in Table 4.5, the reader is directed to some reviews[176–182] of biomarker discovery in saliva. Since saliva is somewhat off the beaten track for most investigations in biomarker research, there are a number of review articles on the composition of this complex fluid.[183–187] Finally, saliva is a medium for the analysis of drugs derived from the circulation.[188–193] There is particular interest in the use of salivary levels of a drug as a measure of free drug as opposed to drug bound to proteins such as albumin in the circulation as well as drugs as abuse.

MISCELLANEOUS BIOLOGICAL FLUIDS/EXCRETORY PRODUCTS AS SOURCES OF BIOMARKERS

Fecal biomarkers are of considerable interest for colorectal cancer and inflammatory bowel disease.[194–198] Exhaled breath is being investigated as a source of biomarkers for pulmonary function[199] and oxidative stress secondary to asbestos or silica-induced lung disease.[200] Sweat serves as a biofluid with chloride (conductance) as a biomarker for the diagnosis of cystic fibrosis[201] and is receiving some attention for the development of biomarkers in diabetes.[202,203]

BIOMARKERS IN TISSUE SAMPLES

A PubMed search for tissue biomarker yielded more than 100,000 citations. Despite this large number of sources, there will be only limited coverage of this material. A tissue sample is generally invasive when compared to a biological fluid and more processing (tissue fixation, preparation of sections, staining, and imaging) is required. However, having said that, demonstration of a biomarker in a tissue section is an unequivocal evidence for the presence of said biomarker. We will confine our discussion to studies that use quantitative immunohistochemistry.[204–218] Some of this work is based on the use of monoclonal antibodies and localization and quantitation with gold particles or by peroxidase systems.[219–222] More recently, there has been use of quantum dot technology.[223–225] Fundamentally, the technology is similar to that in ELISA assays in that there is detection of a specific antigen–antibody complex and mention should be given to the use of ELISPOT assays for assessing biomarkers for cancer vaccine development[226] and for tuberculosis prognosis.[227] This concept is also used in protein microarrays.[228–230] It should be noted that tissue microarrays are of considerable value in screening large numbers of samples.[231–235] It is also possible to use laser microdissection for the identification of biomarkers.[236–240] Technological advances in spectroscopy are permitting the application of Raman scattering and UV-VIS spectroscopy to the direct study of tissues.[241–244] Some examples of the use of quantitative immunohistochemistry for the identification of biomarkers are shown in Table 4.6.

The take-home message is that there are many factors other than the underlying biology that can influence a sample obtained from biological fluids or tissues. It is difficult, if not impossible to eliminate these factors; the best that one can do is to very carefully document the conditions of blood processing. It is absolutely critical that a standard operating procedure (SOP) be established for the process of obtaining blood, urine, saliva, or tissue sample.[256–260] It is only by doing this that one is able to assure reproducibility of samples and to allow some rationale comparison of data from various laboratories. It is also important to do this sooner as opposed to later so early interesting results need not be discarded because of issues related to sample integrity. Likewise, there should be an SOP for the analytical process.[261–268]

Analyte stability in any matrix is of considerable importance because of the clinical use of these materials and the reader is directed to several studies that provide a general review of this area.[269–272] It is generally assumed that freezing a sample provides stability but it is noted that in one situation,[273] does not protect an analyte from degradation. Stability can also be a function of the storage container[81,274–278]

TABLE 4.6
The Use of Quantitative Immunohistochemistry and Gene Expression to Identify Biomarkers in Tissue Samples

Biomarker Study	Reference
Use of quantitative fluorescence image analysis (QFIA) to measure G-actin and DNA as biomarkers in fine needle aspiration aspirates for breast cancer intermediate risk assessment. The technique permitted examination of single cells in archival samples	245
Use of immunohistochemistry and PCR to identify HCCR oncoprotein as diagnostic biomarker for breast cancer. Subsequent work suggests that HCCR oncoprotein is a serological biomarker with sensitivity greater than CA15-3	246
Use of immunohistochemistry and RT-PCR to identify CD63 as a prognostic biomarker in lung cancer	247
Use of quantitative fluorescence imaging analysis for discovery of biomarker in formalin-fixed tissue section. Quantitation of β-catenin in archival samples from prostate cancer correlated with tissue content after extractions and immunological assay. Evaluation of β-catenin as a biomarker for cancer	248
Use of quantitative fluorescence analysis (tissue microarray[a]) to support human kallikrein 7 as prognostic biomarker for ovarian cancer	249
Use of serial analysis of gene expression (SAGE) to identify connective tissue growth expression as a prognostic biomarker in gall bladder cancer	250
Use of RT-PCR and immunohistocytochemistry to identify midkine as a prognostic biomarker (survival rates) in oral squamous cell carcinoma	251
Use of FACS analysis of PMNs to identify CD64 (high-affinity Fc receptor) as biomarker to discriminate between inflammatory bowel disease, infectious entercolitis, and functional intestinal disorders	252
Immunohistochemistry used to measure the expression of TGF-α (transforming growth factor-alpha) in rectal mucosa as prognostic indicator in colorectal cancer	253
Immunohistochemical quantitative determination of p16(INK4A)[b] as a prognostic biomarker for cervical intraepithelial neoplasia. This study also measured human papilloma virus (HPV) by in situ hybridization	254
Immunochemical measurement and RT-PCR measurement of glucagon/insulin ratio for differentiation between pancreatic ductal adenocarcinoma-related diabetes mellitus and diabetes mellitus type 2	255

[a] See Rimm, D.L., Camp, R.L., Charette, L.A. et al., Tissue microarray: A new technology for amplification of tissue resources, *Cancer J.* 7, 24–31, 2001; Camp, R.L., Neumeister, V., and Rimm, D.L., A decade of tissue microarrays: Progress in the discovery and validation of cancer biomarkers, *J. Clin. Oncol.* 26, 5630–5637, 2008.

[b] p16(INK4A) is an inhibitor of cyclin-dependent kinases functioning as tumor suppressor (Shapiro, G.I. and Rollins, B.J., p16INK4A as a human tumor suppressor, *Biochim. Biophys. Acta* 1242, 165–169, 1996) and is important in transformation (Drayton, S., Brookes, S., Rowe, J. et al., The significance of p16INK4a in cell defenses against transformation, *Cell Cycle* 3, 611–615, 2004).

and should be evaluated for the specific analyte as there is differential stability on storage, and antigenic activity may not parallel biological activity.[279] Studies on the stability of blood biomarkers are shown in Table 4.7. Lyophilization is a possibility for sample and standard storage.[292–298] Biobanking of samples is of increasing interest[299–303] and stability is an important consideration. It is possible to use paraffin-embedded formaldehyde (formalin*)-fixed tissues for DNA microarray studies.[304–307]

The majority of studies in this book describe the measurement of analytes in biological fluids and most of those in blood. Most biomarkers of interest are proteins or peptides. Thus, the most reasonable approach for the discovery of biomarkers would be based on the use of proteomic technologies. It is increasingly common to see an investigator state that proteomic technologies were used in their studies. The track record is a bit discouraging; as cited by Kiernan,[308] only one clinical assay based on a protein biomarker has been approved by FDA since 1998.

Proteomics is a somewhat ill-defined area of study with some 23,000 citations on PubMed since 1997; 40,605 records were obtained with SciFinder Scholar® while 18,518 citations were obtained with Web of Science®. SciFinder Scholar® retrieves abstracts from ACS meetings and patent information in addition to literature citations. The term "proteome" dates back to 1995[309] when Wasinger and colleagues defined proteome as "the total protein content of a genome." A consideration of the current literature would suggest that proteomic technologies are various analytical methods such as mass spectrometry, electrophoresis, and liquid chromatography. It would follow that proteomics is the study of the proteome using proteomic technologies where the proteome includes intracellular and extracellular proteins. There appear to be two general types of activities in proteomics, which are closely related to each other and of value to the discovery of biomarkers. Analytical proteomics includes various separation technologies, coupled analytical technologies including mass spectrometry, and various microarray platforms (Chapter 7 is concerned with microarray technology). The second general activity is expression proteomics.[310–314] Expression profiling uses technologies such as isotope-coded affinity tags (ICAT) and stable isotope labeling with amino acid in cell culture (SILAC).[315–321] Expression profiling should provide information similar to that obtained with DNA microarray technology.

There has been limited use of ICAT in the identification of biomarkers. It is a complex technique where two samples are compared in a process involving multiple steps including chemical modification. The goal is the demonstration of differential expression of a protein with respect to a challenge; this usually involves the comparison of two cell culture populations subjected to different challenges[322,323] although it is possible to compare paired human subjects.[324] Zhang and coworkers[325] used ICAT technology to study changes in the proteins of cerebrospinal fluid in Alzheimer's disease and related dementia. There is increasing use[326–331] of SILAC to identify biomarkers in cell culture experiments. Shan and coworkers[327] prepared a stable isotope labeled proteome (SILAP) standard from the culture of human columnar epithelial endocervical-1 cells and vaginal cells for use in the identification of biomarkers for preterm labor in cervicovaginal

---

* Formalin is a generic term used to refer to 37% (w/v) formaldehyde.

TABLE 4.7
Biomarker Stability Studies in Plasma and Serum

Anticoagulant	Study	Reference
EDTA	The stability of soluble adhesion molecules, selectins, and CRP in EDTA-anticoagulated whole blood and plasma at 5°C and 21°C is reported. The soluble adhesion molecules (sVCAM) and CRP are stable in either plasma or blood at either temperature for 3 days. sE-Selectin was stable for 2 days and only sP-selectin was not stable requiring immediate assay	280
	Stability of choline in plasma as a function of anticoagulant, storage temperature, and time	281
EDTA	Progastrin-releasing peptide increased during storage in EDTA plasma while it decreased in serum. The decrease in serum was not prevented by protease inhibitors. There was no significant difference in progastrin-releasing peptide concentration between EDTA plasma and serum immediately after collection	282
EDTA	Stabilization of ascorbic acid in plasma samples by the addition of trichloroacetic acid	283
Not given	Effect of delay of centrifugation on common laboratory analytes; hemolysis did have an effect on some analytes	284
Lithium heparin, potassium EDTA, serum	Storage at 23°C[a] in potassium EDTA plasma resulted in insignificant loss of β-C-telopeptides from type I collagen (β-CTX) immunoreactivity for up to 48 h; there was significant loss of immunoreactivity in lithium heparin plasma or serum with much greater inter-individual variability	61
Serum	Stability of PSA and PSA complexed with α_1-antichymotrypsin in serum has been evaluated as function of time and temperature. The PSA complex is more stable than free PSA under all conditions. Serum samples can be stored up to 8 h at 4°C before assay or stored at −80°C	285
Serum	Specimen preparation conditions for human serum samples including clotting conditions, storage temperature, freeze/thaw cycles	286
Serum	Serum samples collected in serum separator tubes. Markers (α-fetoprotein; unconjugated estriol; total human chorionic gonadotropin; dimeric inhibin A) used in Down syndrome screening were stable for 6 days at 4°C with shipping	287
Urine	Comprehensive study on the stability of urine samples for SELDI-TOF mass spectrometry. Up to five freeze/thaw cycles did not influence analytical results while storage at room temperature resulted in protein degradation, which was only prevented up to 2 h by the addition of proteinase inhibitors (leupeptin, AEBSF, bestatin, aprotinin, pepstatin A)[a]	90
Platelet-free plasma	Atrial natriuretic peptide stable in platelet-free plasma stored at −80°C for at least 12 months	288
Serum	Prostate-specific antigen stable for 7 years at −80°C	289
Saliva	Salivary IgA and salivary lysozyme stable for up to 3 months at −30°C	290
Plasma	Stability of ascorbic acid and dehydroascorbic acid in plasma after acidification with metaphosphoric acid	291

[a] Components of a commercially available protease inhibitor cocktail (Sigma-Aldrich; http://www.sigmaaldrich.com).

fluid. Yu and coworkers[332] used a SILAP prepared from the culture of CAPAN-2 human pancreatic cells as a standard for the identification of biomarkers for pancreatic cancer. Prokhorova and coworkers[333] used SILAC to identify potential biomarkers for human embryonic stem cell differentiation. There has been limited application of chemical proteomics and activity-based proteomics to biomarker discovery.[334–338]

Analytical proteomics has taken the analysis of biological fluids and tissues to a new level. The combination of basic analytical chemistry and separation technologies has permitted the visualization of thousands of analytes in serum samples. Advances in data processing have enabled the organization of this mass of information and the subsequent formation of various databases. There is a large international project to define the human proteome.[339] The common analytical tool is mass spectrometry, while the variable is the processing/prefractionation/separation/purification of the sample material. It must be emphasized that several of the separation techniques can also be considered as free-standing analytical techniques.

The size and dynamic range of the human proteome make a global analytical approach a challenging proposition.[16,340–344] One approach to reducing sample complexity is sample prefractionation.[345–350] The most widely used approach depletes the sample of the high-abundance proteins such as albumin,[351] but there is concern that additional proteins are also removed.[347] The use of a subpopulation such as phosphorylated proteins (phosphoproteome)[352] is also a useful approach. While this is a useful approach for the identification of biomarkers, the addition of a prefractionation step is not useful with a validated clinical assay; however, it is unlikely that the assay technology used for biomarker discovery will be used in the clinical laboratory. Another approach is to define secretomes[353–360] and predict the proteins that could be secreted into the circulation.[360] This basic research can lead to the development of immunoassays, which can be used to determine the validity of putative biomarkers identified through the study of secretomes.[361–363]

The discovery of a new biomarker implies the separation and identification of the analyte from bulk solution. Separation is based on differences in the analytes, which can be exploited by the application of technologies. Proteins differ from each other on the basis of electrical charge (isoelectric point), size (molecular weight), and perhaps shape, and biological properties.[364,365] Proteins can also differ on the basis of post-translational modifications resulting from enzymatic action such as γ-carboxylation, acetylation, or nonenzymatic modification with peroxynitrite or 4-hydroxy-2-nonenal.

The most logical approach is to devise a series of separation steps based on the different characteristics.

The oldest of the proteomic techniques and likely also the most frequently used is two-dimensional (2D) electrophoresis. The first dimension uses isoelectric focusing (separation based on the charge of the protein; isoelectric point) with the second dimension is performed in the presence of sodium dodecyl sulfate (separation is based on molecular size).[366,367] This is a complex system but has been demonstrated to be reproducible to the extent of being able to identify potential biomarkers.[368–383] Depending on the detection system, the number of protein "spots" can be more

TABLE 4.8
The Use of 2D Gel Electrophoresis to Identify Biomarkers[a]

Study	Reference
2D difference gel electrophoresis (2D-DIGE)[b] with subsequent analysis by mass spectrometry used to identify biomarkers in human lung squamous carcinoma (LSC). Microdissection was used to prepare samples from LSC samples from subjects with and without lymph node metastasis. Differences in protein expression might provide biomarkers and further understanding of pathology	411
2D-DIGE coupled with mass spectrometry is used to identify biomarkers in Immunodepleted[c] serum from patients with prostate cancer. The results suggest that pigment epithelium–derived factor is a biomarker for early stage prostate cancer	412
2D gel electrophoresis used to identify biomarkers for endometriosis in endometrial fluid	413
2D gel electrophoresis used to identify biomarkers in patients with Kawasaki disease. Increased fibrinogen-related proteins (fibrinogen, α-1-antitrypsin, clusterin, CD5L) and decreases levels of immunoglobulin free light chains. The demonstration of abnormal fibrinogen proteins may result in useful biomarker and understanding of cardiovascular complications in Kawasaki disease 1-antitrypsin, clusterin, CD5L and decreases levels of immunoglobulin free light chains. The demonstration of abnormal fibrinogen proteins may result in useful biomarker and understanding of cardiovascular complications in Kawasaki disease	414
Use of 2D gel electrophoresis to identify biomarkers in plasma for multiple sclerosis	415
The effect of sample pH on the 2D gel electrophoresis of urine. The results suggest that urine pH does not affect the electrophoretic analysis of urine	118
2D-DIGE coupled with mass spectrometry used to identify biomarkers in serum for head and neck cancer. Patients with recurrent disease showed significant differential expression (overexpression and underexpression) of several proteins	416
2D gel electrophoresis used to identify biomarkers in saliva for aggressive periodontitis	417
2-DIGE used to identify biomarkers in serum for colorectal cancer	418
2D electrophoresis used to identify biomarkers in extracts from tissues obtained from subjects' nasal polyps and chronic sinusitis and compared with extracts from normal nasal mucosal tissues. Differential expression of Cu/Zn superoxide dismutase and PLUNC[d] was observed with nasal polyps and chronic sinusitis and confirmed with immunohistochemistry and may be prognostic biomarkers	419
Use of 2D-DIGE to identify serum biomarkers illustrating differences I individuals before and after tai chi chuan exercise. There was a significant increase in complement factor H with a decrease in C1q esterase inhibitor and complement factor B	420

(continued)

TABLE 4.8 (continued)
The Use of 2D Gel Electrophoresis to Identify Biomarkers[a]

[a] Selected studies were taken from a PubMed search of 2009.

[b] Two-dimensional difference gel electrophoresis (2-DIGE) uses the differential labeling of samples with fluorescent dyes having different spectral characteristics (Cy dyes) and subsequent electrophoresis analysis. This permits the identification of differences between normal and disease samples. See Marouga, R., David, S., and Hawkins, E., The development of the DIGE system: 2D fluorescence difference gel analysis technology, *Anal. Bioanal. Chem.* 382, 669–678, 2005; Hoffman, S.A., Joo, W.A., Echan, L.A. et al., Higher dimensional (Hi-D) separation strategies dramatically improve the potential for cancer biomarker detection in serum and plasma, *J. Chromatogr. B. Analyt. Technol. Biomed. Life Sci.* 849, 43–52, 2007.

[c] A commercial (Aligent) immunoaffinity column was used for the removal of albumin, IgG, IgA, transferrin, haptoglobin, and anti-trypsin.

[d] PLUNC, palate, lung, nasal epithelium clone is an innate immune protein expressed by epithelial tissue and by neutrophils. See Bingle, C.D. and Gorr, S.U., Host defense in oral and airway epithelia: chromosome 20 contributes a new protein family, *Int. J. Biochem. Cell Biol.* 36, 2144–2152, 2004; Bartlett, J.A., Hicks, B.J., Scholmann, J.M. et al., *J. Leukoc. Biol.* 83, 1201–1206, 2008.

than 5000. The identification of biomarkers by 2D electrophoresis is based on the comparison of diseased tissue versus normal tissue. In most studies, the protein "spots" are identified on the electrophoretograms by staining with Coomassie blue, ammoniacal silver or with fluorescent dyes such as SYPRO.[384–394] The developed electrophoretograms are compared and the difference "spots" are excised either manually or with the aid of a robotic process, the proteins either digested in situ with proteolytic enzymes or analyzed directly by mass spectrometry.[395,396] Another option is the derivatization of the protein sample with fluorescent dyes prior to electrophoresis; the selection of different fluorophores allows the more facile comparison of several electrophoretograms.[385,397–410] Table 4.8 provides a partial list of studies useful for the identification of biomarkers.

Multidimensional protein identification technologies (MuDPiT)[421] is quite similar in concept to the 2D electrophoretic approach described above. This approach is described as the use of orthogonal techniques; for example, an ion-exchange column would be the first "phase" and a reverse-phase column would be the second "phase." Mass spectrometry of the intact proteins or digests thereof is used for analysis and identification using the same databases as described above. There has been limited use of this technology in the identification of biomarkers.

Surface-enhanced laser desorption/ionization-time-of-flight mass spectrometry was developed as affinity mass spectrometry.[422] This approach has been commercialized by Cipergen (Fremont, California) as ProteinChip® technology.[423–428] This is a facile technique that will provide a pattern for analysis. It is a high-throughput technology but is complicated by low resolution and difficult reproducibility.[427] Some selected applications of SELDI-TOF-MS for the study of biomarkers are shown in Table 4.9.

TABLE 4.9

Selected Studies on the Use of SELDI-TOF-MS in Identifying Biomarkers[a]

Study	Reference
SELDI-TOF-MS (anion array Q10[b]) was used to identify biomarkers in colon-derived liver metastasis. S100A6 and S100A11 could discriminate between metastases derived from primary colorectal carcinomas and hepatocellular carcinomas	429
SELDI-TOF-MS used for the profiling of cervical mucous proteins. This study optimizes pre-analytical parameters such as extraction reagent, matrix type, and sample fractionation	430
Use of SELDI-TOF-MS (cation exchange, CM10) to identify serum biomarkers for pancreatic adenocarcinoma	431
Use of SELDI-TOF-MS (Q10, anion exchange) to identify biomarkers in malignant and nonmalignant ascites	432
Use of SELDI-TOF MS for profiling protein expression in kidney tissue. This article discusses preparation of sample, array selection and processing, and data processing	433
Use of SELDI-TOF-MS (cation exchange chip[c]) to identify serum biomarkers for HIV-1-associated dementia. Gelsolin and prealbumin were found to be differentially expressed. This observation was confirmed by Western blotting	434
Use of SELDI-TOF-MS (anion exchange and weak cation exchange chips) to identify biomarkers in tissue extracts from microdissected human papillomavirus (HPV)-related oral squamous carcinoma. Thioredoxin and epidermal fatty acid–binding protein were upregulated in HPV-related tumor tissue. This observation was supported by immunohistochemistry	435
Use of SELDI-TOF-MS to identify hepcidin in urine as a biomarker for iron-related disorders. It was found that the CM10 protein chip was superior to the NP20 proteinchip	436
Use of SELDI-TOF-MS to identify serum biomarkers to distinguish between pancreatic cancer, chronic pancreatic, and type 2 diabetes mellitus. Peptide distribution patterns established with SELDI-TOF-MS can be combined with CA-19 to discriminate pancreatic cancer from acute pancreatitis and/or type II diabetes mellitus	437
SELDI-TOF-MS (IMAC-3 copper-treated chip) used to identify serum biomarkers for recurrence of head and neck cancer (prognostic biomarkers). Underexpression and overexpression of specific proteins were observed in the sample from patients with recurring disease	416
SELDI-TOF-MS (IMAC metal affinity matrix) is used to identify serum biomarkers for breast cancer. Ten "clusters" were identified as differences between breast cancer patients and controls. Peak clusters were C3a des-arginine anaphylotoxin, inter-α-trypsin inhibitor heavy chain 4 fragments and a fibrinogen fragment. It is suggested that the heterogeneity of breast cancer makes the selection of breast cancer subgroups for comparison with healthy controls a consideration for improving data	438
Use of SELDI-TOF-MS (cation exchange chip, CM10) for identification of serum biomarkers for renal cell carcinoma. While it is possible to identify two serum profiles to differentiate between renal cell carcinoma and control patients, sensitivity and specificity are not sufficient for screening/diagnostic use, but the profiles might be useful prognostic/predictive biomarkers	439

(continued)

TABLE 4.9 (continued)
Selected Studies on the Use of SELDI-TOF-MS in Identifying Biomarkers[a]

Study	Reference
Use of SELDI-TOF-MS (weak cation exchange, WCX-2) to identify serum biomarkers for myasthenia gravis. There were upregulated and downregulated proteins in serum differing between subjects with myasthenia gravis and control subjects. Some of these proteins are potential biomarkers for myasthenia gravis	440
Use of SELDI-TOF-MS (immobilized metal ion, IMAC; strong anion exchange, Q10) to identify serum biomarker for breast cancer. Serum samples and corresponding tissue samples were compared in this study. Three peaks were observed in breast cancer serum while there were 27 different protein in breast cancer tissue. Two tissue peaks were N-terminal albumin fragments	441
Use of SELDI-TOF-MS (cation exchange, CM10) to identify serum biomarkers in fibrosis. Studies involved serum and nasal epithelial cells from individuals with cystic fibrosis, individuals with asthma, individuals with chronic obstructive pulmonary disease (COPD), and control individuals. Profile patterns differed between the various study groups and may provide biomarkers for cystic fibrosis, asthma, and COPD	442
SELDI-TOF-MS (cation exchange) is used to identify biomarkers in cerebrospinal fluid for Creutzfeld–Jacob disease (CJD). Ubiquitin was identified as a potential biomarker in cerebrospinal fluid for CJD	443
SELDI-TOF-MS (cation exchange, anion exchange, immobilized metal ion) used to identify biomarkers in saliva for oral squamous cell carcinoma. The study also used prefractionation of saliva by anion exchange and cation exchange. A truncated cystatin SA-I was identified as a potential biomarker in saliva for oral squamous cell carcinoma	444
Use of SELDI-TOF-MS to identify serum biomarkers for inflammation in elderly hip fracture patients being treated with heparin	445

[a] ProteinChip® technology is a commercial term for SELDI-TOF-MS as technique for mass spectrometry developed by Ciphergen (Fremont, CA, USA). This technique was originally developed as affinity mass spectrometry at the University of California at Davis (Kuwata, H., Yip, T.T., Yip, C.L. et al., Bactericidal domain of lactoferrin: Detection, quantitation, and characterization of lactoferricin in serum by SELDI affinity mass spectrometry, *Biochem. Biophys. Res. Commun.* 245, 764–772, 1998). There were an earlier study from another institution (Brockman, A.H. and Orlando, R., New immobilization chemistry for probe affinity mass spectrometry, *Rapid Commun. Mass Spectrom.* 10, 1688–1692, 1996).

[b] See Melle, C., Ernst, G., Scheibner, O. et al., Identification of specific protein markers in microdissected hepatocellular carcinoma, *J. Proteome Res.* 6, 306–315, 2007.

[c] See Enose, Y., Destache, C.J., Mack, A.L. et al., Proteomic fingerprints distinguish microglia, bone marrow, and spleen macrophage populations, *Glia* 51, 161–172, 2005.

REFERENCES

1. Kornberg, A., Why purify enzymes?, *Methods Enzymol.* 182, 1–5, 1990.
2. Lasky, F.D. and Boser, R.B., Designing in quality through design control: A manufacturer's perspective, *Clin. Chem.* 43, 866–872, 1997.
3. Powers, D.M. and Greenberg, N., Development and use of analytical quality specifications in the in vitro diagnostics medical device industry, *Scand. J. Clin. Lab. Invest.* 59, 539–543, 1999.
4. Panteghini, M. and Forest, J.C., Standardization in laboratory medicine: New challenges, *Clin. Chim. Acta* 355, 1–12, 2005.
5. Burgess, C., Analytical quality management, in *Analytical Chemistry. A Modern Approach to Analytical Science*, 2nd edn., R. Kellner, J.-M. Mermet, M. Otto, M. Varacel, and H.M. Widmer (eds.), Wiley-VCH, Weinheim, Germany, 2004.
6. Molloy, M.P., Brzezinski, E.E., Hang, J. et al., Overcoming technical variations and biological variations in quantitative proteomics, *Proteomics* 3, 1912–1929, 2003.
7. Tammen, H., Specimen collection and handling: Standardization of blood sample collection, *Methods Mol. Biol.* 428, 35–42, 2008.
8. Favaloro, E.J., Lippi, G., and Adcock, D.M., Preanalytical and postanalytical variables: The leading causes of diagnostic errors in hemostasis?, *Semin. Thromb. Hemost.* 34, 612–634, 2008.
9. Rudež, G., Meijer, P., Spronk, H.M.M. et al., Biological variation in inflammatory and hemostatic markers, *J. Thromb. Haemost.* 7, 1247–1255, 2009.
10. Blomberg, A., Blomberg, L., Norbeck, J. et al., Intralaboratory reproducibility of yeast protein patters analyzed by immobilized pH gradient two-dimensional gel electrophoresis, *Electrophoresis* 16, 1935–1945, 1995.
11. Noonan, K., Kalu, M.E., Holownia, P., and Burrin, J.M., Effect of different storage temperature, sample collection procedures and immunoassay methods on osteocalcin measurements, *Eur. J. Clin. Chem. Clin. Biochem.* 34, 341–344, 1996.
12. Young, D.S. and Bermes, E.W.J., Specimen collection and process: Sources of biological variation, in *Tietz Textbook of Clinical Chemistry*, 3rd edn., C.A. Burris and E.R. Ashwood (eds.), W.B. Saunders, Philadelphia, PA, Chapter 2, pp. 42–72, 1999.
13. Zawda, B., The unexpected result: Fault of the laboratory? Traps in laboratory diagnostics, *Methods Find. Exp. Clin. Pharmacol.* 21, 65–67, 1999.
14. Dufour, D.R., Sources and control of preanalytical variation, in *Clinical Chemistry Theory, Analysis. Correlation*, 4th edn., L.A. Kaplan, A.J. Pesce, and S.C. Kazmierczak (eds.), Mosby, St. Louis, MO, Chapter 3, pp. 64–82, 2003.
15. Ahmad, S., Sundaramoorthy, E., Arora, R. et al., Progressive degradation of serum samples limits proteomic marker discovery, *Anal. Biochem.* 394, 237–242, 2009.
16. Apweiler, R., Aslandix, C., Deufel, T. et al., Approaching clinical proteomics: Current state and future fields of application in cellular proteomics, *Cytometry A* 75A, 816–832, 2009.
17. Turk, M.K., Chan, D.W., Chia, D. et al., Standard operating procedures for serum and plasma collection: Early detection network consensus statement standard operating procedure integration working group, *J. Proteome Res.* 8, 113–117, 2009.
18. Page, I.H., Blood—The circulatory computer tape, *Perspect. Biol. Med.* 15, 219–220, 1972.
19. Koepke, J.A., Specimen collection—Cellular hematology, in *Laboratory Hematology*, Vol. 2, J.A. Koepke (ed.), Churchill Livingstone, New York, Chapter 32, pp. 821–831, 1984.
20. Thompson, J.M., Specimen collection for blood coagulation testing, in *Laboratory Hematology*, Vol. 2, J.A. Keopke (ed.), Churchill Livingstone, New York, Chapter 33, p. 846, 1984.

21. Rommel, K., Koch, K.-D., and Spilker, K., Einfluss der Materialgewinnung auf Klinisch-chemische parameter in blut, plasma, und serum bei patienten mit stabilem und zentral-isiertem Kreislauf, *J. Clin. Chem. Clin. Biochem.* 16, 373–380, 1978.
22. Irjala, K.M. and Grönroos, P.E., Preanalytical and analytical factors affecting laboratory results, *Ann. Med.* 30, 267–272, 1998.
23. Statland, B.E., Bokelund, H., and Winkel, P., Factors contributing to intraindividual variation of serum constituents: 4. Effects of posture and tourniquet application on variation of serum constituents in health subjects, *Clin. Chem.* 20, 1513–1519, 1974.
24. Omote, M., Asakura, H., Takamichi, S. et al., Changes in molecular markers of hemostatic and fibrinolytic activation under various sampling conditions using vacuum tube samples from health volunteers, *Thromb. Res.* 123, 390–395, 2008.
25. Preissner, C.M., Reilly, W.M., Cyr, R.C. et al., Plastic versus glass tubes: Effects on analytical performance of selected serum and plasma hormone assays, *Clin. Chem.* 50, 1245–1246, 2004.
26. Wang, S.H., Ho, V., Roquemore-Goins, A., and Smith, F.A., Effects of blood collection tubes including pediatric devices on 16 common immunoassays, *Clin. Chem.* 52, 892–893, 2006.
27. Morovat, A., James, T.S., Cox, S.D. et al., Comparison of Bayer Advia Centaur® immunoassay results obtained on samples collected in four different Becton Dickinson Vacutainer® tubes, *Ann. Clin. Biochem.* 43, 481–487, 2006.
28. Hayward, R.M., Kirk, M.J., Sproull, M. et al., Post-collection, pre-measurement variables affecting VEGF levels in urine biospecimens, *J. Cell. Mol. Med.* 12, 343–350, 2008.
29. Evans, M.J., Livesey, J.H., Ellis, M.J., and Yandle, T.O., Effect of anticoagulants and storage temperatures on stability of plasma and serum hormones, *Clin. Biochem.* 34, 107–112, 2001.
30. Prisco, D., Panicca, R., Bandinelli, B. et al., Euglobulin lysis time in fresh and stored samples, *Am. J. Clin. Pathol.* 102, 794–796, 1994.
31. Guder, W.G., Who cares about the stability of analytes?, *Eur. J. Clin. Chim. Biochem.* 33, 177, 1995.
32. Heins, M., Heil, W., and Withold, W., Storage of serum or whole blood samples? Effects of time and temperature on 22 serum analytes, *Eur. J. Clin. Chem. Biochem.* 33, 231–238, 1995.
33. Qvist, P., Munk, M., Hoyle, N., and Christianen, C., Serum and plasma fragments of C-telopeptides of type I collagen (CTX) are stable during storage at low temperatures for 3 years, *Clin. Chim. Acta* 350, 167–173, 2004.
34. Kioukia-Fouglia, N., Christofidis, I., and Strantzalis, N., Physicochemical conditions affecting the formation/stability of serum complexes and the determination of prostate-specific antigen (PSA), *Anticancer Res.* 19, 3315–3320, 1999.
35. White, J.G., EDTA-induced changes in platelet structure and function: Clot retraction, *Platelets* 11, 49–55, 2000.
36. Mikaelsson, M.E., Forsman, N., and Oswaldsson, U.M., Human factor VIII: A calcium-linked protein complex, *Blood* 62, 1006–1015, 1982.
37. Green, D., McMahon, B., Foiles, N., and Tien, L., Measurement of hemostatic factors in EDTA plasma, *Am. J. Clin. Pathol.* 130, 811–815, 2008.
38. Palmer, D.S., Rosborough, D., Perkins, H. et al., Characterization of factors affecting the stability of frozen heparinized plasma, *Vox Sang.* 65, 258–270, 1993.
39. Cecchi, F., Ruggiero, M., Cappelletti, R. et al., Improved method for analysis of glycosaminoglycans in glycosaminoglycan/protein mixtures: Application in Cohn-Oncley fractions of human plasma, *Clin. Chim. Acta* 376, 142–149, 2007.
40. Saito, A. and Munakata, H., Analysis of plasma proteins that bind to glycosaminoglycans, *Biochim. Biophys. Acta* 1770, 241–246, 2007.

41. Tucholska, M., Bowden, P., Jacks, K. et al., Human serum proteins fractionated by preparative partition chromatography prior to LC-ESI-MS/MS, *J. Proteome Res.* 8, 1143–1155, 2009.

42. Jin, Y. and Manabe, T., Difference in protein distribution between human plasma preparations, EDTA-plasma and heparin-plasma, analyzed by non-denaturing micro-2-DE and MALDI-MS PMF, *Electrophoresis* 30, 931–938, 2009.

43. Imai, H., Yamada, O., Morita, S. et al., Detection of HIV-1 RNA in heparinized plasma of HIV-1 seropositive individuals, *J. Virol. Methods* 36, 181–184, 1992.

44. Taylor, A.C., Titration of heparinase for removal of the PCR-inhibitory effect of heparin in DNA samples, *Mol. Ecol.* 6, 383–385, 1997.

45. Bai, X., Fischer, S., Keshavjee, S., and Liu, M., Heparin interference with reverse transcriptase polymerase chain reaction of RNA extracted from lungs after ischemia-reperfusion, *Transpl. Int.* 13, 146–150, 2000.

46. Glaum, M.C., Wang, Y., Raible, D.G., and Schulman, E.S., Degranulation influences heparin-associated inhibition of RT-PCR in human lung mast cells, *Clin. Exp. Allergy* 31, 1631–1635, 2001.

47. Nohara, O., Gilchrist, M., Déry, R.E. et al., Reverse transcriptase in situ polymerase chain reaction for gene expression in rat mast cells and macrophages, *J. Immunol. Methods* 226, 147–158, 1999.

48. Vjordjevic, V., Stankovic, M., Nikolic, A. et al., PCR amplification on whole blood samples treated with different commonly used anticoagulants, *Pediatr. Hematol. Oncol.* 23, 517–521, 2006.

49. Poli, F., Cattaneo, R., Crespiatico, L. et al., A rapid and simple method for reversing the inhibitory effect of heparin on PCR for HLA class II typing, *PCR Methods Appl.* 2, 356–358, 1993.

50. Hartman, L.J., Coyne, S.R., and Norwood, D.A., Development of a novel internal positive control for Taqman based assays, *Mol. Cell. Probes* 19, 51–59, 2005.

51. Del Prete, M.J., Vernal, R., Dolznig, H. et al., Isolation of polysome-bound mRNA from solid tissues amenable for RT-PCR and profiling experiments, *RNA* 13, 414–421, 2007.

52. Zimmermann, R., Ringwald, J., and Eckstein, R., EDTA plasma is unsuitable for *in vivo* determinations of platelet-derived angiogenic kinins, *J. Immunol. Methods* 347, 91–92, 2009.

53. Sharma, R., Mahadreswaraswamy, Y.H., Harish Kumar, K. et al., Effect of anticoagulant on plasma hyaluronidase activity, *J. Clin. Lab. Anal.* 23, 29–33, 2009.

54. Dittadi, R., Meo, S., Fabris, F. et al., Validation of blood coagulation procedures for the determination of circulating vascular endothelial growth factor (VEGF) in different blood compartments, *Int. J. Biol. Markers* 16, 87–96, 2001.

55. Ivandic, B.T., Spanuth, E., Hasse, D. et al., Increased plasma concentrations of soluble CD40 ligand I acute coronary syndrome depend on in vitro platelet activation, *Clin. Chem.* 53, 1231–1234, 2007.

56. Weber, M., Rabenau, B., Stanisch, M. et al., Influence of sample type and storage conditions on soluble CD40 ligand assessment, *Clin. Chem.* 52, 888–890, 2006.

57. Nilsson, T.K., Boman, K., Jansson, J.H. et al., Comparison of soluble thrombomodulin, von Willebrand factor, tPA/PAI-1 complex and high sensitivity CRP concentration in serum EDTA plasma, citrated plasma, and acidified citrated plasma (Stabilyte™) stored at −70°C for 8–11 years, *Thromb. Res.* 116, 249–254, 2005.

58. Hsieh, S.-Y., Chen, R.-K., Pan, Y.-H., and Lee, H.-L., Systematical evaluation of the effects of sample collection procedures on low-molecular-weight serum/plasma proteome profiling, *Proteomics* 6, 3189–3198, 2006.

59. Knudsen, L.S., Christensen, I.J., Lottenburger, T. et al., Pre-analytical and biological variability in circulating interleukin 6 in healthy subjects and patients with rheumatoid arthritis, *Biomarkers* 13, 59–78, 2008.

60. Mannello, F., Serum or plasma samples? The "Cinderella" role of blood coagulation procedures preanalytical methodological issues influence the release and activity of circulating matrix metalloproteinase and their tissue inhibitors, hampering diagnostic trueness and leading to misinterpretation, *Arterioscler. Thromb. Vasc. Biol.* 28, 611–614, 2008.

61. Lippi, G., Brocco, G., Salvagno, G.L. et al., Influence of the sample matrix on the stability of β-CTX at room temperature for 24 and 48 hours, *Clin. Lab.* 53, 455–459, 2007.

62. Yi, J., Kim, C., and Gelfand, C.A., Inhibition of intrinsic proteolytic activities moderates preanalytical variability and instability of human plasma, *J. Proteome Res.* 6, 1768–1781, 2007.

63. Tammen, H., Schulte, I., Hess, R. et al., Peptidomic analysis of human blood specimens: Comparison between plasma specimens and serum by differential peptide display, *Proteomics* 5, 3414–3422, 2005.

64. Budtz-Olsen, O.E., *Clot Retraction*, C.C. Thomas, Springfield, IL, 1951.

65. MacFarlane, R.G., A single method for maximum clot retraction, *Lancet* I, 236, 1199–1201, 1939.

66. Luque-Garcia, J.L. and Neubert, T.A., Sample preparation for serum/plasma profiling and biomarker identification by mass spectrometry, *J. Chromatogr. A* 1153, 259–276, 2007.

67. Govorukhina, N.I., de Vries, M., Reijmers, T.H. et al., Influence of clotting time on the protein composition of serum samples based on LC-MS data, *J. Chromatogr. A* 877, 1281–1291, 2009.

68. Timms, J.F., Arslan-Low, E., Gentry-Maharaj, A. et al., Preanalytical influence of sample handling on SELDI-TOF serum protein profiles, *Clin. Chem.* 53, 645–656, 2007.

69. Ferry, J.D., Protein gels, *Adv. Protein Chem.* 4, 1–78, 1948.

70. Scheraga, H.A. and Laskowski Jr. M., The fibrinogen-fibrin conversion, *Adv. Protein Chem.* 12, 1–131, 1957.

71. Faulk, W.P., Torny, D.S., and McIntyre, J.A., Effects of serum versus plasma on agglutination of antibody-coated indicator cells by human rheumatoid factor, *Clin. Immunol. Immunopathol.* 46, 169–176, 1988.

72. Adkins, J.N., Varnum, S.M., Auberry, K.J. et al., Toward a human blood serum proteome. Analysis by multidimensional separation coupled with mass spectrometry, *Mol. Cell. Proteomics* 1, 947–955, 2002.

73. Lum, G. and Gambino, S.R., A comparison of serum *versus* heparinized plasma for routine clinical tests, *Am. J. Clin. Pathol.* 61, 108–113, 1974.

74. Ladenson, J.H., Tsai, L.-M., Michael, J.M., Kessler, G., and Joist, J.H., Serum *versus* heparinized plasma for eighteen common chemistry tests, *Am. J. Clin. Pathol.* 62, 545–552, 1974.

75. George, J.N., Thai, L.L., McManus, L.M., and Reiman, T.A., Isolation of human platelet membrane microparticles from plasma serum, *Blood* 60, 834–840, 1982.

76. Levine, R.B. and Rebellino, E.M., Platelet glycoprotein-IIB and glycoprotein-IIIa associated with blood monocytes are derived from platelets, *Blood* 67, 207–219, 1986.

77. Lindemann, S., Tolley, N.D., Dixon, P.A. et al., Activated platelets mediate inflammatory signaling by regulated interleukin 1beta synthesis, *J. Cell. Biol.* 154, 485–490, 2001.

78. Benoy, I., Salgado, R., Colpaert, C. et al., Serum interleukin 6, plasma VEGF, serum VEGF, and VEGF platelet load in breast cancer patients, *Clin. Breast Cancer* 2, 311–315, 2002.

79. Spence, G.M., Graham, A.N., Mulholland, K. et al., Vascular endothelial growth factor levels in serum and platelets following esophageal cancer resection—Relationship to platelet count, *Int. J. Biol. Markers* 17, 119–124, 2002.

80. Boyanton Jr. B.L. and Blick, K.E., Stability studies of twenty-four analytes in human plasma and serum, *Clin. Chem.* 48, 2242–2247, 2002.

81. Daves, M., Trevisan, D., and Cemin, R., Different collection tubes in cardiac biomarkers detection, *J. Clin. Lab. Anal.* 22, 391–394, 2008.
82. Dominici, R., Infusino, I., Valente, C. et al., Plasma or serum samples: Measurement of cardiac troponin T and of other analytes compared, *Clin. Chem. Lab. Med.* 42, 945–951, 2004.
83. Kingdon, H.S. and Lundblad, R.L., Factors affecting the evolution of factor XIa during blood coagulation, *J. Lab. Clin. Med.* 85, 826–831, 1976.
84. Heppinstall, S., The use of a chelating ion-exchange resin to evaluate the effects of the extracellular calcium concentration on adenosine diphosphate induced aggregation of human blood platelets, *Thromb. Haemost.* 36, 208–220, 1976.
85. Venelinov, T.I., Davies, I.M., and Beattie, J.H., Dialysis-Chelex method for determination of exchangeable copper in human plasma, *Anal. Bioanal. Chem.* 379, 777–780, 2004.
86. Bierau, J., Bakker, J.A., Lindhout, M., and van Gennip, A.H., Determination of ITPase in erythrocyte lysates obtained for determination of TPMT activity, *Nucleosides Nucleotides Nucleic Acids* 25, 1129–1132, 2006.
87. Dardenne, M., Pléau, J.M., Nabarra, B. et al., Contribution of zinc and other metals to the biological activity of the serum thymic factor, *Proc. Natl. Acad. Sci. USA* 79, 5370–5373, 1982.
88. Bertolero, F., Kaighn, M.E., Camalier, R.F., and Saffiotti, U., Effects of serum and serum-derived factors of growth and differentiation of mouse keratinocytes, *In Vitro Cell. Dev. Biol.* 22, 423–428, 1986.
89. Carpentieri, U., Myers, J., Daeschner III W., and Haggard, M.E., Observations on the use of a cation exchange resin for the preparation of metal-depleted media for lymphocyte culture, *J. Biochem. Biophys. Methods* 14, 93–100, 1987.
90. Papale, M., Pedicillo, M.C., Thatcher, B.J. et al., Urine profiling by SELDI-TOF/MS: Monitoring of the critical steps in sample collection, handling and analysis, *J. Chromatogr. B* 856, 205–213, 2007.
91. Wechuck, J.B., Goins, W.F., Glorioso, J.C. et al., Effect of protease inhibitors on yield of HSF-1-based viral vectors, *Biotechnol. Prog.* 16, 493–496, 2000.
92. Price, C.P., Newall, R.G., and Boyd, J.C., Use of protein: Creatinine ratio measurements on random urine samples for prediction of significant proteinuria: A systematic review, *Clin. Chem.* 51, 1577–1586, 2005.
93. Lamb, E., Newman, D.J., and Price, C.P., Kidney function tests, in *Tietz Textbook of Clinical Chemistry and Molecular Diagnostics*, 4th edn., C.A. Burtis, E.R. Ashwood, and D.E. Burns (eds.), Saunders, St. Louis, MO, Chapter 24, pp. 797–835, 2006.
94. McIntyre, N.J. and Taal, M.W., How to measure proteinuria?, *Curr. Opin. Nephrol. Hypertens.* 17, 600–603, 2008.
95. Arndt, T., Urine-creatinine concentrations as a marker of urine dilution: Reflections using a cohort of 45,000 samples, *Forensic Sci. Int.* 186, 48–51, 2009.
96. Kuusimäki, L., Peltonen, Y., Mutanen, P. et al., Urinary hydroxy-metabolites of naphthalene, phenanthrene and pyrene as markers of exposure to diesel exhaust, *Int. Arch. Occup. Environ. Health* 77, 23–30, 2004.
97. Meeker, J.D., Sathyanarayana, S., and Swan, S.H., Phthalates and other additives in plastics: Human exposure and associated health outcomes, *Philos. Trans. R. Soc. Lond. B. Biol. Sci.* 364, 2097–2113, 2009.
98. Basilicata, P., Miraglia, N., Pieri, M. et al., Application of the standard addition approach for the quantification of urinary benzene, *J. Chromatogr. B. Analyt. Technol. Biomed. Life Sci.* 818, 293–299, 2005.
99. Creely, K.S., Hughson, G.W., Cocker, J., and Jones, K., Assessing isocyanate exposures in polyurethane industry sectors using biological and air monitoring methods, *Ann. Occup. Hyg.* 50, 609–621, 2006.

100. Fustinoni, S., Mercadante, R., and Campo, L., Self-collected urine sampling to study the kinetics of urinary toluene (and *o*-cresol) and define the best sampling time for biomonitoring, *Int. Arch. Occup. Environ. Health* 82, 703–713, 2009.

101. Sobus, J.R., McLcean, M.D., Herrick, R.F. et al., Comparing urinary biomarkers of airborne and dermal exposure to polycyclic aromatic compounds in asphalt-exposed workers, *Ann. Occup. Hyg.* 53, 561–571, 2009.

102. Lowe, F.J., Gregg, E.O., and McEwan, M., Evaluation of biomarkers of exposure and potential harm in smokers, former smokers and never-smokers, *Clin. Chem. Lab. Med.* 47, 311–320, 2009.

103. Cobanoglu, N., Kiper, N., Dilber, E. et al., Environmental tobacco smoke exposure and respiratory morbidity in children, *Inhal. Toxicol.* 19, 779–785, 2007.

104. Ducos, P., Berode, M., Francin, J.M. et al., Biological monitoring of exposure to solvents using the chemical itself in urine: Application to toluene, *Int. Arch. Occup. Environ. Health* 81, 273–284, 2007.

105. Craig, A., Cloarec, O., Holmes, E. et al., Scaling and normalization effects in NMR spectroscopic metanomic data sets, *Anal. Chem.* 78, 2262–2267, 2006.

106. Beckonert, O., Keun, H.C., Ebbels, T.M. et al., Metabolic profiling, metabolomic and metabonomic procedures for NMR spectroscopy of urine, plasma, serum and tissue extracts, *Nat. Protoc.* 2, 2692–2703, 2007.

107. Kim, K., Aronov, P., Zakarkin, S.O. et al., Urine metabolomics analysis for kidney cancer detection and biomarker discovery, *Mol. Cell. Proteomics* 8, 558–570, 2009.

108. Walsh, M.C., Brennan, L., Malthouse, J.P. et al., Effect of acute dietary standardization on the urinary, plasma, and salivary metabolomic profiles of healthy humans, *Am. J. Clin. Nutr.* 84, 531–539, 2006.

109. Smith, S.M., Dillon, E.L., DeKerlegand, D.E., and Davis-Street, J.E., Variability of collagen crosslinks: Impact of sample collection period, *Calcif. Tissue Int.* 74, 336–341, 2004.

110. Gerber, L.M., Schwartz, J.E., and Pickering, T.G., Albumin-to-creatinine ratio predicts change in ambulatory blood pressure in normotensive persons: A 7.5 year prospective study, *Am. J. Hypertens.* 19, 220–226, 2006.

111. Gansevoort, R.T., Brinkman, J., Bakker, S.J. et al., Evaluation of measures of urinary albumin excretion, *Am. J. Epidemiol.* 164, 725–727, 2006.

112. Cai, T., Mazzoli, S., Meacci, F., Interleukin-6/10 ratio as a prognostic marker of recurrence in patients with intermediate risk urothelial bladder cancer, *J. Urol.* 178, 1906–1911, 2007.

113. Hegedus, C.M., Skibola, C.F., Warner, M. et al., Decreased urinary beta-defensin-1 expression as a biomarker of response to arsenic, *Toxicol. Sci.* 106, 74–82, 2008.

114. Trachtenberg, F., Barregard, L., and McKinlay, S., The influence of urinary flow rate in children on excretion of markers used for assessment of renal damage: Albumin, γ-glutamyl transpeptidase, *N*-acetyl-β-D-glucosaminidase, and alpha1-microglobulin, *Pediatr. Nephrol.* 23, 445–456, 2008.

115. Pickering, A., Carter, J., Hanning, I., and Townend, W., Emergency department measurement of urinary S100B in children following head injury: Can extracranial injury confound findings?, *Emerg. Med. J.* 25, 88–89, 2008.

116. Jamshidian, H., Kor, K., and Djalali, M., Urine concentration of nuclear matrix protein 22 for diagnosis of transitional cell carcinoma of bladder, *Urol. J.* 5, 243–247, 2008.

117. Quintana, L.F., Solé-Gonzalez, A., Kalko, S.G. et al., Urine proteomics to detect biomarkers for chronic allograft dysfunction, *J. Am. Soc. Nephrol.* 20, 236–235, 2009.

118. Thongboonkerd, V., Mungdee, S., and Chiangjong, W., Should urine pH be adjusted prior to gel-based proteome analysis?, *J. Proteome Res.* 8, 3206–3211, 2009.

119. Palmer, S.C., Endre, Z.H., Richards, A.M., and Yandle, T.G., Characterization of NT-proBNP in human urine, *Clin. Chem.* 55, 1126–1134, 2009.

120. Machtejevas, E., Marko-Varga, G., Lindberg, C. et al., Profiling of endogenous peptides by multidimensional liquid chromatography: On-line automated sample cleanup for biomarker discovery in human urine, *J. Sep. Sci.* 32, 2223–2232, 2009.

121. Siew, E.D., Ware, L.B., Gebretsadik, T. et al., Urine neutrophil gelatinase-associated lipocalin moderately predicts acute kidney injury in critically ill adults, *J. Am. Soc. Nephrol.* 20, 1823–1832, 2009.

122. Theodorescu, D. and Mischak, H., Mass spectrometry based proteomics in urine biomarker discovery, *World J. Urol.* 25, 435–443, 2007.

123. Thongboonkerd, V., Practical points in urinary proteomics, *J. Proteome Res.* 6, 3881–3890, 2007.

124. Decramer, S., Gonzalez de Peredo, A., Breuil, B. et al., Urine in clinical proteomics, *Mol. Cell. Proteomics* 7, 1850–1862, 2008.

125. Vaezzadeh, A.R., Steen, H., Freeman, M.R., and Lee, R.S., Proteomics and opportunities for clinical translation in urological disease, *J. Urol.* 182, 835–843, 2009.

126. Wu, L.L., Chiou, C.C., Chang, P.Y., and Wu, J.T., Urinary 8-OHdG: A marker of oxidative stress to DNA and a risk factor for cancer, atherosclerosis and diabetes, *Clin. Chim. Acta* 339, 1–9, 2004.

127. O'Driscoll, L., Extracellular nucleic acids and their potential as diagnostic, prognostic and predictive biomarkers, *Anticancer Res.* 27, 1257–1265, 2007.

128. Lodde, M. and Fradet, Y., The detection of genetic markers of bladder cancer in urine and serum, *Curr. Opin. Urol.* 18, 499–503, 2008.

129. Simpson, R.J., Lim, J.W., Moritz, R.L., and Mathivanan, S., Exosomes: Proteomic insights and diagnostic potential, *Expert Rev. Proteomics* 6, 267–283, 2009.

130. Ben-Aryeh, H., Miron, D., Szargel, R., and Gutman, D., Whole-saliva secretion rates in old and young healthy subjects, *J. Dent. Res.* 63, 1147–1148, 1984.

131. Ben-Aryeh, H., Shalev, A., Szargel, R. et al., The salivary flow rate and composition of whole and parotid resting and stimulated saliva in young and old healthy subjects, *Biochem. Med. Metab. Biol.* 36, 260–265, 1986.

132. Shern, R.J., Fox, P.C., and Li, S.H., Influence of age of the secretory rates of the human minor salivary glands and whole saliva, *Arch. Oral Biol.* 38, 755–761, 1993.

133. Bots, C.P., Brand, H.S., Veerman, E.C. et al., Preferences and saliva stimulation of eight different chewing gums, *Int. Dent. J.* 54, 143–148, 2004.

134. Aframian, D.J., Helcer, M., Livni, D., and Markitzui, A., Pilocarpine for the treatment of salivary glands' impairment caused by radioiodine therapy for thyroid cancer, *Oral Dis.* 12, 297–300, 2006.

135. Sevón, L., Laine, M.A., Karjalainen, S. et al., Effect of age on flow-rate, protein and electrolyte composition of stimulated whole saliva in healthy, non-smoking women, *Open Dent. J.* 2, 89–92, 2008.

136. Yamamoto, K., Kurihara, M., Matsusue, Y. et al. Whole saliva flow rate and body profile in healthy young adults, *Arch. Oral Biol.* 54, 464–467, 2009.

137. Rayment, S.A., Liu, B., Soares, R.V. et al., The effects of duration and intensity of stimulation on total protein and mucin concentration in resting and stimulated whole saliva, *J. Dent. Res.* 80, 1584–1587, 2001.

138. Curtis, M.A., Gilllett, I.R., Griffiths, G.S. et al., Detection of high-risk groups and individuals for periodontal diseases: Laboratory markers from analysis of gingival crevicular fluid, *J. Clin. Periodont.* 16, 1–11, 1989.

139. Ozmeric, N., Advances in periodontal disease markers, *Clin. Chim. Acta* 343, 1–16, 2004.

140. Taba Jr. M., Kinney, J., Kim, A.S., and Giannobile, W.V., Diagnostic biomarkers for oral and periodontal diseases, *Dent. Clin. North Am.* 49, 551–571, 2005.

141. Lamster, I.B. and Ahlo, J.K., Analysis of gingival crevicular fluid as applied to the diagnosis of oral and systemic diseases, *Ann. N. Y. Acad. Sci.* 1098, 216–229, 2007.

142. Garito, M.L., Prihoda, T.J., and McManus, L.M., Salivary PAF levels correlate with the severity of periodontal inflammation, *J. Dent. Res.* 74, 1048–1056, 1995.
143. Lamster, I.B., Kaufman, E., Grbic, J.T. et al., β-Glucuronidase activity in saliva: Relationship to clinical periodontal parameters, *J. Periodontol.* 74, 353–359, 2003.
144. Miller, C.S., King Jr. C.P., Langub, M.C. et al., Salivary biomarkers of existing periodontal disease: A cross-sectional study, *J. Am. Dent. Assoc.* 137, 322–329, 2006.
145. Inzitari, R., Cabras, T., Pisano, E. et al., HPLC-ESI-MS analysis of oral human fluids reveals that gingival crevicular fluid is the main source of oral thymosins β_4 and β_{10}, *J. Sep. Sci.* 32, 57–63, 2009.
146. Mandel, I.D. and Wotman, S., The salivary secretions in health and disease, *Oral Sci. Rev.* 8, 25–47, 1976.
147. Schaeffer, M.E., Rhodes, M., Prince, S. et al., A plastic intraoral device for the collection of human parotid saliva, *J. Dent. Res.* 56, 728–733, 1977.
148. Ericson, T. and Nordlund, A., A new device of collection of parotid saliva, *Ann. N. Y. Acad. Sci.* 694, 274–275, 1977.
149. Veerman, E.C., van den Keybus, P.A., Vissink, A., and Nieuw Amerogen, A.V., Human glandular salivas: Their separate collection and analysis, *Eur. J. Oral. Sci.* 104, 346–352, 1996.
150. Fischer, D. and Ship, J.A., Effect of age on variability of parotid salivary gland flow rates over time, *Age Ageing* 28, 557–561, 1999.
151. Ghezzi, E.M., Lange, L.A., and Ship, J.A., Determination of variation of stimulated salivary flow rates, *J. Dent. Res.* 79, 1874–1878, 2000.
152. Nagler, R.M., Hershkovich, O., Lischinsky, S. et al., Saliva analysis in the clinical setting: Revisiting an underused diagnostic tool, *J. Investig. Med.* 50, 214–225, 2002.
153. Kaufman, E. and Lamster, I.B., The diagnostic applications of saliva—A review, *Crit. Rev. Oral Biol. Med.* 13, 197–212, 2002.
154. Jenzano, J.W., Courts, N.F., Timko, D.A., and Lundblad, R.L., Levels of glandular kallikrein in whole saliva obtained from patients with solid tumors remote from the oral cavity, *J. Dent. Res.* 65, 67–70, 1986.
155. Tarui, H. and Nakamura, A., Saliva cortisol: A good indicator for acceleration stress, *Aviat. Space Environ. Med.* 58, 573–575, 1987.
156. Kirschbaum, C. and Hellhammer, D.H., Salivary cortisol in psychobiological research: An overview, *Neuropsychobiology* 22, 150–169, 1989.
157. Hellhammer, D.H., Wüst. S., and Kudielka, B.M., Salivary cortisol as a biomarker in stress research, *Psychoneuroendocrinology* 34, 163–171, 2009.
158. Aggreger, A.L., Cardoso, E.M., Tumilasci, O., and Contreras, L.N., Diagnostic value of salivary cortisol in end stage renal disease, *Steroids* 73, 77–82, 2008.
159. Walker, R.F., Wilson, D.W., Read, G.F., and Riad-Fahmy, D., Assessment of testicular function by the radioimmunoassay of testosterone in saliva, *Int. J. Androl.* 3, 105–120, 1980.
160. Navarro, M.A., Aquiló, F., Villabona, C.M. et al., Salivary testosterone in prostatic carcinoma, *Br. J. Urol.* 63, 306–308, 1989.
161. Granger, D.A., Shirtcliff, E.A., Booth, A. et al., The "trouble" with salivary testosterone, *Psychoneuroendocrinology* 29, 1229–1240, 2004.
162. Aggreger, A.L., Contreras, L.N., Tumilasci, O.R. et al., Salivary testosterone: A reliable approach to the diagnosis of male hypogonadism, *Clin. Endocrinol.* 67, 656–662, 2007.
163. Ferrari, M., Negri, S., Zadra, P. et al., Saliva as an analytical tool to measure occupational exposure to toluene, *Int. Arch. Occup. Environ. Health* 81, 1021–1028, 2008.
164. Qvarnstrom, M., Janket, S., Jones, J.A. et al., Salivary lysozyme and prevalent hypertension, *J. Dent. Res.* 87, 480–484, 2008.
165. Bassim, C.W., Redman, R.S., DeNucci, D.J. et al., Salivary procalcitonin and periodontitis in diabetes, *J. Dent. Res.* 87, 630–634, 2008.

166. Cardoso, E.M., Arregger, A.L., Tumilasci, O.R. et al., Assessment of salivary urea as a less invasive alternative to serum determinations, *Scand. J. Clin. Lab. Invest.* 69, 330–334, 2009.

167. Nater, U.M. and Rohleder, N., Salivary α-amylase as a non-invasive biomarker for the sympathetic nervous system: Current state of research, *Psychoneuroendocrinology* 34, 486–496, 2009.

168. Schapher, M., Wendler, O., Gröschl, M. et al., Salivary leptin as a candidate diagnostic marker in salivary gland tumors, *Clin. Chem.* 55, 914–922, 2009.

169. He, H., Chen, G., Zhou, L., and Liu, Y., A joint detection of CEA and CA-50 levels in saliva and serum of patients with tumors in oral region and salivary gland, *J. Cancer Res. Clin. Oncol.* 135, 1315–1321, 2009.

170. Osaka, I., Kurihara, Y., Tanaka, K. et al., Endocrinology evaluations of brief hand masses in palliative care, *J. Altern. Complement Med.* 15, 981–985, 2009.

171. Sapan, C.V., Lundblad, R.L., and Price, N.C., Colorimetric protein assay techniques, *Biotechnol. Appl. Biochem.* 29, 99–108, 1999.

172. Jenzano, J.W., Hogan, S.L., Noyes, C.M. et al., Comparison of five techniques for the determination of protein content in mixed human saliva, *Anal. Biochem.* 159, 370–376, 1986.

173. Eppel, G.A., Nagy, S., Jenkins, M.A. et al., Variability of standard clinical protein assays in the analysis of a model urine solution of fragmented albumin, *Clin. Biochem.* 33, 487–494, 2000.

174. Greive, K.A., Balazs, N.D., and Comper, W.D., Protein fragments in urine have been considerably underestimated by various protein fragments in urine have been considerably underestimated by various protein assays, *Clin. Chem.* 47, 1717–1719, 2001.

175. Miller, W.G., Bruns, D.E., Hortin, G.L. et al., Current issues in measurement and reporting urinary albumin excretion, *Clin. Chem.* 55, 24–38, 2009.

176. Schipper, R.G., Silletti, E., and Vingerhoeds, M.H., Saliva as research material: Biochemical, physicochemical and practical aspects, *Arch. Oral Biol.* 52, 1114–1135, 2007.

177. Hu, S., Loo, J.A., and Wong, D.T., Human saliva proteome analysis and disease biomarker discovery, *Expert Rev. Proteomics* 4, 531–538, 2007.

178. Hu, S., Yu, T., Xie, Y. et al., Discovery of oral fluid biomarkers for human oral cancer by mass spectrometry, *Cancer Genomics Proteomics* 4, 55–64, 2007.

179. Segal, A. and Wong, D.T., Salivary diagnostics: Enhancing disease detection and making medicine better, *Eur. J. Dent. Educ.* 12 (Suppl 1), 22–29, 2008.

180. Hu, S., Arellano, M., Bootheung, P. et al., Salivary proteomics for oral cancer biomarker discovery, *Clin. Cancer Res.* 14, 6246–6252, 2008.

181. Al-Tarawneh, S.K. and Bencharit, S., Applications of surface-enhanced laser desorption/ionization time-of-flight (SELDI-TOF) mass spectrometry in defining salivary proteomic profiles, *Open Dent. J.* 3, 74–79, 2009.

182. Zhang, L., Xiao, H., and Wong, D.T., Salivary biomarkers for clinical applications, *Mol. Diagn. Ther.* 13, 245–259, 2009.

183. Edgar, W.M., Saliva: Its secretion, composition and functions, *Br. Dent. J.* 172, 305–312, 1992.

184. Schenkels, L.C.P.M., Veerman, E.C.I., and Nieuw Amerongen, A.V., Biochemical composition of human saliva in relation to other mucosal fluids, *Crit. Rev. Oral Biol. Med.* 6, 161–175, 1995.

185. Humphrey, S.P. and Williamson, R.T., A review of saliva: Normal composition, flow, and function, *J. Prosthet. Dent.* 85, 162–169, 2001.

186. Aps, J.K. and Martens, L.C., Review: The physiology of saliva and transfer of drugs into saliva, *Forensic Sci. Int.* 150, 119–131, 2005.

187. de Almeida Pdel, V., Grégio, A.M., Machado, M.A. et al., Saliva composition and functions: A comprehensive review, *J. Contemp. Dent. Pract.* 9, 72–80, 2008.

188. Skopp, G. and Pötsch, L., Perspiration versus saliva—Basic aspects concerning their use in roadside drug testing, *Int. J. Legal Med.* 112, 213–221, 1999.
189. Dasgupta, A., Clinical utility of free drug monitoring, *Clin. Chem. Lab. Med.* 40, 986–993, 2002.
190. Crouch, D.J., Oral fluid collection: The neglected variable in oral fluid testing, *Forensic Sci. Int.* 150, 165–173, 2005.
191. Cone, E.J. and Huestis, M.A., Interpretation of oral fluid tests for drugs of abuse, *Ann. N. Y. Acad. Sci.* 1098, 51–103, 2007.
192. Drummer, O.H., Introduction and review of collection techniques and applications of drug testing of oral fluid, *Ther. Drug. Monit.* 30, 203–206, 2008.
193. Gallardo, E. and Queiroz, J.A., The role of alternative specimens in toxicological analysis, *Biomed. Chromatogr.* 22, 795–821, 2008.
194. Sutherland, A.D., Gearry, R.B., and Frizelle, F.A., Review of fecal biomarkers in inflammatory bowel disease, *Dis. Colon. Rectum* 51, 1283–1291, 2008.
195. Foell, D., Wittkowski, H., and Roth, J., Monitoring disease activity by stool analyses: From occult blood to molecular markers of intestinal inflammation and damage, *Gut* 58, 859–868, 2009.
196. Pearson, J.R., Gill, C.I., and Rowland, I.R., Diet, fecal water, and colon cancer—Development of a biomarker, *Nutr. Rev.* 67, 509–526, 2009.
197. Dal Pont, E., D'Incà, R., Caruso, A., and Sturniolo, G.C., Non-invasive investigation in patients with inflammatory joint disease, *World J. Gastroenterol.* 15, 2463–2468, 2009.
198. Babu, S., Mohapatra, S., Zubkov, L. et al., A PMMA microcapillary quantum dot linked immunosorbent assay (QLISA), *Biosens. Bioelectron.* 24, 3467–3474, 2009.
199. Hunt, J., Exhaled breath condensate: An overview, *Immunol. Allergy Clin. North Am.* 27, 587–596, 2007.
200. Syslová, K., Kačer, P., Kuzma, M. et al., Rapid and easy method for monitoring oxidative stress markers in body fluids of patients with asbestos or silica-induced lung diseases, *J. Chromatogr. B* 877, 2477–2486, 2009.
201. Webster, H.L., Laboratory diagnosis of cystic fibrosis, *Crit. Rev. Clin. Lab. Sci.* 18, 313–338, 1983.
202. Asahina, M., Yamanaka, Y., Akaogi, Y. et al., Measurements of sweat response and skin vasomotor reflex for assessment of autonomic dysfunction in patients with diabetes, *J. Diabetes Complications* 22, 278–283, 2008.
203. Petrofsky, J.S. and McLellan, K., Galvanic skin resistance—A marker for endothelial damage in diabetes, *Diabetes Technol. Ther.* 11, 461–467, 2009.
204. Bosman, F.T., Some recent developments in immunocytochemistry, *Histochem. J.* 15, 189–200, 1983.
205. Fritz, P., Multhaupt, H., Hoenes, J. et al., Quantitative immunohistochemistry. Theoretical background and its application in biology and surgical pathology, *Prog. Histochem. Cytochem.* 24, 1–53, 1992.
206. Goldschmidt, D., Decaestecker, C., Berthe, J.V. et al., The contribution of image cytometry and artificial intelligence-related methods of numerical data analysis for adipose tumor histopathologic classification, *Lab. Invest.* 75, 295–306, 1996.
207. Rabouille, C., Quantitative aspects of immunogold labeling in embedded and nonembedded sections, *Methods Mol. Biol.* 117, 125–144, 1999.
208. Santella, R.M., Immunological methods for detection of carcinogen-DNA damage in humans, *Cancer Epidemiol. Biomarkers Prev.* 8, 733–739, 1999.
209. Raina, A.K., Petty, G., Nunomura, A. et al., Histochemical and immunocytochemical approaches to the study of oxidative stress, *Clin. Chem. Lab. Med.* 38, 93–97, 2000.
210. Hanna, W., Testing for HER2 status, *Oncology* 61(Suppl 2), 22–30, 2001.
211. Lewis, F., Jackson, P., Lane, S. et al., Testing for HER2 in breast cancer, *Histopathology* 45, 207–217, 2004.

212. Taylor, C.R. and Levenson, R.M., Quantification of immunohistochemistry—Issues concerning methods, utility and semiquantitative assessment II, *Histopathology* 49, 411–424, 2006.
213. Cregger, M., Berger, A.J., and Rimm, D.L., Immunohistochemistry and quantitative analysis of protein expression, *Arch. Pathol. Lab. Med.* 130, 1026–1030, 2006.
214. D'Amico, F. and Skarmoutsou, E., Quantifying immunogold labelling in transmission electron microscopy, *J. Microsc.* 230, 9–15, 2008.
215. Peterson, R.A., Krull, D.L., and Butler, L., Applications of laser scanning cytometry in immunohistochemistry and routine histopathology, *Toxicol. Pathol.* 36, 117–132, 2008.
216. Tholouli, E., Sweeney, E., Barrow, E. et al., Quantum dots light up pathology, *J. Pathol.* 216, 275–285, 2008.
217. Sullivan, C.A. and Chung, G.G., Biomarker validation: In situ analysis of protein expression using semiquantitative immunohistochemistry-based techniques, *Clin. Colorectal Cancer* 7, 172–177, 2008.
218. Mayhew, T.M. and Lucocq, J.M., Developments in cell biology for quantitative immunoelectron microscopy based on thin sections: A review, *Histochem. Cell Biol.* 130, 299–313, 2008.
219. Chi, V. and Chandy, K.G., Immunohistochemistry: Paraffin sections using the Vectastain ABC kit from vector labs, *J. Vis. Exp.* 8, 308, 2007.
220. Kiernan, J.A., Indigogenic substrates for detection and localization of enzymes, *Biotech. Histochem.* 82, 73–103, 2007.
221. Hopkins, C., Gibson, A., Stinchcombe, J. et al., Chimeric molecules employing horseradish peroxidase as reporter enzyme for protein localization in the electron microscope, *Methods Enzymol.* 327, 35–45, 2000.
222. Fritz, P., Wu, X., Tuczek, H. et al., Quantitation in immunohistochemistry. A research method or a diagnostic tool in surgical pathology?, *Pathologica* 87, 300–309, 1995.
223. Sweeney, E., Ward, T.H., Gray, N. et al., Quantitative multiplexed quantum dot immunohistochemistry, *Biochem. Biophys. Res. Commun.* 374, 181–186, 2008.
224. Li, R., Dai, H., Wheeler, T.M. et al., Prognostic value of Akt-1 in human prostate cancer: A computerized quantitative assessment with quantum dot technology, *Clin. Cancer Res.* 15, 3568–3573, 2009.
225. Snyder, E.L., Bailey, D., Shipitsin, M. et al., Identification of CD44v6+/CD24-breast carcinoma cells in primary human tumors by quantum dot-conjugated antibodies, *Lab. Invest.* 89, 857–866, 2009.
226. Hogrefe, W.R., Biomarkers and assessment of vaccine responses, *Biomarkers* 10 (Suppl 1), S50–S57, 2005.
227. Bakir, M., Millington, K.A., Soysal, A. et al., Prognostic value of a T-cell-based, interferon-gamma biomarker in children with tuberculosis contact, *Ann. Int. Med.* 149, 777–787, 2009.
228. Elia, G., Silacci, M., Scheurer, S. et al., Affinity-capture reagents for protein arrays, *Trends Biotechnol.* 20 (12 Suppl), S19–S22, 2002.
229. Becker, K.F., Schott, C., Hipp, S. et al., Quantitative protein analysis from formalin-fixed tissues: Implications for translational clinical research and nanoscale molecular diagnostics, *J. Pathol.* 211, 370–378, 2007.
230. Chergui, F., Chrétien, A.S., Bouali, S. et al., Validation of a phosphoprotein array assay for characterization of human tyrosine kinase receptor downstream signaling in breast cancer, *Clin. Chem.* 55, 1327–1336, 2009.
231. Aguilar-Maheca, A., Hassan, S., Ferrario, C., and Basik, M., Microarrays as validation strategies in clinical samples: Tissue and protein microarrays, *OMICS* 10, 311–326, 2006.
232. Bayani, J. and Squire, J.A., Application and interpretation of FISH in biomarker studies, *Cancer Lett.* 249, 97–109, 2007.

233. Spisak, S., Tulassay, Z., Molnar, B., and Guttman, A., Protein microchips in biomedicine and biomarker discovery, *Electrophoresis* 28, 4261–4273, 2007.
234. Das, K., Mohd Omar, M.F., Ong, C.W. et al., TRARESA: A tissue microarray-based hospital system for biomarker validation and discovery, *Pathology* 40, 441–449, 2008.
235. Osunkoya, A.O., Yin-Goen, Q., Phan, J.H. et al., Diagnostic biomarkers for renal cell carcinoma: Selection using novel bioinformatics systems for microarray data analysis, *Hum. Pathol.* 40, 1671–1678, 2009.
236. Gozal, Y.M., Cheng, D., Duong, D.M. et al., Merger of laser capture microdissection and mass spectrometry: A window into the amyloid plaque proteome, *Methods Enzymol.* 412, 77–93, 2006.
237. Hashimoto, A., Matsui, T., Tanaka, S. et al., Laser-mediated microdissection for analysis of gene expression in synovial tissue, *Mod. Rheumatol.* 17, 185–190, 2007.
238. Cheng, A.L., Huang, W.G., Chen, Z.C. et al., Identification cathepsin D as a biomarker for differentiation and prognosis of nasopharyngeal carcinoma by laser capture microdissection and proteomic analysis, *J. Proteome Res.* 7, 2415–2426, 2008.
239. Li, X.M., Huang, W.G., and Yi., H., Proteomic analysis to identify cytokeratin 18 as a novel biomarker of nasopharyngeal carcinoma, *J. Cancer Res. Clin. Oncol.* 135, 1763–1775, 2009.
240. Liu, Y.F., Xiao, Z.Q., Li, M.X. et al., Quantitative proteome analysis reveals annexin A3 as a novel biomarker in lung adenocarcinoma, *J. Pathol.* 217, 54–64, 2009.
241. Keller, M.D., Kanter, E.M., Lieber, C.A. et al., Detecting temporal and spatial effects of epithelial cancers with Raman spectroscopy, *Dis. Markers* 25, 323–337, 2008.
242. Pawlak, A.M., Glenn, J.V., Beattie, J.R. et al., Advanced glycation as a basis for understanding retinal aging and noninvasive risk prediction, *Ann. N. Y. Acad. Sci.* 1126, 59–65, 2008.
243. Krafft, C., Steiner, G., Beleites, C. et al., Disease recognition by infrared and Raman spectroscopy, *J. Biophotonics* 2, 13–28, 2009.
244. Brown, J.Q., Vishwanath, K., Palmer, G.M. et al., Advances in quantitative UV-visible spectroscopy for clinical and pre-clinical application in cancer, *Curr. Opin. Biotechnol.* 20, 119–131, 2009.
245. Rao, J.Y., Apple, S.K., Hemstreet, G.P. et al., Single cell multiple biomarker analysis in archival breast fine-needle aspiration specimens: Quantitative fluorescence image analysis of DNA content, p53, and G-actin as breast cancer biomarkers, *Cancer Epidemiol. Biomarkers Prev.* 7, 1027–1033, 1998.
246. Jung, S.S., Park, H.S., Lee, I.J. et al., The HCCR oncoprotein as a biomarker for human breast cancer, *Clin. Cancer Res.* 11, 7700–7708, 2005.
247. Kwon, M.S., Shin, S.H., Yin, S.H. et al., CD63 as a biomarker for predicting the clinical outcomes in adenocarcinoma of the lung, *Lung Cancer* 57, 46–53, 2007.
248. Huang, D., Casale, G.P., Tian, J. et al., Quantitative fluorescence imaging analysis for cancer biomarker discovery: Application to β-catenin in archived prostate specimens, *Cancer Epidemiol. Biomarkers Prev.* 16, 1371–1381, 2007.
249. Psyrri, A., Kountourakis, P., Scorilas, A. et al., Human tissue kallikrein 7, a novel biomarker for advanced ovarian carcinoma using a novel *in situ* quantitative method of protein expression, *Ann. Oncol.* 19, 1271–1277, 2008.
250. Alvarez, H., Corvalan, A., Roa, J.C. et al., Serial analysis of gene expression identifies connective tissue growth factor expression as a prognostic biomarker in gallbladder cancer, *Clin. Cancer Res.* 14, 2631–2638, 2008.
251. Ota, K., Fujimori, H., Ueda, M. et al., Midkine as a prognostic biomarker in oral squamous cell carcinoma, *Br. J. Cancer* 99, 655–662, 2008.
252. Tillinger, W., Jilich, R., Jilma, B. et al., Expression of the high-affinity IgG receptor FcRI (CD64) in patients with inflammatory bowel disease: A new biomarker for gastroenterologic diagnostics, *Am. J. Gastroenterol.* 104, 102–109, 2009.

253. Daniel, C.R., Bostick, R.M., Flanders, W.D. et al., TGF-α expression as a potential bio-marker of risk within the normal-appearing colorectal mucosa of patients with and without incident sporadic adenoma, *Cancer Epidemiol. Biomarkers Prev.* 18, 65–73, 2009.

254. Mirasoli, M., Guardigli, M., Simoni, P. et al., Multiplex chemiluminescence micro-scope imaging of p16(INK4A) and HPV DNA as biomarker of cervical neoplasia, *Anal. Bioanal. Chem.* 394, 981–987, 2009.

255. Kolb, A., Rieder, S., Born, D. et al., Glucagon/insulin ratio as a potential biomarker for pancreatic cancer in patients with new-onset diabetes mellitus, *Cancer Biol. Ther.* 8, 1527–1533, 2009.

256. Mager, S.R., Oomen, M.H., Morente, M.M. et al., Standard operating procedure for the collection of fresh frozen tissue samples, *Eur. J. Cancer* 43, 828–834, 2007.

257. Vaught, J.B., Blood collection, shipment, processing, and storage, *Cancer Epidemiol. Biomarkers Prev.* 15, 1582–1584, 2006.

258. Lomholt, A.F., Frederiksen, C.B., Christensen, I.J. et al., Plasma tissue inhibitor of metalloproteinases-1 as a biological marker? Preanalytical considerations, *Clin. Chim. Acta* 380, 128–132, 2007.

259. Smith, C.L., Dickinson, P., Forster, T. et al., Quantitative assessment of human whole blood RNA as a potential biomarker for infectious disease, *Analyst* 132, 1200–1209, 2007.

260. Ahmann, G.J., Chng, W.J., Henderson, K.J. et al., Effect of tissue shipping on plasma cell isolation, viability, and RNA integrity in the context of a centralized good labora-tory practice-certified tissue banking facility, *Cancer Epidemiol. Biomarkers Prev.* 17, 666–673, 2008.

261. van der Schans, G.P., Mars-Groenendijk, R., de Jong, L.P. et al., Standard operating pro-cedure for immunuslotblot assay for analysis of DNA/sulfur mustard adducts in human blood and skin, *J. Anal. Toxicol.* 28, 316–319, 2004.

262. Gray, A.C., Malton, J., and Clothier, R.H., The development of a standardized protocol to measure squamous differentiation in stratified epithelia, using the fluorescein cadava-rine incorporation technique, *Altern. Lab. Anim.* 32, 91–100, 2004.

263. Schabacker, D.S., Stefanovska, I., Gain, I. et al., Protein array staining methods for undefined protein content, manufacturing quality control, and performance validation, *Anal. Biochem.* 359, 84–93, 2006.

264. Bewarder, N., Abmeier, F., Bergmann, S. et al., Multicenter evaluation of performance of C-reactive protein analysis over a one-year period, *Clin. Lab.* 52, 639–654, 2006.

265. Lamoreaux, L., Roederer, M., and Koup, R., Intracellular cytokine optimization and standard operating procedure, *Nat. Protoc.* 1, 1507–1516, 2006.

266. Wingren, C. and Borrebaeck, C.A., Antibody microarray analysis of directly labelled complex proteomes, *Curr. Opin. Biotechnol.* 19, 55–61, 2008.

267. Popa-Burke, I., Lupotsky, B., Boyer, J. et al., Establishing quality assurance criteria for serial dilution operations on liquid-handling equipment, *J. Biomol. Screen.* 14, 1017–1030, 2009.

268. Sukurmaran, D.K., Garcia, E., Hua, J. et al., Standard operating procedure for metabo-lomics studies of blood serum and plasma samples using ¹H-NMR micro-flow probe, *Magn. Reson. Chem.* 47(Suppl 1), S81–S85, 2009.

269. José, M., Curtu, S., Gajardo, R., and Jorquera, J.I., The effect of storage at different temperatures on the stability of hepatitis C virus RNA in plasma samples, *Biologicals* 31, 1–8, 2003.

270. Fura, A., Harper, T.W., Zhang, H., Fung, L., and Shyu, W.C., Shift in pH of biological fluids during storage and processing: Effect on bioanalysis, *J. Pharm. Biomed. Anal.* 32, 513–522, 2003.

271. Bel-Tal, O., Zwang, E., Eichel, R., Badalber, T., and Hareuveni, M., Vitamin K-dependent coagulation factors and fibrinogen levels in FPP remain stable upon freezing and thawing, *Transfusion* 43, 873–877, 2003.

272. Lewis, M.R., Callas, P.W., Jenny, N.S., and Tracy, R.P., Longitudinal stability of coagulation, fibrinolysis, and inflammation factors in stored plasma samples, *Thromb. Haemost.* 86, 1495–1500, 2001.

273. Rouy, D., Ernens, I., Jeanty, C., and Wagner, D.R., Plasma storage at –80°C does not protect matrix metalloproteinase-9 from degradation, *Anal. Biochem.* 338, 294–298, 2005.

274. Anderson, N.R., Chatha, K., Holland, M.R., and Gama, R., Effect of sample tube type and time to separation on *in vitro levels* of C-reactive protein, *Br. J. Biomed. Sci.* 60, 164–165, 2003.

275. Tanner, M., Kent, N., Smith, B. et al., Stability of common biochemical analytes in serum gel tubes subjected to various storage temperatures and time pre-centrifugation, *Ann. Clin. Biochem.* 45, 375–379, 2008.

276. Sulik, A., Wojtkowska, M., and Oldak, E., Preanalytical factors affecting the stability of matrix metalloproteinase-2 concentrations in cerebrospinal fluid, *Clin. Chim.* 392, 73–75, 2008.

277. Ostroff, R., Foreman, T., Keeney, T.R. et al., The stability of the circulating human proteome to variations in sample collection and handling procedures measure with an aptamers-based proteomics array, *J. Proteomics* 73, 649–666, 2009.

278. Chance, J., Berube, J., Vandersmissen, M., and Blanckaert, N., Evaluation of the BD Vacutainer® PST™ II Blood Collection Tube for special chemistry analytes, *Clin. Chem. Lab. Med.* 47, 358–361, 2009.

279. Betsou, F., Roussel, B., Guillaume, N., and Lefrère, J.-J., Long-term stability of coagulation variables: Protein S as a biomarker for preanalytical storage-related variations in human plasma, *Thromb. Haemost.* 101, 1172–1175, 2009.

280. Hartweg, J., Gunter, M., Perera, R. et al., Stability of soluble adhesion molecules, selectins, and C-reactive protein at various temperatures: Implications for epidemiological and large-scale clinical studies, *Clin. Chem.* 53, 1858–1860, 2007.

281. Yue, B., Pattison, E., and Roberts, W.L., Choline in whole blood and plasma: Sample preparation and stability, *Clin. Chem.* 54, 590–593, 2008.

282. Nordlund, M.J., Bjerner, J., Warren, D.J. et al., Progastrin-releasing peptide: Stability in plasma/serum and upper reference limit, *Tumour Biol.* 29, 204–210, 2008.

283. Salminen, I. and Alfthan, G., Plasma ascorbic acid preparation and storage for epidemiological studies using TCA precipitation, *Clin. Biochem.* 41, 723–727, 2008.

284. Devgun, M.S., Delay in centrifugation and measurement of serum constituents in normal subjects, *Clin. Physiol. Biochem.* 7, 189–197, 1989.

285. Jung, K., Lein, M., Brux, B. et al., Different stability of free and complexed prostate-specific antigen in serum in relation to specimen handling and storage conditions, *Clin. Chem. Lab. Med.* 38, 1271–1275, 2000.

286. West-Nielsen, M., Hogdall, E.V., and Marchiori, E., Sample handling for mass spectrometric proteomic investigations of human sera, *Anal. Chem.* 77, 5114–5123, 2005.

287. Lambert-Messerlian, G.M., Eklund, E.E., Malone, F.D. et al., Stability of first- and second-trimester serum markers after storage and shipment, *Prenat. Diagn.* 26, 17–21, 2006.

288. Zolty, R., Bauer, C., Allen, P. et al., Atrial natriuretic peptide stability, *Clin. Biochem.* 41, 1255–1258, 2008.

289. Reed, A.B., Ankerst, D.P., Leach, R.J. et al., Total prostate specific antigen stability confirmed after long-term storage of serum at –80°C, *J. Urol.* 180, 534–538, 2008.

290. Ng, V., Koh, D., Fu, Q. et al., Effects of storage time on stability of salivary immunoglobulin A and lysozyme, *Clin. Chim. Acta* 338, 131–134, 2003.

291. Lykkesfeldt, J., Loft, S., and Poulsen, H.E., Determination of ascorbic acid and dehydroascorbic acid in plasma by high-performance liquid chromatography with coulometric detection—Are they reliable biomarkers of oxidative stress?, *Anal. Biochem.* 229, 329–335, 1995.

292. Hymas, W., Stevenson, J., Taggart, E.W. et al., Use of lyophilized standards for the calibration of a newly developed real time PCR assay for human herpes type six (HHV6) variants A and B, *J. Virol. Methods* 128, 143–150, 2005.
293. Satterfield, M.B. and Welch, M.J., Comparison by LC-MS and MALDI-MS of prostate-specific antigen from five commercial sources with certified reference material 613, *Clin. Biochem.* 38, 166–174, 2005.
294. Carpentier, S.C., Dens, K., Van den houve, I. et al., Lyophilization, a practical way to store and transport tissues prior to protein extraction for 2DE analysis, *Practical Proteomics* 7(Suppl 1), 64–69, 2007.
295. Bach, P.R. and Larson, J.W., Stability of standard curves prepared for EMIT homogeneous enzyme immunoassay kits stored at room temperature after reconstitution, *Clin. Chem.* 26, 652–654, 1980.
296. Nishimura, M., Iwanaga, T., Ohkaru, Y. et al., Change in immunoreactive human hepatic triglyceride (HTGL) mass and the shelf-life of the HTGL ELISA kit in long-term storage, *J. Immunoassay Immunochem.* 27, 89–102, 2006.
297. Caserman, S., Menart, V., Gaines Das, R. et al., Thermal stability of the WHO International Standard of interferon alpha 2b (IFN-α 2b): Application of new reporter gene assay for IFN-α 2b potency determinations, *J. Immunol. Methods* 319, 6–12, 2007.
298. Chaigneau, C., Calbloch, T., Beaumont, K., and Betsou, F., Serum biobank certification and the establishment of quality controls for biological fluids: Examples of serum biomarker stability after temperature variation, *Clin. Chem. Lab. Med.* 45, 1390–1395, 2007.
299. Riegman, P.H., Dinjens, W.N., and Oosterhuis, J.W., Biobanking for interdisciplinary clinical research, *Pathobiology* 74, 239–244, 2007.
300. Schrohl, A.S., Würtz, S., Kohn, E. et al., Banking of biological fluids for studies of disease-associated protein biomarkers, *Mol. Cell. Proteomics* 7, 2061–2066, 2008.
301. Day, J.G. and Stacey, G.N., Biobanking, *Mol. Biotechnol.* 40, 202–213, 2008.
302. Shevde, L.A. and Riker, A.I., Current concepts in biobanking: Development and implementation of a tissue repository, *Front. Biosci.* 1, 188–193, 2009.
303. Lipworth, W., Morrell, B., Irvine, R., and Kerridge, I., An empirical reappraisal of public trust in biobanking research: Rethinking restrictive consent requirements, *J. Law Med.* 17, 119–132, 2009.
304. Lumachi, F., Marino, F., Varotto, S., and Basso, U., Oligonucleotide probe array for p53 gene alteration analysis in DNA from formalin-fixed paraffin-embedded breast cancer tissues, *Ann. N. Y. Acad. Sci.* 1175, 89–92, 2009.
305. Schwers, S., Reifenberger, E., Gehrmann, M. et al., A high-sensitivity, medium-density, and target amplification-free planar waveguide microarray system for gene expression analysis of formalin-fixed and paraffin-embedded tissue, *Clin. Chem.* 55, 1995–2003, 2009.
306. Toussaint, J., Sieuwerts, A.M., Haibe-Kains, B. et al., Improvement of the clinical applicability of the Genome Grade Index through a qRT-PCR test performed on frozen and formalin-fixed paraffin-embedded tissues, *BMC Genomics* 10, 424, 2009.
307. Koh, S.S., Open, M.L., Wei, J.P. et al., Molecular classification of melanomas and nevi using gene expression microarray signatures and formalin-fixed and paraffin-embedded tissue, *Mod. Pathol.* 22, 538–546, 2009.
308. Kiernan, U.A., Biomarker rediscovery in diagnostics, *Expert Opin. Med. Diagn.* 2, 1391–1400, 2008.
300. Wasinger, V., Cordwell, S., Cerpa-Poljak, A. et al., Progress with gene-product mapping of the Mollicutes: *Mycoplasma genitalium*, *Electrophoresis* 16, 1090, 1995.
310. Miles, A.K., Matharoo-Ball, B., Li, G. et al., The tumour antigens: Current status and future developments, *Cancer Immunol. Immunother.* 55, 996–1003, 2006.
311. Azad, N.S., Rasool, N., Annuziata, C.M. et al., Proteomics in clinical trials and practice: Present uses and future promise, *Mol. Cell. Proteomics* 5, 1819–1829, 2006.

312. Zubarev, R.A., Nielsen, M.L., Fung, E.M. et al., Identification of dominant signaling pathways from proteomics expression data, *J. Proteomics* 71, 89–96, 2008.
313. Bhattacharjee, M., Botting, C.H., and Sillanpää, M.J., Bayesian biomarker identification based on marker-expression proteomics data, *Genomics* 92, 384–392, 2008.
314. Gast, M.C., Schllens, J.H., and Beijnen, J.H., Clinical proteomics in breast cancer, a review, *Breast Cancer Res. Treat.* 116, 17–29, 2009.
315. Gygi, S.P., Rist, B., Gerber, S.A., Turecek, F., Gelb, M.H., and Aebersold, R., Quantitative analysis of complex protein mixtures using isotope-coded affinity tags, *Nat. Biotechnol.* 17, 994, 1999.
316. Ong, S.E., Blagoev, B., Kratchmarova, I. et al., Stable isotope labeling by amino acids in cell culture, SILAC, as a simple and accurate approach to expression proteomics, *Mol. Cell. Proteomics* 1, 376, 2002.
317. Haqqani, A.S., Kelly, J.F., and Stanimirovic, D.B., Quantitative protein profiling by mass spectrometry using label-free proteomics, *Methods Mol. Biol.* 439, 241–256, 2008.
318. Himeda, C.L., Ranish, J.A., and Hauschka, S.D., Quantitative proteomic identification of MAZ as a transcriptional regulator of muscle-specific genes in skeletal and cardiac myocytes, *Mol. Cell. Biol.* 28, 6521–6535, 2008.
319. Alex, P., Gucek, M., and Li, X., Applications of proteomics in the study of inflammatory bowel diseases: Current status and future directions with available technologies, *Inflamm. Bowel Dis.* 15, 616–629, 2009.
320. Lin, B., Cheng, L., White, J.T. et al., Quantitative proteomics analysis integrated with microarray data reveals that extracellular matrix proteins, catenins, and p53 binding protein 1 are important for chemotherapy response in ovarian cancers, *OMICS* 13, 345–354, 2009.
321. Zamò, A. and Cecconi, D., Proteomic analysis of lymphoid and hematopoietic neoplasms: There's more than biomarker discovery, *J. Proteomics* 73, 508–520, 2009.
322. Turecek, F., Mass spectrometry in coupling with affinity capture-release and isotope-coded affinity tags for quantitative protein analysis, *J. Mass Spectrom.* 37, 1–14, 2002.
323. Xiao, Z. and Veenstra, T.D., Comparison of protein expression by isotope-coded affinity tag labeling, *Methods Mol. Biol.* 428, 181–192, 2008.
324. Pawlik, T.M., Hawke, D.H., Liu, Y. et al., Proteomic analysis of nipple aspirate fluid from women with early-stage breast cancer using isotope-coded affinity tags and tandem mass spectrometry reveals differential expression of vitamin D binding protein, *BMC Cancer* 6, 68, 2006.
325. Zhang, J., Goodlett, D.R., Quinn, J.F. et al., Quantitative proteomics of cerebrospinal fluid from patients with Alzheimer disease, *J. Alzheimers Dis.* 9, 81–88, 2006.
326. Dhungana, S., Merrick, B.A., Tomer, K.B., and Fessler, M.B., Quantitative proteomics analysis of macrophage rafts reveals compartmentalized activation of the proteosome and proteosome-mediated ERK activation in response to lipopolysaccharide, *Mol. Cell. Proteomics* 8, 201–213, 2009.
327. Shan, S.J., Yu, K.H., Sangar, V. et al., Identification and quantification of preterm birth biomarkers in human cervicovaginal fluid by liquid chromatography/tandem mass spectrometry, *J. Proteome Res.* 8, 2407–2417, 2009.
328. Aggelis, V., Craven, R.A., Peng, J. et al., Proteomic identification of differentially expressed plasma membrane proteins in renal cell carcinoma by stable isotope labelling of a von Hippel-Lindau transfectant cell line model, *Proteomics* 9, 2118–2130, 2009.
329. Lund, R., Leth-Larsen, R., Jensen, O.N., and Ditzel, H.J., Efficient isolation and quantitative proteomic analysis of cancer cell plasma membrane proteins for identification of metastasis-associated cell surface markers, *J. Proteome Res.* 8, 3078–3090, 2009.
330. Kawase, H., Fujii,K., Miyamoto, M. et al., Differential LC-MS-based proteomics of surgical human cholangiocarcincoma tissues, *J. Proteome Res.* 8, 4092–4103, 2009.

331. Rangiah, K., Tippornwong, M., Sangar, V. et al., Differential secreted proteome approach in murine model for candidate biomarker discovery in colon cancer, *J. Proteome Res.* 8, 5153–5164, 2009.

332. Yu, K.H., Barry, C.G., Austin, D. et al., Stable isotope dilution multidimensional liquid chromatography-tandem mass spectrometry for pancreatic cancer serum biomarker discovery, *J. Proteome Res.* 8, 1565–1576, 2009.

333. Prokhorova, T.A., Rigbolt, K.T., Johansen, P.T. et al., Stable isotope labeling by amino acids in cell culture (SILAC) and quantitative comparison of the membrane proteomes of self-renewing and differentiating human embryonic stem cells, *Mol. Cell. Proteomics* 8, 959–970, 2009.

334. Berger, A.B., Vitorino, P.M., and Bogyo, M., Activity-based protein profiling: Applications to biomarker discovery, in vivo imaging and drug discovery, *Am. J. Pharmacogenomics* 4, 371–381, 2004.

335. Pullela, P.K., Chiku, T., Carvan III M.J., and Sem, D.S., Fluorescence-based detection of thiols *in vitro* and *in vivo* using dithiol probes, *Anal. Biochem.* 352, 265–273, 2006.

336. Castronovo, V., Waltregny, D., Kischel, P. et al., A chemical proteomics approach for identification of accessible antigens in human kidney cancer, *Mol. Cell. Proteomics* 5, 2083–2091, 2006.

337. Birner-Gruenberger, R. and Hermetter, A., Activity-based proteomics of lipolytic enzymes, *Curr. Drug. Discov. Technol.* 4, 1–11, 2007.

338. Paulick, M.G. and Bogyo, M., Application of activity-based probes to the study of enzymes involved in cancer progression, *Curr. Opin. Genet. Dev.* 18, 97–106, 2008.

339. Human Proteome Organisation; http://www.hupo.org/research/hppp/

340. Righetti, P.G., Castagna, A., Antonucci, F. et al., The proteome: Anno Domini 2002, *Clin. Chem. Lab. Med.* 41, 425–438, 2002.

341. Anderson, N.L. and Anderson, N.G., The human plasma proteome: History, character, and diagnostics prospects, *Mol. Cell. Proteomics* 1, 845–867, 2003.

342. Nielsen, M.L., Savitski, M.M., and Zubarev, R.A., Extent of modifications in human proteome samples and their effect on dynamic range of analysis in shotgun proteomics, *Mol. Cell Proteomics* 5, 2384–2391, 2006.

343. Addona, T.A., Abbatiello, S.E., Schilling, B. et al., Multi-site assessment of the precision and reproducibility of multiple reaction monitoring-based measurements proteins in plasma, *Nat. Biotechnol.* 27, 633–641, 2009.

344. Hortin, G.L. and Sviridov, D., The dynamic range problem in the analysis of the plasma proteome, *J. Proteomics* 73, 629–636, 2010.

345. Cho, S.Y., Lee, E.Y., Lee, J.S. et al., Efficient prefractionation of low-abundance proteins in human plasma and construction of a two-dimensional map, *Proteomics* 5, 3386–3396, 2005.

346. Barnea, E., Sorkin, R., Ziv, T. et al., Evaluation of prefractionation methods as a preparatory step for multidimensional based chromatography of serum proteins, *Proteomics* 5, 3367–3375, 2005.

347. Pernemalm, M., Orre, L.M., Lengqvist, J. et al., Evaluation of three principally different intact protein prefractionation methods for plasma biomarker discovery, *J. Proteome Res.* 7, 2712–2722, 2008.

348. Jmeian, Y. and El Rassi, Z., Multicolumn separation platform for simultaneous depletion and prefractionation prior to 2-DE for facilitating in-depth serum proteomics profiling, *J. Proteome Res.* 8, 4592–4603, 2009.

349. Joo, W.A. and Speicher, D., Prefractionation using microscale solution IEF, *Method Mol. Biol.* 591, 291–304, 2009.

350. Pernemalm, M., Lewensohn, R., and Lehtiö, J., Affinity prefractionation for MS-based plasma proteomics, *Proteomics* 9, 1420–1427, 2009.

351. Siegmund, R., Kiehntopf, M., and Deufel, T., Evaluation of two different albumin depletion strategies for improved analysis of human CSF by SELDI-TOF-MS, *Clin. Biochem.* 42, 1136–1143, 2009.

352. McNulty, D.E. and Annan, R.S., Hydrophilic interaction chromatography for fractionation and enrichment of the phosphoproteome, *Methods Mol. Biol.* 527, 93–105, 2009.

353. Grønborg, M., Kristiansen, T.Z., Iwahori, A. et al., Biomarker discovery from pancreatic cancer secretome using as a differential proteomic approach, *Mol. Cell Proteomics* 5, 157–171, 2006.

354. Xue, H., Lu, B., and Lai, M., The cancer secretome: A reservoir of biomarkers, *J. Transl. Med.* 6, 52, 2008.

355. Faca, V.M. and Hanash, S.M., In-depth proteomics to define the cell surface and secretome of ovarian cancer cells and processes of protein shedding, *Cancer Res.* 69, 728–730, 2009.

356. Mutch, D.M., Rouault, C., Keophiphath, M. et al., Using gene expression to predict the secretome of differentiating human preadipocytes, *Int. J. Obes.* 33, 354–363, 2009.

357. Gundacker, N.C., Haudek, V.J., Wimmer, H. et al., Cytoplasmic proteome and secretome profiles of differently stimulated human dendritic cells, *J. Proteome Res.* 8, 2799–2811, 2009.

358. Chenau, J., Michelland, S., de Fraipont, F. et al., The cell line secretome, a suitable tool for investigating proteins released in vivo by tumors: Application to the study of p53-modulated proteins secreted in lung cancer cells, *J. Proteome Res.* 8, 4579–4591, 2009.

359. Tunica, D.G., Yin, X., Sidibe, A. et al., Proteomic analysis of the secretome of human umbilical vein endothelial cells using a combination of free-flow electrophoresis and nanoflow LC-MS/MS, *Proteomics* 9, 4991–4996, 2009.

360. Cui, J., Liu, Q., Puett, D., and Xu, Y., Computational prediction of human proteins that can be secreted into the bloodstream, *Bioinformatics* 24, 2370–2375, 2008.

361. Chen, Y., Zhang, H., Zu, A. et al., Elevation of serum l-lactate dehydrogenase B correlated with the clinical stage of lung cancer, *Lung Cancer* 54, 95–102, 2006.

362. Sarkissian, G., Fergelot, P., and Lamy, P.J., Identification of pro-MMP-7 as a serum marker for renal cell carcinoma by use of proteomic analysis, *Clin. Chem.* 54, 574–581, 2008.

363. Weng, L.P., Wu, C.C., Hsu, B.L. et al., Secretome-based identification of Mac-2 binding protein as a potential oral cancer marker involved in cell growth and motility, *J. Proteome Res.* 7, 3765–3775, 2008.

364. Misek, D.E., Kuick, R., Wang, H. et al., A wide range of protein isoforms in serum and plasma uncovered by a quantitative intact protein analysis system, *Proteomics* 5, 3343–3352, 2005.

365. Hoffman, S.A., Joo, W.A., Echan, L.A. et al., Higher dimensional (Hi-D) separation strategies dramatically improve the potential for cancer biomarker detection in serum and plasma, *J. Chromatogr. B. Analyt. Technol. Biomed. Life Sci.* 849, 43–52, 2007.

366. O'Farrell, P.H., High resolution two-dimensional electrophoresis of proteins, *J. Biol. Chem.* 250, 4007–4021, 1975.

367. Görg, A., Obermaier, C., Boguth, G. et al., The current status of two-dimensional electrophoresis with immobilized pH gradients, *Electrophoresis* 21, 1037–1053, 2000.

368. Choe, L.H. and Lee, K.H., Quantitative and qualitative measure of intralaboratory two-dimensional protein gel reproducibility and the effects of sample preparation, sample load, and image analysis, *Electrophoresis* 24, 3500–3507, 2003.

369. Challapalli, K.K., Zabel, C., Schuchhardt, C. et al., High reproducibility of large-gel two-dimensional electrophoresis, *Electrophoresis* 25, 3040–3047, 2004.

370. Gustafsson, J.S., Ceasar, R., Glasbey, C.A. et al., Statistical exploration of variation in quantitative two-gel electrophoresis, *Proteomics* 4, 3791–3799, 2004.

371. Aikkokallio, T., Salmi, J., Nyman, T.A., and Nevalainen, O.S., Geometrical distortions in two-dimensional gels: Applicable correction methods, *J. Chromatogr. B. Analyt. Technol. Biomed. Life Sci.* 815, 25–37, 2005.

372. Li, Z.B., Flint, P.W., and Boluyt, M.O., Evaluation of several two-dimensional gel electrophoresis in cardiac proteomics, *Electrophoresis* 26, 3572–3585, 2005.

373. Rowell, C., Carpenter, M., and Lamartiniere, C.A., Modeling biological variability in 2-D gel proteomic carcinogenesis experiments, *J. Proteome Res.* 4, 1619–1627, 2005.

374. Hoorn, E.J., Hoffert, J.D., and Knepper, M.A., The application of DIGE-based proteomics to renal physiology, *Nephron Physiol.* 104, 61–72, 2006.

375. Damodaran, S. and Rabin, R.A., Minimizing variability in two-dimensional electrophoresis gel image analysis, *OMICS* 11, 225–230, 2007.

376. Tannu, N.S. and Hemby, S.E., Two-dimensional fluorescence difference gel electrophoresis for comparative proteomics profiling, *Nat. Protoc.* 1, 1732–1742, 2006.

377. Valcu, C.M. and Valcu, M., Reproducibility of two-dimensional gel electrophoresis at different replication levels, *J. Proteome Res.* 6, 4677–4683, 2007.

378. Winkler, W., Zellner, M., Diestinger, M. et al., Biological variation of the platelet proteome in the elderly population and its implication for biomarker research, *Mol. Cell. Proteomics* 7, 193–203, 2008.

379. Bandow, J.E., Baker, J.D., Berth, M. et al., Improved image analysis workflow for 2-D gels enables large-scale 2-D gel-based proteomic studies—COPD biomarker discovery study, *Proteomics* 8, 3030–3041, 2008.

380. Karp, N.A., Feret, R., Rubtsov, D.V., and Lilly, K.S., Comparison of DIGE and post-stained gel electrophoresis with both traditional and SameSpots analysis for quantitative proteomics, *Proteomics* 8, 948–960, 2008.

381. Malm, C., Hadrevi, J., Bergstrom, S.A. et al., Evaluation of 2-D DIGE for skeletal muscle: Protocol and repeatability, *Scand. J. Clin. Lab. Invest.* 68, 793–800, 2008.

382. Schroder, S., Brandmuller, A., Deng, X. et al., Improved precision in gel electrophoresis by stepwisely decreasing variance components, *J. Pharm. Biomed. Anal.* 50, 320–327, 2009.

383. Kondo, T. and Hirohashi, S., Application of 2D-DIGE in cancer proteomics toward personalized medicine, *Methods Mol. Biol.* 577, 135–154, 2009.

384. Miura, K., Imaging technologies for the detection of multiple stains in proteomics, *Proteomics* 3, 1097–1108, 2003.

385. White, I.T., Pickford, R., Wood, J. et al., A statistical comparison of silver and SYPRO Ruby staining for proteomic analysis, *Electrophoresis* 25, 3048–3054, 2004.

386. Lanne, B. and Panfilov, O., Protein staining influences the quality of mass spectra obtained by peptide mass fingerprinting after separation on 2-D gels. A comparison of staining with Coomassie brilliant blue and Sypro Ruby, *J. Proteome Res.* 4, 175–179, 2005.

387. Chevalier, F., Centeno, D., Rofidal, V. et al., Different impact of staining procedures using visible stains and fluorescent dyes for large-scale investigation of proteomics by MALDI-TOF mass spectrometry, *J. Proteome Res.* 5, 512–520, 2006.

388. Harris, L.R., Churchward, M.A., Butt, R.H., and Coorssen, J.R., Assessing detection methods for gel-based proteomic analyses, *J. Proteome Res.* 6, 1418–1425, 2007.

389. Ball, M.S. and Karuso, P., Mass spectral compatibility of four proteomics stains, *J. Proteome Res.* 6, 4313–4320, 2007.

390. Nock, C.M., Bass, M.S., White, I.R. et al., Mass spectrometric compatibility of Deep Purple and SYPRO Ruby total protein stains for high-throughput proteomics using large-format two-dimensional gel electrophoresis, *Rapid Commun. Mass Spectrom.* 22, 881–886, 2008.

391. Galesio, M., Vieira, D.V., Rial-Otero, R. et al., Influence of the protein staining in the fast ultrasonic sample treatment for protein identification through peptide mass fingerprint and matrix-assisted laser desorption ionization time of flight mass spectrometry, *J. Proteome Res.* 7, 2097–2106, 2008.

392. Chakravarti, B., Ratanaprayul, W., Dalal, N., and Chakravarti, D.N., Comparison of SYPRO Ruby and Deep Purple using commonly available UV transilluminator: Widescale application in proteomic research, *J. Proteome Res.* 7, 2797–2802, 2008.

393. Cong, W.T., Hwang, S.Y., Jin, L.T., and Choi, J.K., Sensitive fluorescent staining for proteomic analysis of proteins in 1-D and 2-D SDS-PAGE and its comparison with SYPRO Ruby by PMF, *Electrophoresis* 29, 4304–4315, 2008.

394. Chiangjog, W. and Thongboonkerd, V., A comparative study of different dyes for the detection of proteomes derived from *Escherichia coli* and MDCK cells: Sensitivity and selectivity, *J. Chromatogr. B. Analyt. Technol. Biomed. Life Sci.* 877, 1433–1439, 2009.

395. Doherty, N.S., Littman, B.H., Reilly, K. et al., Analysis of changes in acute-phase plasma proteins in an acute-phase plasma proteins in an acute inflammatory response and in rheumatoid arthritis using two-dimensional gel electrophoresis, *Electrophoresis* 20, 355–363, 1998.

396. Krah, A., Schmidt, F., Becher, D. et al., Analysis of automatically generated peptide mass fingerprints of cellular proteins and antigens form *Helicobacter pylori* 26695 separated by two-dimensional electrophoresis, *Mol. Cell. Proteomics* 2, 1271–1283, 2003.

397. Tonge, R., Shaw, J., Middleton, B. et al., Validation and development of fluorescence two-dimensional differential gel electrophoresis proteomics, *Proteomics* 1, 377–396, 2001.

398. Hrebicek, T., Duerrschmid, K., Auer, N. et al., Effect of CyDye minimum labeling in differential gel electrophoresis on the reliability of protein identification, *Electrophoresis* 28, 1161–1169, 2007.

399. Sellers, K.F., Miecznikowski, J., Viswanathan, S. et al., Lights, camera, action! Systematic variation in 2-D difference gel electrophoresis images, *Electrophoresis* 28, 3324–3332, 2007.

400. Krogh, M., Liu, Y., Waldermarson, S. et al., Analysis of DIGE data using a linear mixed model allowing for protein-specific dye effects, *Proteomics* 7, 4235–4244, 2007.

401. Pretzer, E. and Wiktorowicz, J.E., Saturation fluorescence labeling of proteins for proteomic analysis, *Analyt. Biochem.* 374, 250–262, 2008.

402. Kvach, M.V., Ustinov, A.V., Stepanova, I.A. et al., A convenient synthesis of cyanine dyes: Reagents for the labeling of biomolecules, *Eur. J. Org. Chem.* 12, 2107–2117, 2008.

403. van den Broeck, H.G., America, A.H.P., Smulders, M.J.M. et al., Staining efficiency of specific proteins depends on the staining method: Wheat gluten proteins, *Proteomics* 8, 1880–1884, 2008.

404. Dupont, A., Chwastyniak, M., Beseme, O. et al., Application of saturation dye 2D-DIGE proteomics to characterize proteins modulated by oxidized low density lipoprotein treatment of human macrophages, *J. Proteome Res.* 7, 3572–3582, 2008.

405. Riederer, B.M., Non-covalent and covalent protein labeling in two-dimensional gel electrophoresis, *J. Proteomics* 71, 231–244, 2008.

406. Timms, J.F. and Cramer, R., Difference gel electrophoresis, *Proteomics* 8, 4886–4897, 2008.

407. Bruschi, M., Grilli, S., and Candiano, G., New iodo-acetamido cyanines for labeling cysteine thiol residues. A strategy for evaluating plasma proteins and their oxido-redox status, *Proteomics* 9, 460–469, 2009.

408. Tsolakos, N., Techanukui, T., Wallington, A. et al., Comparison of two combinations of cyanine dyes for prelabelling and gel electrophoresis, *Proteomics* 9, 1727–1730, 2009.

409. Van den Bergh, G. and Arckens, L., High resolution protein display by two-dimensional electrophoresis, *Curr. Analyt. Chem.* 5, 106–115, 2009.

410. Heinemeyer, J., Scheibe, B., Schmitz, U.K., and Braun, H.P., Blue native DIGE as a tool for comparative analysis of protein complexes, *J. Proteomics* 72, 539–544, 2009.

411. Yao, H., Zhang, Z., Ziao, Z. et al., Identification of metastasis associated proteins in human lung squamous carcinoma using two-dimensional difference gel electrophoresis and laser capture microdissection, *Lung Cancer* 65, 41–48, 2009.

412. Byrne, J.C., Downes, M.R., O'Donoghue, N. et al., 2D-DIGE as a strategy to identify serum markers for the progression of prostate cancer, *J. Proteome Res.* 8, 942–957, 2009.

413. Ametzazurra, A., Matorras, R., Garcia-Velasco, J.A. et al., Endometrial fluid is a specific and non-invasive biological sample for protein biomarker identification in endometriosis, *Hum. Reprod.* 24, 954–965, 2009.

414. Yu, H.R., Kuo, H.C., Sheen, J.M. et al., A unique plasma proteomic profiling with imbalanced fibrinogen cascade in patients with Kawaski disease, *Pediatr. Allergy Immunol.* 20, 699–707, 2009.

415. Rithidech, K.N., Honikel, L., Milazzo, M. et al., Protein expression profiles in pediatric multiple sclerosis: Potential biomarkers, *Mult. Scler.* 15, 455–464, 2009.

416. Gourin, C.G., Zhi, W., and Adam, B.L., Proteomic identification of serum biomarkers for head and neck cancer surveillance, *Laryngoscope* 119, 1291–1302, 2009.

417. Wu, Y., Shu, R., Luo, L.J. et al., Initial comparison of proteomic profiles of whole unstimulated saliva obtained from generalized aggressive periodontitis patients and healthy control subjects, *J. Periodont. Res.* 44, 636–644, 2009.

418. Ma, Y., Peng, J., Huang, L. et al., Searching for tumor markers for colorectal cancer using a 2-D DIGE approach, *Electrophoresis* 30, 2591–2599, 2009.

419. Min-man, W., Hong, S., Zhi-giang, X. et al., Differential proteomic analysis of nasal polyps, chronic sinusitis, and normal nasal mucosal tissues, *Otolaryngol. Head Neck Surg.* 141, 364–368, 2009.

420. Yang, K.D., Chang, W.C., Chuang, H. et al., Increased complement factor H with decreased factor B determined by proteomic differential displays as a biomarker of tai chi chuan exercise, *Clin. Chem.* 56, 127–131, 2010.

421. Yates, J.R., Ruse, C.I., and Nakorschevsky, A., Proteomics by mass spectrometry: Approaches, advances, and applications, *Annu. Rev. Biomed. Eng.* 11, 49–79, 2009.

422. Hutchins, T.W. and Yip, T.-T., New desorption strategies for the mass spectrometric analysis of macromolecules, *Rapid Commun. Mass Spectrom.* 7, 576, 1993.

423. Merchant, M. and Weinberg, S.R., Recent advancements in surface-enhanced laser desorption/ionization time-of-flight mass spectrometry, *Electrophoresis* 21, 1164, 2000.

424. Tang, N., Tornatore, P., and Weinberger, S.R., Current developments in SELDI affinity technology, *Mass Spectrom. Rev.* 23, 34, 2004.

425. Woolley, J.F. and Al-Rubeai, M., The application of SELDI-TOF mass spectrometry to mammalian cell culture, *Biotechnol. Adv.* 27, 177–184, 2009.

426. Garrisi, V.M., Abbate, I., Quarante, M. et al., Serum proteomics and breast cancer: Which perspective?, *Expert Rev. Proteomics* 5, 799–785, 2008.

427. Callesen, A.K., Madsen, J.S., Vach, W. et al., Serum protein profiling by solid phase extraction and mass spectrometry: A future diagnostics tool?, *Proteomics* 9, 1428–1441, 2009.

428. Merrigan, T.L., Hunniford, C.A., Timson, D.J. et al., Development of a novel mass spectrometric technique for studying DNA damage, *Biochem. Soc. Trans.* 37, 905–909, 2009.

429. Melle, C., Ernst, G., Schimmel, B. et al., Colon-derived liver metastasis, colorectal carcinoma, and hepatocellular carcinoma can be discriminated by the Ca^{2+}-binding proteins S100A6 and S100A11, *PLoS One* 3, e3767, 2008.

430. Panicker, G., Lee, D.R., and Unger, E.R., Optimization of SELDI-TOF protein profiling for analysis of cervical mucous, *J. Proteomics* 71, 637–646, 2009.

431. Liu, D., Cao, L., Yu, J. et al., Diagnosis of pancreatic adenocarcinoma using protein chip technology, *Pancreatology* 9, 127–135, 2009.

432. Braunschweig, T., Krieg, R.C., Bar-Or, R. et al., Protein profiling of non-malignant ascites by SELDI-TOF MS: Proof of principle, *Int. J. Mol. Med.* 23, 3–8, 2009.
433. Giannakis, E., Samuel, C.S., Boon, W.M. et al., SELDI-TOF mass spectrometry-based protein profiling of kidney tissue, *Methods Mol. Biol.* 466, 237–249, 2009.
434. Wiederin, J., Rozek, W., Duan, F., and Ciborowski, P., Biomarkers of HIV-1 associated dementia: Proteomic investigation of sera, *Proteome Sci.* 7, 8, 2009.
435. Melle, C., Ernst, G., Winkler, R. et al., Proteomic analysis of human papillomavirus-related oral squamous cell carcinoma: Identification of thioredoxin and epidermal-fatty acid binding protein as upregulated protein markers in microdissected tumor tissue, *Proteomics* 9, 2193–2201, 2009.
436. Altamura, S., Kiss, J., Blattman, C. et al., SELDI-TOF MS detection of urinary hepcidin, *Biochemie* 91, 1335–1338, 2009.
437. Navaglia, F., Fogar, P., Basso, D. et al., Pancreatic cancer biomarkers discovery by surface-enhanced laser desorption and ionization time-of-flight mass spectrometry, *Clin. Chem. Lab. Med.* 47, 713–723, 2009.
438. Gast, M.C., Van Gils, C.H., Wessels, L.F. et al., Serum protein profiling for diagnosis of breast cancer using SELDI-TOF MS, *Oncol. Rep.* 22, 205–213, 2009.
439. Engwegen, J.Y., Mehra, N., Haanen, J.B. et al., Identification of two new serum profiles for renal cell carcinoma, *Oncol. Rep.* 22, 401–408, 2009.
440. Cheng, C., Wu, G., Yeung, S.C. et al., Serum protein profiles in myasthenia gravis, *Ann. Thorac. Surg.* 88, 1118–1123, 2009.
441. Gast, M.C., van Dulken, E.J., van Loenen, T.K. et al., Detection of breast cancer by surface-enhanced laser desorption/ionization time-of-flight mass spectrometry tissue and serum protein profiling, *Int. J. Biol. Markers* 24, 130–141, 2009.
442. Gomes-Alves, P., Imrie, M., Gray, R.D. et al., SELDI-TOF biomarker signatures for cystic fibrosis, asthma, and chronic obstructive pulmonary disease, *Clin. Biochem.* 46, 168–177, 2009.
443. Steinacker, P., Rist, W., Swiatek-de-Lange, M. et al., Ubiquitin as potential cerebrospinal fluid marker of Creutzfeld-Jakob disease, *Proteomics* 10, 81–89, 2010.
444. Shintani, S., Hamakawa, H., Ueyama, Y. et al., Identification of a truncated cystatin SA-I as a saliva biomarker for oral squamous cell carcinoma using the SELDI ProteinChip platform, *Int. J. Oral Maxillofac. Surg.* 39, 68–74, 2009.
445. Knesek, M.J., Litinas, E., Adiguzel, C. et al., Inflammatory biomarker profiling in elderly patients with acute hip fracture treated with heparin, *Clin. Appl. Thromb. Hemost.* 16, 42–50, 2009.

5 Modified Proteins, Oligosaccharides, and Oligonucleotides as Biomarkers

The vast majority of biomarkers are proteins and many proteins are subject to post-translational modification such as glycosylation, sulfation, phosphorylation, and γ-carboxylation.[1] Abnormalities in these processes can result in biomarkers distinct from the parent protein. This is particularly true of differences in glycosylation.[2–6] Some selected biomarkers based on changes in the glycosylation of proteins are shown in Table 5.1. There has been particular interest in the fucosylation of proteins as biomarker for oncology,[16,25–29] but fucosylated proteins are also suggested as biomarkers for liver cirrhosis and fibrosis.[4] Studies of fucosylation are enabled by the use of a specific lectin from *Lotus tetragonolobus*, which is also known as the lotus lectin.[25,27] Other lectins specific for fucosylation are available,[30] as well as several lectin microarrays.[31,32]

Biological polymers are also modified by nonenzymatic chemical modification resulting in the formation of biomarkers. One example is the modification of proteins by glycation, which is process of the reaction of the aldehyde function of a sugar with the ε-amino group of lysine although reaction can also occur with nucleic acids.[33,34] This results in advanced glycation end products (AGE) (Figure 5.1), which have specific receptors (RAGE) and are discussed in Chapter 2. Another nonenzymatic modification of proteins that can serve as a biomarker is the reaction with lipid oxidation such as 4-hydroxynonenal (4-HNE) (Figure 5.2).[35,36] 4-Hydroxynonenal is formed from the peroxidation of lipid[37] and is thought to be biomarker for oxidative stress, which can modify proteins and nucleic acids.[38,39] Various examples of 4-HNE and 4-HNE-modified biopolymers as biomarkers are listed in Table 5.2. Proteins modified with 4-HNE by mass spectrometry[62,63] or by immunological methods.[64] The identification of 4-HNE as a biomarker provides another example of the wide spectrum of biomarkers. There are mechanisms for the removal of 4-HNE-modified proteins.[65]

The nitration/nitrosylation of biopolymers (Figure 5.3) provides another source of biomarkers.[66–69] Nitrosylation is usually the result of the reaction of nitric oxide with protein sulfhydryl groups as part of cellular regulation[66,70,71] and the nitrosylated proteins are not usually considered to be biomarkers. The nitration of proteins at tyrosine residues has provided the bulk of biomarkers. Nitrotyrosine (Figure 5.3)

TABLE 5.1

Changes in Glycosylation as Biomarker

Study	References
Glyco-isoforms of apolipoprotein C used a biomarker for obesity and predictive for bariatric surgery, chronic hepatitis, alcoholic cirrhosis, sepsis, and graft-vs-host disease	7
Use of lectins to identity abnormal glycoproteins in cancer patients	6
Use of a lectin microarray to identify biomarker glycoproteins. Target protein is concentrated by immunoprecipitation and differences in glycosylation studied by antibody overlay lectin microarray. This technique permits identification at the sub-picomole level	8
Carbohydrate-deficient transferrin is a biomarker for alcohol abuse	9–11
Patients with colorectal cancer had higher levels of sialylation and fucosylation compared to normal controls. Further analysis suggested that complement C3, histidine-rich glycoprotein, and kininogen 1 had elevated sialylation and fucosylation and were candidate biomarkers to distinguish colorectal cancer from adenoma and normal subjects	12
Fucosylation as a biomarker for cancer	13
Use of multilectin affinity chromatography to isolate glycoproteins from cancer patient serum	14
A glycoform of acetylcholinesterase in cerebrospinal fluid (CSF) is a potential biomarker for Alzheimer's disease; alone a sensitivity of 80% but can be improved by combining with other biomarkers	15
Fucosylated proteins as biomarkers for primary hepatocellular cancer. The combination of fucosylated kininogen, α-fetoprotein, and GP73 yielded a sensitivity of 95%	16, 17
Use of lectin (Jacalin, SNA)-coupled protein chips for SELDI-TOF-MS identification of different glycoforms in serum from cancer patients	18
DNA sequencer–based carbohydrate analytical profiling technology to identify N-glycan structures associated with aging and hepatocellular cancer	19
Ascitic fluid glycoforms may be biomarkers that differentiate between malignant and benign disease	20
Differential N-glycosylation of kallikrein 6 is a biomarker for ovarian cancer. Differences in the molecular weight kallikrein 6 isoforms are due to glycosylation at a single site	21
A core disialyl-Le(x) hexasaccharide as biomarker for colon cancer	22
Increase in a trisialylated triantennary glyan containing α-1,3-linked fucose, which forms part of the sialyl Lewis x epitope in breast cancer	23
Increase in fucosylation in liver fibrosis and cirrhosis	4
Glycoforms as biomarkers	24

biomarkers are usually determined by immunological analysis[72,73] although mass spectrometry is also used.[74,75] Immunohistochemical techniques can determine the presence of nitrated protein in tissues.[76,77] While most of the studies focus on the use of protein-bound nitrotyrosine as a biomarker, there are some studies on free 3-nitrotyrosine in plasma and urine.[78–83] Some examples of the use of nitration of biopolymers as biomarkers are listed in Table 5.3.

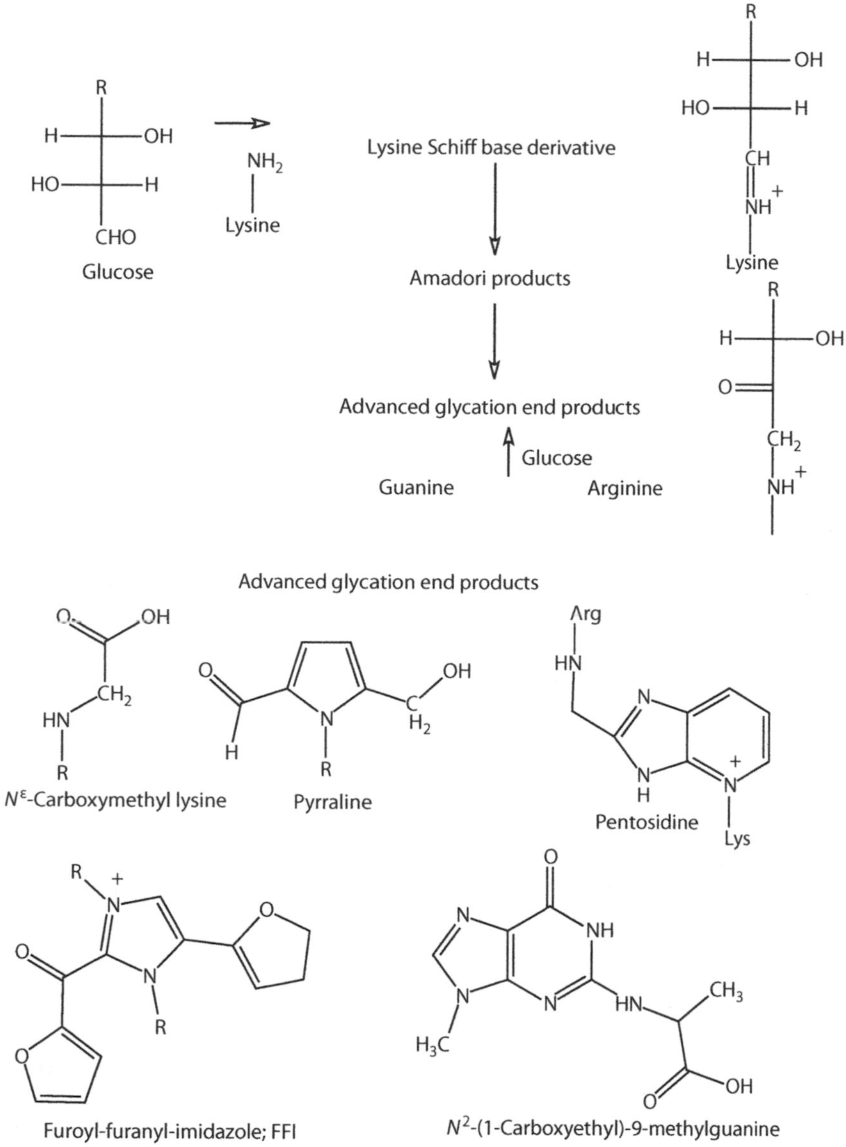

FIGURE 5.1   The glycation of proteins. Shown is the reaction of glucose with lysine to form the Amadori product. A similar reaction occurs with other nitrogen nucleophiles in proteins to yield a variety of derivative referred to as advanced glycation endproducts (AGE). See al-Abed, Y., Schleicher, E., Voelter, W. et al., Identification of N^2-(1-carboxymethyl)guanine (CMG) as a guanine advanced glycation end product, *Bioorg. Med. Chem. Lett.* 8, 2109–2110, 1998; Bierhaus, A., Hofmann, M.A., Ziegler, R., and Nawroth, P.P., AGEs and their interaction with AGE-receptors in vascular disease and diabetes mellitus. I. The AGE concept, *Cardiovasc. Res.* 37, 586–600, 1998; Cho, S.-J., Roman, G., Yeboath, F., and Konishi, Y., The road to advanced glycation end products: A mechanistic perspective, *Curr. Med. Chem.* 14, 1653–1671, 2007.

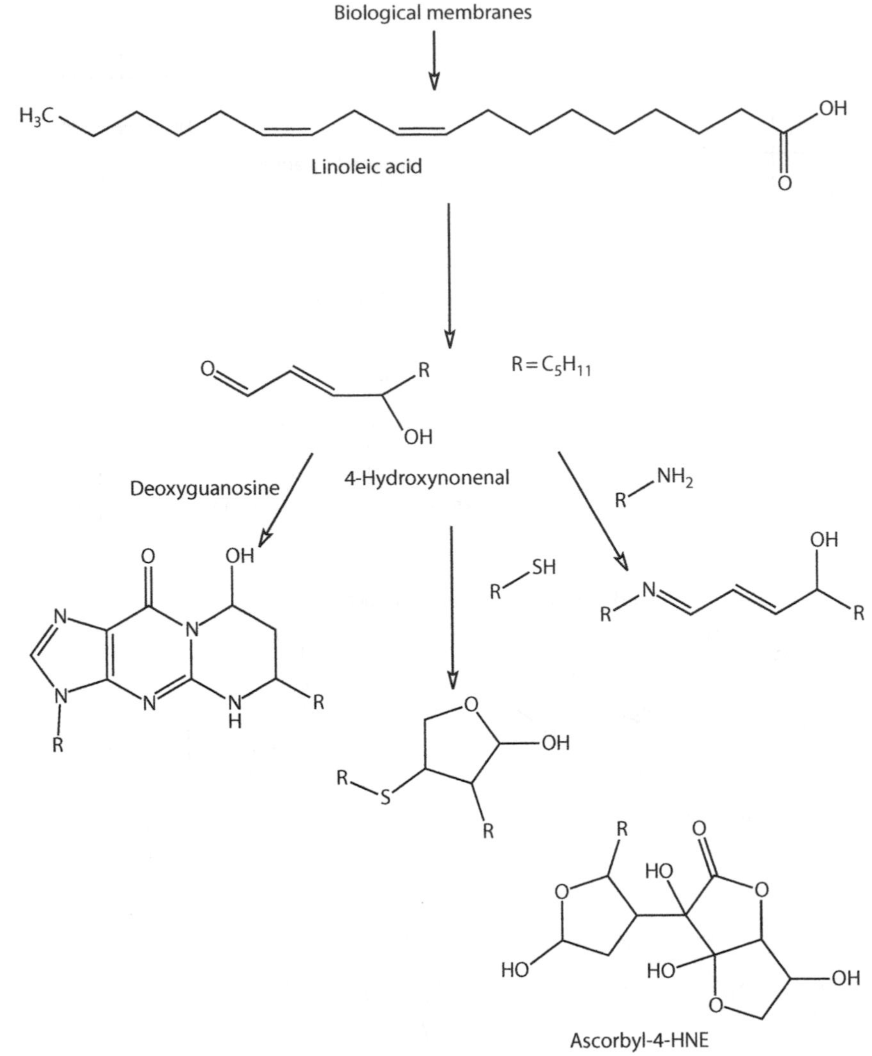

FIGURE 5.2 The chemistry of 4-hydroxy-2-nonenal (4-HNE) as a biomarker. 4-HNE is derived from the peroxidation of lipids. It is a reactive species that can react via the aldehyde function to form a Schiff base, and with amino and sulfhydryl functions via a Michael addition to form cyclic derivatives. See Esterbauer, H., Schaur, R.J., and Zollner, H., Chemistry and biochemistry of 4-hydroxynonenal, malonaldehyde and related aldehydes, *Free Radic. Biol. Med.* 11, 81–128, 1991; Schaur, R.J., Basic aspects of the biochemical reactivity of 4-hydroxynonenal, *Mol. Aspects Med.* 24, 149–159, 2003; Petersen, D.R. and Doorn, J.A., Reactions of 4-hydroxynonenal with proteins and cellular targets, *Free Radic. Biol. Med.* 37, 937–945, 2004; Minko, I.G., Kozekov, I.D., Harris, T.M. et al., Chemical and biology of DNA containing 1,*N*²-deoxyguanosine adducts of the α,β-unsaturated aldehydes acrolein, crotonaldehyde, and 4-hydroxynonenal, *Chem. Res. Toxicol.* 22, 759–778, 2009; Sowell, J., Conway, H.M., Bruno, R.S. et al., Ascorbylated 4-hydroxy-2-nonenal as a potential biomarker of oxidative stress response, *J. Chromatogr. B* 827, 139–145, 2005.

TABLE 5.2
4-Hydroxynonenal as a Biomarker

Biomarker	References
Immunohistochemical identification of carbonyl modification products ("carbonyl stress") in diabetic glomerular lesions as a biomarker for increased oxidative stress	40
Measurement of 4-hydroxynonenal (4-HNE) in urine as a biomarker for in vivo lipid peroxidation occurring during oxidative stress	41
Exocyclic DNA adducts as biomarkers for familial adenomatous polyposis. The exocyclic DNA adducts derived from lipid peroxidation	42
4-HNE as a biomarker in serum for acute pancreatitis. The elevation of 4-HNE is associated with decrease in sulfhydryl groups	43
Serum 4-HNE is a biomarker for oxidative stress	44
4-HNE as histochemical biomarker in kidney for methamphetamine toxicity	45
4-HNE as oxidative stress biomarker determined with immunohistochemistry. It is suggested that proteins modified with 4-HNE may be involved in etiology of chorioamnionitis	46
4-HNE in exhaled breath condensate (EBC) as a biomarker for oxidative stress. Other biomarker of oxidative stress such as hydrogen peroxide and malondialdehyde are also found in EBC	47
Hemoglobin modified with 4-HNE as a biomarker of oxidative stress	48
Elevated 4-HNE in brain tissue from subjects with mild cognitive impairment and early Alzheimer's disease. It is suggested that 4-HNE is a biomarker for the risk of Alzheimer's disease	49, 50
Ascorbylated 4-HNE as a biomarker for oxidative stress	51
Use of 1,4-dihydroxynonane-mercapturic acid, a metabolite of 4-HNE, as a biomarker in urine	52–54
4-HNE as a biomarker together with protein oxidation (carbonyl formation) and DNA damage as clinical oxidation parameters of aging	55
4-HNE-protein adducts as potential biomarkers for coronary artery disease	56
Exocyclic DNA adducts as biomarkers for precancerous lesions	57
4-HNE-protein adducts as biomarker of lipid oxidation in liver disease	58, 59
4-HNE as a predictive biomarker for the use of N-acetylcysteine to reduce oxidative stress in type 2 diabetes mellitus	60
Elevation of 4-HNE in brain arterovenous malformation	61

There are a variety of posttranslational modifications that influence protein function.[1] In addition to the covalent modifications of structure such as carboxylation, phosphorylation, and acylation (e.g., farnesylation), modifications such as limited proteolysis and splicing are included in this category. The absence of a required modification such as the γ-carboxylation of the vitamin K-dependent proteins also results in the formation of biomarkers such as protein induced by vitamin K absence (PIVKA) proteins.[92,93] PIVKA was the term used to describe proteins, mainly acarboxyprothrombin, which was synthesized in the presence of vitamin K antagonists such as the coumarin derivatives. PIVKA has received more attention recently as a biomarker for hepatocellular cancer.[94,95] PIVKA-II (undercarboxylated prothrombin)

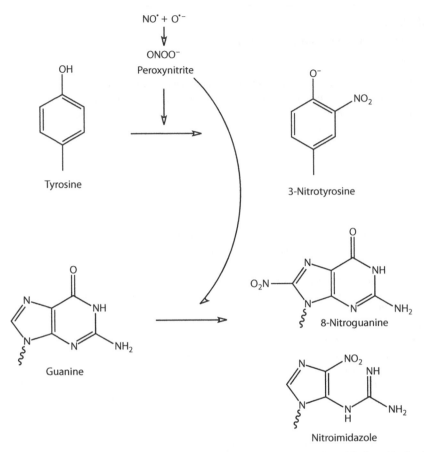

FIGURE 5.3 The reaction of peroxynitrite with protein and nucleic acid. See Abello, N., Kerstjens, H.A., Postma, D.S. et al., Protein tyrosine nitration: Selectivity, physicochemical and biological consequences, denitration, and proteomics methods of the identification of tyrosine-nitrated proteins, *J. Proteome Res.* 8, 3222–3238, 2009; Son, J., Pang, B., McFaline, J.L. et al., Surveying the damage: The challenges of developing nucleic acid biomarkers of inflammation, *Mol. Biosyst.* 4, 902–908, 2008.

was also observed in ICU patients.[96] The citrullination of proteins is a posttranslational modification event used as biomarker for rheumatoid arthritis[97,98] (see Chapter 2). Some other posttranslational modifications that serve as biomarkers are listed in Table 5.4.

Nucleic acids can also serve as biomarkers[109–113] where detection of trace amounts is made possible by use of the polymerase chain reaction.[114] Most of the interest is focused on tumor-derived nucleic acid.[109–120] Some selected nucleic acid biomarkers are listed in Table 5.5. DNA concentration is higher in serum than plasma and there is a division of opinion as to the most reliable source for evaluation of biomarkers.[131–135]

TABLE 5.3
The Use of Nitration as Biomarker

Study	Reference
Urinary nitrotyrosine as biomarker of oxidative stress. Urinary free nitrotyrosine determined (after derivatization) by gas chromatography. Urinary nitrotyrosine may be predictive of antioxidant therapy	82
Nitrotyrosine (protein bound detected by immunoassay) in diabetic plasma as biomarker of oxidative stress	84
Nitration of tyrosine-99 in calmodulin as a cellular biomarker of oxidative stress	85
Nitration of protein as biomarker for exhaustive oxidative stress. Urinary nitrotyrosine and protein carbonyl measured in subjects after a 4 day super marathon	86
Nitrotyrosine as biomarker for peroxynitrite-mediated damage in dopamine neuronal system	87
Use of modified tau protein (including nitrated tau) in cerebrospinal fluid as diagnostic biomarker for Alzheimer's disease	88
Protein-bound nitrotyrosine in plasma is decreased in patients with Raynaud's syndrome	89
Plasma nitrotyrosine as biomarker for endothelial dysfunction in obese individuals	90
Immunohistochemistry demonstrates protein nitration in brain as biomarker for transition from mild cognitive impairment to Alzheimer's disease	91

TABLE 5.4
Posttranslational Modifications as Biomarker

Study	References
Lipid-modified protein as biomarker for cardiovascular disease (review)	99
Oxidized transthyretin as a biomarker in cerebral spinal fluid for Alzheimer's disease	100
Review of posttranslational protein modifications as biomarkers for cancer	101
Phosphorylation of fibrinogen as biomarker for ovarian cancer	102
Posttranslational modifications of transthyretin as biomarkers in serum for mycosis fungoids, *Neoplasia* 9, 254–259, 2007	103
Succinylation of cysteine as a biomarker for mitochondrial stress in metabolic stress syndrome	104, 105
Immunohistochemical identification of trimethylated histone (trimethylated histone 3 lysine 27, H3K27triMe) as biomarker for esophageal squamous cell carcinoma	106
Isomerization of aspartate as a biomarker in osteoarthritis	107
Arginine methylation as a biomarker for obstructive coronary artery disease	108

TABLE 5.5

Nucleic Acid Biomarkers in Blood

Biomarker/Study	References
Microsatellite DNA markers (PCR analysis) in serum as biomarker for tumor burden in head and neck cancer. Microsatellite alternations in serum DNA matched those in primary tumors in 6 of 21 patients	121
Tumor-derived mRNA can be extracted and amplified using RT-PCR in patients with breast cancer	122
Epstein–Barr virus (EBV) in serum and plasma as biomarker of EBV-associated malignancies	123
Circulating DNA as a result of tumor cell death (reviews)	124, 125
Plasma hnRNP B1 mRNA as a diagnostic biomarker (RT-PCR) in plasma for lung cancer	126
Circulating DNA in plasma or serum as diagnostic and/or prognostic biomarker in lung cancer	127
EBV DNA in plasma as biomarker for nasopharyngeal cancer	128
Serum microRNA (miR141) is a biomarker for prostate cancer	129
Longer DNA fragments derived from non-apoptotic cells in breast cancer are the major contributors to increased DNA levels during adjuvant systemic chemotherapy	130
Increased concentration of circulating DNA in plasma as biomarker for endometriosis. Circulating DNA was much higher in serum than plasma but there was no significant difference between individuals with endometriosis and the controlled population	131

REFERENCES

1. Walsh, C.T., Garneau-Tsodikova, S., and Gatto, G.J. Jr., Protein posttranslational modifications: The chemistry of proteome diversifications, *Angew. Chem. Int. Ed. Engl.* 44, 7342–7372, 2005.
2. Kobata, A. and Amano, J., Altered glycosylation of proteins produced by malignant cells and application for the diagnosis and immunotherapy of tumours, *Immunol. Cell Biol.* 83, 429–439, 2005.
3. Cole, L.A., Hyperglycosylated hCG, *Placenta* 28, 977–986, 2007.
4. Mehta, A. and Block, T.M., Fucosylated glycoproteins as markers of liver disease, *Dis. Markers* 25, 259–265, 2008.
5. Wei, X. and Li, L., Comparative glycoproteomics: Approaches and applications, *Brief Funct. Genomic Proteomic* 8, 104–113, 2009.
6. Kim, Y.S., Yoo, H.S., and Ko, J.H., Implications of aberrant glycosylation in cancer and use of lectin for cancer biomarker discovery, *Protein Pept. Lett.* 16, 499–507, 2009.
7. Harvey, S.B., Zhang, Y., Wilson-Grady, J. et al., *O*-Glycoside biomarker of apolipoprotein C3: Responsiveness to obesity, bariatric surgery, and therapy with metformin, to chronic or severe disease and to mortality in severe sepsis and graft vs host disease, *J. Proteome Res.* 8, 603–612, 2009.
8. Kuno, A., Kato, Y., Matsuda, A. et al., Focused differential glycan analysis with the platform antibody-assisted lectin profiling for glycan-related biomarker verification, *Mol. Cell. Proteomics* 9, 99–108, 2009.
9. Oberrauch, W., Bergman, A.C., and Helander, A., HPLC and mass spectrometric characterization of a candidate reference material for the alcohol biomarker carbohydrate-deficient transferring (CDT), *Clin. Chim. Acta* 395, 142–145, 2008.

10. Bergstrom, J.P. and Helander, A., Influence of alcohol use, ethnicity, age, gender, BMI and smoking on the serum transferrin glycoform pattern: Implications for use of carbohydrate-deficient transferring (CDT) as alcohol biomarker, *Clin. Chim. Acta* 388, 59–67, 2008.

11. Golka, K. and Wiese, A., Carbohydrate-deficient transferring (CDT)—A biomarker for long-term alcohol consumption, *J. Toxicol. Environ. Health B: Crit. Rev.* 7, 319–327, 2004.

12. Qiu, Y., Patwa, T.H., Xu, L. et al., Plasma glycoprotein profiling for colorectal cancer biomarker identification by lectin glycoarray and lectin blot, *J. Proteome Res.* 7, 1693–1703, 2008.

13. Ueda, K., Katagiri, T., Shimada, T. et al., Comparative profiling of serum glycoproteome by sequential purification of glycoproteins and 2-nitrobenzenesulfonyl (NBS) stable isotope labeling: A new approach for the novel biomarker discovery for cancer, *J. Proteome Res.* 6, 3475–3483, 2007.

14. Yang, Z., Harris, L.E., Palmer-Toy, D.E., and Hancock, W.S., Multilectin affinity chromatography for characterization of multiple glycoprotein biomarker candidates in serum from breast cancer patients, *Clin. Chem.* 52, 1897–1905, 2006.

15. Sáez-Valero, J., Mok, S.S., and Small, D.H., An unusually glycosylated form of acetylcholinesterase is a CSF biomarker for Alzheimer's disease, *Acta Neurol. Scand. Suppl.* 176, 49–52, 2000.

16. Wang, M., Long, R.E., Comunale, M.A. et al., Novel fucosylated biomarkers for the early detection of hepatocellular carcinoma, *Cancer Epidemiol. Biomarkers Prev.* 18, 1914–1921, 2009.

17. Comunale, M.A., Wang, M., Hafner, J. et al., Identification and development of fucosylated glycoproteins as biomarkers of primary hepatocellular carcinoma, *J. Proteome Res.* 8, 595–602, 2009.

18. Ueda, K., Fukase, Y., Katagiri, T. et al., Targeted serum glycoproteomics for the discovery of lung cancer-associated glycosylation disorders using lectin-coupled ProteinChip arrays, *Proteomics* 9, 2182–2192, 2009.

19. Vanhooren, V., Liu, X.E., Franceschi, C. et al., *N*-glycan profiles as tools in diagnosis of hepatocellular carcinoma and prediction of healthy human ageing, *Mech. Ageing Dev.* 130, 92–97, 2009.

20. Radziejewska, I., Borzym-Kluczyk, M., Kisiel, D.G. et al., Characterisation of glycoforms of ascitic fluids in benign and malignant diseases, *Clin. Biochem.* 42, 72–77, 2009.

21. Kuzmanov, U., Jiang, N., Smith, C.R. et al., Differential *N*-glycosylation of kallikrein 6 derived from ovarian cancer cells or the central nervous system, *Mol. Cell. Proteomics* 8, 791–798, 2009.

22. Robbe-Masselot, C., Hermann, A., Maes, E. et al., Expression of a core 3 disialyl-Le(x) hexasaccharide in human colorectal cancer: A potential marker of malignant transformation in colon, *J. Proteome Res.* 8, 702–711, 2009.

23. Abd Hamid, U.M., Royle, L., Saldova, R. et al., A strategy to reveal potential glycan markers from serum glycoproteins associated with breast cancer progression, *Glycobiology* 18, 1105–1108, 2008.

24. Alavi, A. and Axford, J.S., Glyco-biomarkers: Potential determinants of cellular physiology and pathology, *Dis. Markers* 25, 193–205, 2008.

25. Thompson, S., Guthrie, D., and Turner, G.A., Fucosylated forms of α-1-antitrypsin that predict unresponsiveness to chemotherapy in ovarian cancer, *Br. J. Cancer* 58, 589–593, 1988.

26. Aoyagi, Y., Suzuki, Y., Isemura, M. et al., The fucosylation index of α-fetoprotein and its usefulness in the early diagnosis of hepatocellular carcinoma, *Cancer* 61, 769–774, 1988.

27. Yan, L., Wilkins, P.P., Alvarez-Manilla, G. et al., Immobilized *Lotus tetragonolobus* agglutinin bind oligosaccharides containing the Lc(x) determinant, *Glycoconj. J.* 14, 45–55, 1997.
28. Naitoh, A., Aoyagi, Y., and Asakura, H., Highly enhanced fucosylation of serum glycoproteins in patients with hepatocellular carcinoma, *J. Gastroenterol. Hepatol.* 14, 436–445, 1999.
29. Miyoshi, E. and Nakano, M., Fucosylated haptoglobin is a novel marker for pancreatic cancer: Detailed analyses of oligosaccharide structures, *Proteomics* 8, 3257–3262, 2008.
30. Matsumura, K., Higashida, K., Hata, Y. et al., Comparative analysis of oligosaccharide specificities of fucose-specific lectins from *Aspergillus oryzae* and *Aleuria aurantia* using frontal affinity chromatography, *Anal. Biochem.* 386, 217–221, 2009.
31. Zhao, J., Patwa, T.H., Qui, W. et al., Glycoprotein microarrays with multilectin detection: Unique lectin binding patterns as a tool for classifying normal chronic pancreatitis and pancreatic cancer sera, *J. Proteome Res.* 6, 1864–1874, 2007.
32. Ito, H., Kuno, A., Sawaki, H. et al., Strategy for glycoproteomics: Identification of glycol-alteration using multiple glycan profiling tools, *J. Proteome Res.* 8, 1358–1367, 2009.
33. Meerwaldt, R., van der Vaart, M.G., van Dam, G.M. et al., Clinical relevance of advanced glycation endproducts for vascular surgery, *Eur. J. Vasc. Endovasc. Surg.* 36, 125–131, 2008.
34. Peppa, M. and Raptis, S.A., Advanced glycation end products and cardiovascular disease, *Curr. Diab. Rep.* 4, 92–100, 2008.
35. Onorato, J.M., Thorpe, S.R., and Baynes, J.W., Immunohistochemical and ELISA assays for biomarkers of oxidative stress in aging and disease, *Ann. N. Y. Acad. Sci.* 28, 1745–1750, 2000.
36. Uchida, K., 4-Hydroxy-2-nonenal: A product and mediator of oxidative stress, *Prog. Lipid Res.* 42, 318–343, 2003.
37. Schneider, C., Porter, N.A., and Brash, A.R., Routes to 4-hydroxynonenal: Fundamental issues in the mechanisms of lipid peroxidation, *J. Biol. Chem.* 283, 15539–15543, 2009.
38. Petersen, D.R. and Doorn, J.A., Reactions of 4-hydroxynonenal with proteins and cellular targets, *Free Radic. Biol. Med.* 37, 937–945, 2004.
39. Minko, I.G., Kozekov, I.D., Harris, T.M. et al., Chemistry and biology of DNA containing $1,N^2$-deoxyguanosine adducts of the α,β-unsaturated aldehydes acrolein, crotonaldehyde, and 4-hydroxynonenal, *Chem. Res. Toxicol.* 22, 759–778, 2009.
40. Suzuki, D., Miyata, T., Saotome, N. et al., Immunohistochemical evidence for an increased oxidative stress and carbonyl modification of proteins in diabetic glomerular lesions, *J. Am. Soc. Nephrol.* 10, 822–832, 1999.
41. Meagher, E.A. and FitzGerald, G.A., Indices of lipid peroxidation in vivo: Strengths and limitations, *Free Radic. Biol. Med.* 28, 1745–1750, 2000.
42. Bartsch, H., Nair, J., and Owen, R.W., Exocyclic DNA adducts as oxidative stress markers in colon carcinogenesis: Potential role of lipid peroxidation, dietary fat and antioxidants, *Biol. Chem.* 383, 915–921, 2002.
43. Wereszcczynska-Siemiatkowska, U., Dabrowska, A., Siemiatkowski, A. et al., Serum profiles of E-selectin, interleukin-10, and interleukin-6 and oxidative stress parameters in patients with acute pancreatitis and nonpancreatic acute abdominal pain, *Pancreas* 26, 144–152, 2003.
44. Zarkovic, N., 4-Hydroxynonenal as a bioactive marker of pathophysiological processes, *Mol. Aspects Med.* 24, 281–291, 2003.
45. Ishigami, A., Tokunaga, I., Gotohda, T., and Kubo, S., Immunohistochemical study of myoglobin and oxidative injury-related markers in the kidney of methamphetamine abusers, *Leg. Med.* (Tokyo) 5, 42–48, 2003.

46. Temma, K., Shimoya, K., Zhang, Q. et al., Effects of 4-hydroxy-2-nonenal, a marker of oxidative stress, on the cyclooxygenase-2 of human placenta in chorioamnionitis, *Mol. Hum. Reprod.* 10, 167–171, 2004.

47. Rahman, I. and Biswas, S.K., Non-invasive biomarkers of oxidative stress: Reproducibility and methodological issues, *Redox Rep.* 9, 125–143, 2004.

48. Yocum, A.K., Oe, T., Yergey, A.L., and Blair, I.A., Novel lipid hydroperoxide-derived hemoglobin histidine adducts as biomarkers of oxidative stress, *J. Mass Spectrom.* 40, 754–764, 2005.

49. Williams, T.I., Lynn, B.C., Markesbery, W.R., and Lovell, M.A., Increased levels of 4-hydroxynonenal and acrolein, neurotoxic markers of lipid peroxidation, in the brain in Mild Cognitive Impairment and early Alzheimer's disease, *Neurobiol. Aging* 27, 1094–1099, 2006.

50. Reed, T., Perluigi, M., Sultana, R. et al., Redox proteomic identification of 4-hydroxy-2-nonenal-modified brain proteins in amnestic mild cognitive impairment: Insight into the role of lipid peroxidation in the progression and pathogenesis of Alzheimer's disease, *Neurobiol. Dis.* 30, 107–120, 2008.

51. Sowell, J., Conway, H.M., Bruno, R.S. et al., Ascorbylated 4-hydroxy-2-nonenal as a potential biomarker of oxidative stress response, *J. Chromatogr. B* 827, 139–145, 2005.

52. Guéraud, F., Peiro, G., Bernard, H. et al., Enzyme immunoassay for a urinary metabolite of 4-hydroxynonenal as a marker of lipid peroxidation, *Free Radic. Biol. Med.* 40, 54–62, 2006.

53. Peiro, G., Alary, J., Cravedi, J.P. et al., Dihydroxynonene mercapturic acid, a urinary metabolite of 4-hydroxynonenal, as a biomarker of lipid peroxidation, *Biofactors* 24, 89–96, 2005.

54. Kuiper, H.C., Miranda, C.L., Sowell, J.D., and Stevens, J.F., Mercapturic acid conjugates of 4-hydroxy-2-nonenal and 4-oxo-2-nonenal metabolites are in vivo markers of oxidative stress, *J. Biol. Chem.* 283, 17131–17138, 2008.

55. Voss, P. and Siems, W., Clinical oxidation parameters of aging, *Free Radic. Res.* 40, 1339–1349, 2006.

56. Sottero, B., Pozzi, R., Leonarduzzi, G. et al., Lipid peroxidation and inflammatory molecules as markers of coronary artery disease, *Redox Rep.* 12, 81–85, 2007.

57. Nair, U., Bartsch, H., and Nair, J., Lipid peroxidation-induced DNA damage in cancer-prone inflammatory diseases: A review of published adduct types and levels in humans, *Free Radic. Biol. Med.* 43, 1109–1120, 2007.

58. Poli, G., Biasi, F., and Leonarduzzi, G., 4-Hydroxynonenal-protein adducts: A reliable biomarker of lipid oxidation in liver diseases, *Mol. Aspects Med.* 29, 67–71, 2008.

59. Thiele, G.M., Klassen, L.W., and Tuma, D.J., Formation and immunological properties of aldehyde-derived protein adducts following alcohol consumption, *Methods Mol. Biol.* 447, 235–257, 2008.

60. Masha, A., Brocato, L., Dinatale, S. et al., N-Acetylcysteine is able to reduce the oxidation status and the endothelial activation after a high-glucose content meal in patients with Type 2 diabetes mellitus, *J. Endocrinol. Invest.* 32, 352–356, 2009.

61. Chen. Y., Hao, Q., Kim, H. et al., Soluble endoglobulin modulates aberrant cerebral vascular remodeling, *Ann. Neurol.* 66, 19–27, 2009.

62. Carini, M., Aldini, G., and Facino, R.M., Mass spectrometry for detection of 4-hydroxy-trans-2-nonenal (HNE) adducts with peptides and proteins, *Mass Spectrom. Rev.* 23, 281–305, 2004.

63. Aldini, G., Gamberoini, L., Orioli, M. et al., Mass spectrometric characterization of covalent modification of human serum albumin by 4-hydroxy-trans-2-nonenal, *J. Mass Spectrom.* 41, 1149–1161, 2006.

64. Uchida, K., Itakura, K., Kawakishi, S. et al., Characterization of epitopes recognized by 4-hydroxy-2-nonenal specific antibodies, *Arch. Biochem. Biophys.* 324, 241–248, 1995.

65. Ethen, C.M., Reilly, C., Feng, X. et al., Age-related macular degeneration and retinal protein modification by 4-hydroxy-2-nonenal, *Invest. Ophthalmol. Vis. Sci.* 48, 3469–3479, 2007.

66. Mannick, J.B. and Schonhoff, C.M., Nitrosylation: The next phosphorylation?, *Arch. Biochem. Biophys.* 408, 1–6, 2002.

67. Niles, J.C., Wishnok, J.S., and Tannebaum, S.R., Peroxynitrite-induced oxidation and nitration products of guanine and 9-oxoguanine: Structures and mechanisms of product formation, *Nitric Oxide* 14, 109–121, 2006.

68. Son, J., Pang, B., McFaline, J.L. et al., Surveying the damage: The challenges of developing nucleic acid biomarkers of inflammation, *Mol. Biosyst.* 4, 902–908, 2008.

69. Souza, J.M., Peluffo, G., and Radi, R., Protein tyrosine nitration—Functional alteration or just a biomarker?, *Free Radic. Biol. Med.* 45, 357–366, 2008.

70. Torta, F., Ursuelli, V., Malgaroli, A., and Bachi, V., Proteomic analysis of protein S-nitrosylation, *Proteomics* 8, 4484–4494, 2008.

71. Urshio-Fukai, M., Vascular signaling through G protein-coupled receptors: New concepts, *Curr. Opin. Nephrol. Hypertens.* 18, 153–159, 2009.

72. Herce-Pagliai, C., Kotecha, S., and Shuker, D.E., Analytical methods for 3-nitrotyrosine as a marker of exposure to reactive nitrogen species: A review, *Nitric Oxide* 2, 324–336, 1998.

73. Safinowski, M., Wilhelm, B., Reimer, T. et al., Determination of nitrotyrosine concentrations in plasma samples of diabetes mellitus patients by four different immunoassays leads to contradictive results and disqualifies the majority of the tests, *Clin. Chem. Lab. Med.* 47, 483–488, 2009.

74. Murray, J., Oquendo, C.E., Willis, J.H. et al., Monitoring oxidative and nitrative modification of cellular proteins: A paradigm for identifying key disease related markers of oxidative stress, *Adv. Drug. Deliv. Rev.* 60, 1497–1503, 2008.

75. Nuriel, T., Deeb, R.S., Haller, D.P., and Gross, S.S., Protein 3-nitrotyrosine in complex biological samples: Quantification by high-performance liquid chromatography/electrochemical detection and emergence of proteomic approaches for unbiased identification of modification sites, *Methods Enzymol.* 441, 1–17, 2008.

76. Raina, A.K., Perry, G., Nunomora, A. et al., Histochemical and immunocytochemical approaches to the study of oxidative stress, *Clin. Chem. Lab. Med.* 38, 93–97, 2000.

77. Sultana, R. and Butterfield, D.A., Slot-blot analysis of 3-nitrotyrosine-modified brain proteins, *Methods Enzymol.* 440, 309–316, 2008.

78. Gaut, J.P., Byun, J., Tran, H.D. et al., Artifact-free quantification of free 3-chlorotyrosine, 3-bromotyrosine, and 3-nitrotyrosine in human plasma by electron capture-negative chemical ionization gas chromatography mass spectrometry and liquid chromatography-electrospray ionization tandem mass spectrometry, *Anal. Biochem.* 300, 252–259, 2002.

79. Elfatih, A., Anderson, N.R., Mansoor, S. et al., Plasma nitrotyrosine in reversible myocardial ischaemia, *J. Clin. Pathol.* 58, 95–96, 2005.

80. Zhang, W.Z., Lang, C., and Kaye, D.M., Determination of plasma free 3-nitrotyrosine and tyrosine by reversed-phase liquid chromatography with 4-fluoro-7-nitrobenzofurazan derivatization, *Biomed. Chromatogr.* 21, 273–278, 2007.

81. Chen, H.J. and Chiu, W.L., Simultaneous detection and quantification of 3-nitrotyrosine and 3-bromotyrosine in human urine by stable isotope dilution liquid chromatography tandem mass spectrometry, *Toxicol. Lett.* 181, 31–39, 2008.

82. Schwemmer, M., Fink, B., Köckerbauer, R., and Bassenge, E., How urine analysis reflects oxidative stress—Nitrotyrosine as a potential marker, *Clin. Chim. Acta* 297, 207–216, 2000.

83. Tsikas, D., Mitschke, A., Suchy, M.T. et al., Determination of 3-nitrotyrosine in human urine at the basal state by gas chromatography-tandem mass spectrometry and evaluation of the excretion after oral intake, *J. Chromatogr. B* 827, 146–156, 2005.

84. Ceriello, A., Mercuri, F., Quagliaro, L. et al., Detection of nitrotyrosine in the diabetic plasma: Evidence of oxidative stress, *Diabetologia* 44, 834–838, 2001.
85. Smallwood, H.S., Galeva, N.A., Bartlett, R.K. et al., Selective nitration of Tyr99 in calmodulin as a marker of cellular conditions of oxidative stress, *Chem. Res. Toxicol.* 16, 95–102, 2003.
86. Radák, Z., Ogonovszky, H., Dubecz, J. et al., Super-marathon race increases serum and urinary nitrotyrosine and carbonyl levels, *Eur. J. Clin. Invest.* 33, 726–730, 2003.
87. Kuhn, D.M., Sakowski, S.A., Sadidi, M., and Geddes, T.J., Nitrotyrosine as a marker for peroxynitrite-induced neurotoxicity: The beginning or the end of dopamine neurons?, *J. Neurochem.* 89, 529–536, 2004.
88. Gong, C.X., Liu, E., Grundke-Igbal, I., and Iqbal, K., Post-translational modifications of tau protein in Alzheimer's disease, *J. Neural Transm.* 112, 813–838, 2005.
89. Kingdon, E.J., Mani, A.R., Frost, M.T. et al., Low plasma protein nitrotyrosine levels distinguish primary Raynaud's phenomenon from scleroderma, *Ann. Rheum. Dis.* 65, 952–954, 2006.
90. Rector, R.S., Turk, J.R., Sun, G.Y. et al., Short-term lifestyle modification alters circulating biomarkers of endothelial health in sedentary, overweight adults, *Appl. Physiol. Nutr. Metab.* 31, 512–517, 2006.
91. Butterfield, D.A., Reed, T.T., Perluigi, M. et al., Elevated levels of 3-nitrotyrosine in brain from subjects with amnestic mild cognitive impairment: Implications for the role of nitration in the progress of Alzheimer's disease, *Brain Res.* 1148, 243–248, 2007.
92. Hemker, H.C. and Muller, A.D., Kinetic aspects of the interaction of blood-clotting enzymes. VI. Localization of the site of blood-coagulation inhibition by the protein induced by vitamin K absence (PIVKA), *Thromb. Diath. Haemorrh.* 20, 78–87, 1968.
93. Veltkamp, J.J., Detection and clinical significance of PIVKA, *Mayo Clin. Proc.* 49, 923–924, 1974.
94. Zhou, L., Liu, J., and Luo, F., Serum tumor markers for detection of hepatocellular carcinoma, *World J. Gastroenterol.* 12, 1175–1181, 2006.
95. Gomaa, A.I., Khan, S.A., Leen, E.L. et al., Diagnosis of hepatocellular carcinoma, *World J. Gastroenterol.* 15, 1301–1314, 2009.
96. O'Shaughnessy, D., Allen, C., Woodcock, T. et al., Echis time, under-carboxylated prothrombin and vitamin K status is intensive care patients, *Clin. Lab. Hematol.* 25, 397–404, 2003.
97. Nijenhuis, S., Zendman, A.J., Vossenaar, E.R. et al., Autoantibodies to citrullinated proteins in rheumatoid arthritis: Clinical performance and biochemical aspects of an RA-specific marker, *Clin. Chim. Acta* 350, 17–34, 2004.
98. Chang, X., Yamada, R., Suzuki, A. et al., Citrullination of fibronectin in rheumatoid arthritis synovial tissue, *Rheumatology* 44, 1374–1382, 2005.
99. Ferri, N., Paoletti, R., and Corsini, A., Lipid-modified proteins as biomarkers for cardiovascular disease: A review, *Biomarkers* 10, 219–237, 2005.
100. Biroccio, A., Del Boccio, P., Panella, M. et al., Differential post-translational modifications of transthyretin in Alzheimer's disease: A study of the cerebral spinal fluid, *Proteomics* 6, 2305–2313, 2006.
101. Krueger, K.E. and Srivastava, S., Posttranslational protein modifications: Current implications for cancer detection, prevention and therapeutics, *Mol. Cell. Proteomics* 5, 1799–1810, 2006.
102. Ogata, Y., Heppelmann, C.J., Charlesworth, M.C. et al., Elevated levels of phosphorylated fibrinogen-α-isoforms and differential expression of other post-translationally modified proteins in the plasma of ovarian cancer patients, *J. Proteome Res.* 5, 3318–3325, 2006.
103. Escher, N., Kaatz, M., Melle, C. et al., Posttranslational modifications of transthretin are serum markers in patients with mycosis fungoides, *Neoplasia* 9, 254–259, 2007.

104. Nagai, R., Brock, J.W., Blatnik, M. et al., Succination of protein thiols during adipocyte maturation: A biomarker of mitochondrial stress, *J. Biol. Chem.* 282, 34219–34228, 2007.

105. Frizzell, N., Rajesh, M., Jepson, M.J. et al., Succination of thiol groups in adipose tissue proteins in diabetes: Succination inhibits polymerization and secretion of adiponectin, *J. Biol. Chem.* 284, 25772–25781, 2009.

106. Tzao, C., Tung, H.J., Jin, J.S. et al., Prognostic significance of global histone modifications in resected squamous cell carcinoma of the esophagus, *Mod. Pathol.* 22, 252–260, 2009.

107. Catterall, J.B., Barr, D., Bolognesi, M. et al., Post-translational aging of proteins in osteoarthritic cartilage and synovial fluid as measured by isomerized aspartate, *Arthritis Res. Ther.* 11, R55, 2009.

108. Wang, Z., Tang, W.H., Cho, L. et al., Targeted metabolomic evaluation of arginine methylation and cardiovascular risks: Potential mechanisms beyond nitric oxide synthase inhibition, *Arterioscler. Thromb. Vasc. Biol.* 29, 1383–1391, 2009.

109. Bremnes, R.M., Sirera, R., and Camps, C., Circulating tumour-derived DNA and RNA markers in blood: A tool for early detection, diagnostics, and follow-up?, *Lung Cancer* 49, 1–12, 2005.

110. Gormally, E., Caboux, E., Vineis, P., and Hainaut, P., Circulating free DNA in plasma or serum as biomarker of carcinogenesis: Practical aspects and biological significance, *Mutat. Res.* 635, 105–117, 2007.

111. Tsang, J.C. and Lo, Y.M., Circulating nucleic acids in plasma/serum, *Pathology* 39, 197–207, 2007.

112. Gahan, P.B., Circulating nucleic acids in plasma and serum: Roles in diagnosis and prognosis in diabetes and cancer, *Infect. Disord. Drug Targets* 8, 100–108, 2008.

113. Holdenrieder, S., Nagel, D., Shalhorn, A. et al., Clinical relevance of circulating nucleosomes in cancer, *Ann. N. Y. Acad. Sci.* 1137, 180–189, 2008.

114. Raj, G.V., Moreno, J.G., and Gomella, L.G., Utilization of polymerase chain reaction technology in the detection of solid tumors, *Cancer* 82, 1419–1442, 1998.

115. Zitt, M., Müller, H.M., Rochel, M. et al., Circulating cell-free DNA in plasma of locally advanced rectal cancer patients undergoing preoperative chemoradiation: A potential diagnostic tool for therapy monitoring, *Dis. Markers* 25, 159–165, 2008.

116. Cheng, C., Omura-Minamisawa, M., Kang, Y. et al., Quantification of circulating cell-free DNA in the plasma of cancer patients during radiation therapy, *Cancer Sci.* 100, 303–309, 2009.

117. Yoon, K.A., Park, S., Lee, S.H. et al., Comparison of circulating plasma DNA levels between lung cancer patients and healthy controls, *J. Mol. Diagn.* 11, 182–185, 2009.

118. Hohaus, S., Giachelia, M., Massini, G. et al., Cell-free circulating DNA in Hodgkin's and non-Hogkin's lymphomas, *Ann. Oncol.* 20, 1408–1413, 2009.

119. Chen, Z., Feng, J., Buzin, C.H. et al., Analysis of cancer mutation signatures in blood by a novel ultra-sensitive assay: Monitoring of therapy or recurrence in non-metastatic breast cancer, *PLoS One*, 4, e7220, 2009.

120. van der Drift, M.A., Hol, B.E., Klaassen, C.H. et al., Circulating DNA is a non-invasive prognostic factor for survival in non-small cell lung cancer, *Lung Cancer* 68, 283–287, 2010.

121. Narroz, H., Koch, W., Anker, P. et al., Microsatellite alterations in serum DNA of head and neck cancer patients, *Nat. Med.* 2, 1035–1037, 1996.

122. Chen, X.Q., Bonnefoi, H., Pelte, M.F. et al., Telomerase RNA as a detection marker in the serum of breast cancer patients, *Clin. Cancer Res.* 6, 3823–3826, 2000.

123. Lo, Y.M., Quantitative analysis of Epstein-Barr virus DNA in plasma and serum: Applications to tumor detection and monitoring, *Ann. N. Y. Acad. Sci.* 945, 68–72, 2001.

124. Ziegler, A., Zangemeister-Wittke, U., and Stahel, R.A., Circulating DNA: A new diagnostic gold mine?, *Cancer Treat. Rev.* 28, 255–271, 2002.

125. Goebel, G., Zitt, M., Zitt, M., and Müller, H.M., Circulating nucleic acids in plasma or serum (CNAPS) as prognostic and predictive markers in patients with solid neoplasias, *Dis. Markers* 21, 105–120, 2005.
126. Sueoka, E., Sueoka, N., Iwanaga, K. et al., Detection of plasma hnRNP B1 mRNA, a new cancer biomarker, in lung cancer patients by quantitative real-time polymerase chain reaction, *Lung Cancer* 48, 77–83, 2005.
127. Pathak, A.K., Bhutani, M., Kumar, S. et al., Circulating cell-free DNA in plasma/serum of lung cancer patients as a potential screening and prognostic tool, *Clin. Chem.* 52, 1833–1842, 2006.
128. Leung, S.F., Zee, B., Ma, B.B. et al., Plasma Epstein-Barr viral deoxyribonucleic acid quantitation complements tumor-node-metastasis staging prognostication in nasopharyngeal carcinoma, *J. Clin. Oncol.* 24, 5414–5418, 2006.
129. Mitchell, P.S., Parkin, R.K., Kroh, E.M. et al., Circulating microRNAs as stable blood-based markers for cancer detection, *Proc. Natl. Acad. Sci. USA* 105, 10513–10518, 2008.
130. Deligezer, U., Eralp, Y., Akisik, E.Z. et al., Effect of adjuvant chemotherapy on integrity of free serum DNA in patients with breast cancer, *Ann. N. Y. Acad. Sci.* 1137, 175–179, 2008.
131. Zachariah, R., Schmid, S., Radpour, R. et al., Circulating cell-free DNA as a potential biomarker for minimal and mild endometriosis, *Reprod. Biomed. Online* 18, 407–411, 2009.
132. Thijssen, M.A., Swinkels, D.W., Ruers, T.J., and de Kok, J.B., Difference between free circulating plasma and serum DNA in patients with colorectal liver metastases, *Anticancer Res.* 22, 421–425, 2002.
133. Umetani, M., Hiamatsu, S., and Hoon, D.S., Higher amount of free circulating DNA in serum than in plasma is not mainly caused by contaminated extraneous DNA during separation, *Ann. N. Y. Acad. Sci.* 1075, 299–307, 2006.
134. Zanetti-Daellenbach, R., Wright, E., Fan, A.X.C. et al., Positive correlation of cell-free DNA in plasma/serum in patients with malignant and benign breast disease, *Anticancer Res.* 28, 921–925, 2008.
135. Board, R.E., Williams, V.X., Knight, L. et al., Isolation and extraction of circulating tumor DNA from patients with small cell lung cancer, *Ann. N. Y. Acad. Sci.* 1137, 98–107, 2008.

6 Complex Biomarkers: Cells, Cell Membrane Proteins, and Cell Fragments as Biomarkers

At one time, it seemed like most of the biomarkers were simple molecules such as proteins or nucleic acids or molecular pathways (metabolomics). The conversion of laboratory analytes into biomarkers has been discussed earlier. The consideration of the literature reveals that cellular phenomena, complex physiological phenomena such as respiration and heart rate, and structures obtained by imaging can be considered as biomarkers (see Chapter 2). This chapter will be concerned with cells, cell membrane proteins, and cell fragments as biomarkers. The author observes that the common clinical laboratory hematology parameters such as hemoglobin, hematocrit, and white cell counts are biomarkers with the accepted definition. Erythrocyte sedimentation rate is a well-accepted biomarker for rheumatoid arthritis (see Chapter 2).

Endothelial cells are involved in vascular biology and with the development of stem-cell research there is considerable interest in endothelial progenitor cells.[1–12] Circulating endothelial cells[3,6] have been used as a biomarker for vascular lesions for almost 40 years,[13] and are distinguished from endothelial progenitor cells by the expression of CD146[6,14,15] as well as other surface markers.[15] Some examples of circulating endothelial as biomarkers are listed in Table 6.1.

Endothelial progenitor cells are mononuclear cells with enhanced potential for differentiation and are distinct from mature endothelial cells on the basis of the expression of both hematopoietic cell markers and endothelial cell markers.[1,11] Endothelial progenitor cells are involved in neovascularization[30–32] and in vascular repair.[10,33,34] The measurement of circulating endothelial progenitor cells[35] is suggested as a biomarker to measure regenerative capacity following vascular damage[12] and for bone repair[36] creating interest in the pharmacological intervention.[37] The measurement of endothelial progenitor cells is not trivial[35] with flow cytometry providing quantitative information and cell culture providing qualitative information. The flow cytometry has issues with respect to specificity and the cell culture method is not trivial and is time consuming.[35] The majority of studies use flow cytometry.[19,38–40] Circulating endothelial cells show diurnal variation[41] and are decreased in major depression[42] and in erectile dysfunction.[43] No change was observed in runners immediately following a marathon when there was an increase in inflammatory markers.[44] Some examples of the use of circulating endothelial progenitor cells as biomarkers are listed in Table 6.2. Circulating endothelial

TABLE 6.1

Circulating Endothelial Cells as Biomarkers

Study	References
Development of monoclonal antibody for measurement of circulating endothelial cells providing biomarker for the evaluation of vascular injury	16
Circulating endothelial cells as biomarkers for systemic lupus erythematosus	17
Circulating endothelial cells as biomarkers for systemic vasculitis (ANCA-associated small-vessel vasculitis). Dynabeads™ coated with antibody used to isolate the circulating endothelial cells for assay	18
Use of flow cytometry for the assay of circulating endothelial cells for biomarkers in vascular disease	19
Review of circulating endothelial cells as biomarkers for vascular disease, cardiovascular disease and neoplastic disease	3
Circulating endothelial cells as biomarkers for tumors; elevated in subjects with advanced malignancies	20
Circulating endothelial cells as predictive biomarkers for anti-VEGF therapy in oncology	21–23
Circulating endothelial cells as prognostic biomarkers in prostate cancer	24
Use of Dynabeads coated with antibody to CD146 used isolate and assay circulating endothelial cells as biomarkers for endothelial dysfunction in vasculitis related to systemic lupus erythematosus	25
Circulating endothelial cells biomarkers for irreversible pulmonary hypertension secondary to congenital heart disease. Circulating endothelial cells measured by binding to CD146 antibody bound to Dynabeads and by flow cytometry	26
Circulating endothelial cells as biomarkers in organ damage resulting from sickle cell anemia	27
CD105-positive circulating endothelial cells as biomarkers derived from damaged normal endothelium	28
Circulating endothelial cells as biomarkers for portal hypertension	29

progenitor cells are associated with a variety of situations involving angiogenesis and can be used as a biomarker for anti-VEGF therapy in oncology.

Circulating tumor cells are defined as tumor cells in the peripheral blood derived either from primary tumors or from metastasis and have considerable diagnostic and prognostic potential.[70–72] The detection of circulating tumor cells, however, is challenging as such cells are quite rare with as few as one circulating tumor cells in 10[6] leukocytes.[73] However, technology has been developed permitting the detection of one circulating tumor cell in 7.5 mL of peripheral blood[74] using CellSearch technology, which combines an immunomagnetic separation step followed by detection with immunofluorescence.[75,76] Current methods for the detection of circulating tumor cells include microfluidic platforms.[77–79] Attempts to use the sensitive techniques of molecular diagnostics such as RT-PCR have been unsuccessful.[80]

There is considerable interest in the development of circulating tumor cells as biomarkers. Tan and colleagues[81] discuss the value of circulating tumor cells as biomarkers in the early clinical trials in oncology. In order for circulating tumor cells to serve as biomarkers, there must be considerable work prior to the start of phase I

TABLE 6.2
Circulating Endothelial Progenitor Cells as Biomarkers

Study	Reference
Review of role of circulating endothelial progenitor cells as biomarkers in vascular dysfunction	45
Circulating endothelial progenitor cells as prognostic biomarkers in sepsis. Increased circulating endothelial progenitor cells is associated with a positive outcome	46
Circulating endothelial progenitor cells as biomarkers for regeneration and neovascularization follow acute myocardial infarction	47
Circulating endothelial cells as prognostic biomarkers for coronary artery disease. Decreased number (flow cytometry) and function (cell culture; ex vivo cell colony-forming units) associated with decreased coronary endothelial function (quantitative coronary angiography)	48
Circulating endothelial progenitor cells may be prognostic biomarkers for neovascular age-related macular degeneration	49
Increased dose of EPO required to maintain target hemoglobin in maintenance dialysis subjects associated with decrease in circulating endothelial progenitor cells	50
Circulating endothelial progenitor cell function as prognostic biomarker in stroke. Colony-forming units and tube formation (Matrigel) reduced in patients with acute stroke compared to subjects with chronic stroke or healthy subjects	51
Circulating progenitor endothelial cells are decreased in obesity while intima-media thickness is increased; weight loss associated with an increase in circulating progenitor endothelial cells and a decrease in intima-media thickness	52
Circulating endothelial progenitor cells as biomarkers in idiopathic pulmonary hypertension; treatment with sildenafil (phosphodiesterase inhibitor, Viagra™) increased circulating endothelial progenitor cells	53
Circulating endothelial progenitor cells may be diagnostic biomarkers for atherosclerosis. The regulation of CD34+ subsets seem to differ between coronary endothelial dysfunction and normal coronary endothelium	54
Circulating endothelial progenitor cells as biomarkers for endothelial dysfunction (and subsequent capillary loss and cardiac complications) in systemic scleroderma	55
Decrease in circulating endothelial progenitor cells following endarterectomy	56
Circulating endothelial progenitor cells were elevated in head and neck squamous cell cancer and remained elevated after treatment. The sample size was small and it is suggested that circulating endothelial progenitor cells may be a surrogate biomarker for anti-angiogenic therapy	57
Circulating endothelial progenitor cells as prognostic biomarkers for patients on chronic hemodialysis. It suggested that the circulating endothelial progenitor cells are a surrogate for vascular risk in these patients	58
Expression of osteocalcin (costaining of CD34+/KDR+ and CD34+/CD133+/KDR+ cell for osteocalcin in flow cytometry analysis) in circulating endothelial progenitor cells as biomarker for coronary atherosclerosis	59
Circulating endothelial progenitor cells as possible biomarkers for peripheral artery disease. Different methods of assay yielded different results decreased cells of CD34+ and CD133+ but increased proliferative activity in culture	60

(continued)

TABLE 6.2 (continued)
Circulating Endothelial Progenitor Cells as Biomarkers

Study	Reference
Circulating endothelial progenitor cells as prognostic biomarkers for cardiovascular disease in type D personality	61
Circulating endothelial progenitor cells as prognostic biomarkers for diabetic nephropathy; decrease in number of CD34+ cells with progressive diabetic nephropathy	62
Use of circulating endothelial progenitor cells as biomarker for effect of VEGF inhibitors in renal cell carcinoma	23
Circulating endothelial progenitor cells (CD133+/VEGF-2+) as surrogate biomarkers for tumor angiogenicity in glioblastoma multiforme	63
Circulating endothelial progenitor cells as biomarkers for efficacy of sunitinib in hepatocellular carcinoma	64
Transient increase in the number of circulating endothelial progenitor cells following implantation of ventricular assist device	65
Circulating endothelial progenitor cells as biomarkers for risk of cardiovascular disease in inflammatory bowel disease	66
Elevated levels of circulating endothelial progenitor cells as biomarkers for metastatic disease in pediatric solid tumor patients	67
Circulating endothelial progenitor cells as prognostic biomarkers in lung cancer	68
Circulating endothelial progenitor cells as prognostic biomarkers for metabolic syndrome; low CD34+ predicted cardiovascular outcome, morbidity and mortality	69

clinical trials.[81–84] A search of PubMed for "circulating tumor cells" yielded slightly more than 15,000 (15,274) citations; combination with "biomarker" decreased the number to somewhat more that 3500 (3562). The amount of work on the use of circulating tumor cells as biomarkers has markedly increased in the past several years as a result of advances in detection technologies. Table 6.3 contains a list of selected studies on the use of circulating tumor cells as biomarkers. It is clear that circulating tumor cells are of increasing importance as prognostic biomarkers and that the identification of subtypes of circulating tumor cells will increase that value.[103]

Microparticles in the context of biomarkers can be defined as small particles derived from cells and containing some characteristics of the parent cell as will platelet microparticles.[104] The term "microparticle" refers to size rather than function and thus there are microparticles that are developed for use in drug delivery.[105,106] Cell-derived microparticles are membrane fragments formed from cells (primarily platelets, leukocytes, endothelial cells, erythrocytes) in response to stimuli and should be considered as potential biomarkers.[107–109] Issues on the measurement of microparticles with flow cytometry are discussed by Shah and coworkers.[110] The majority of microparticles are found in blood but there is interest in microparticles in other biological fluids[107] such as urine.[111] It has been suggested that microparticles represent an "enriched" course for discovery of biomarkers with proteomic technology.[112,113] A partial list of microparticles proposed as biomarkers is shown in Table 6.4. As a note of caution, it is not clear that microparticles represent a homogeneous population and thus are more likely representative of cellular events rather than a separate physiological process.

TABLE 6.3
Circulating Tumor Cells as Biomarkers[a]

Study	Reference
Circulating tumor cells are prognostic biomarkers in metastatic breast cancer patients. The persistence of circulating tumor cells significantly correlated with shorter overall survival	85
Circulating tumor cell count is a prognostic biomarker in breast cancer and may be useful as surrogate endpoint in clinical trials	86
Use of RT-PCR in the detection of circulating tumor cells in gastric cancer for use as prognostic/predictive biomarker	87
Circulating tumor cells (CellSearch) as prognostic and predictive biomarkers in metastatic prostate cancer	88
Use of negative selection for the enrichment of circulating tumor cells in head and neck cancer patients. The process involves the lysis of red cell following and subsequent depletion of CD45+ cells	89
Circulating tumor cells as prognostic biomarkers for newly diagnosed metastatic breast cancer	90
Free cancer cells in venous drainage of colorectal cancer as prognostic indicator. Emphasize on site of sample collection in venous drainage as collection in peripheral circulation obtains samples after passage through the liver (post portal venous system)	91
Circulating tumor cells are recruited to the pulmonary bed during allergic inflammation in breast cancer patients	92
Identification of target genes for RT-PCR for the characterization of circulating tumor cells as prognostic biomarkers for colorectal cancer	93
Invasive circulating tumor cells (in vitro cell invasion assay using a cell adhesion matrix) as prognostic biomarker in ovarian cancer	94
Circulating tumor cells as predictive and prognostic biomarkers in lung cancer	95
Circulating tumor cells in prostate cancer enriched with a functional collagen matrix; circulating tumors cells as diagnostic and prognostic biomarkers in castration-resistant prostate cancer	96
Circulating tumor cells as prognostic biomarkers in progressive, castration-resistant prostate cancer; surrogate biomarker for treatment with chemotherapy. Changes in circulating tumor cells closely associated with risk as was LDH; PSA had weak association	97
Detection of circulating tumor cells expressing E6/E7 HR-HPV oncogenes (immunoselection; RT-PCR amplification; immunostaining with EpCAM, endothelial cell adhesion molecule).[b] EpCAM reactivity as a prognostic biomarker in cervical cancer	98
Circulating tumor cells were obtained by immunomagnetic selection (cytokeratin 19, CK19) and the extracellular domain of HER-2 protein (HER-2/ECD) measured. HER-2/ECD is associated with CK19 and has a productive/predictive role (trastuzumab) in metastatic breast cancer patients. It is suggested that circulating tumor cells serve as surrogate biomarkers for immunotherapy of metastatic breast cancer	99
The chemokine receptor CXCR4 is found on pan-cytokeratin positive circulating tumor cells as prognostic biomarkers in non-small-cell lung cancer	100

(continued)

TABLE 6.3 (continued)
Circulating Tumor Cells as Biomarkers[a]

Study	Reference
Circulating tumor cells as surrogate biomarkers for treatment of small cell lung cancer patients with chemotherapy	101
Multicenter study of the use of circulating tumor cells as surrogate biomarkers for the treatment of metastatic breast cancer	102

[a] The studies were selected published manuscripts in the past year as derived from a PubMed search.
[b] See Munz, M., Baeuerle, P.A., and Gires, O., The emerging role of EpCAM in cancer and stem cell signaling, *Cancer Res.* 69, 5627–5629, 2009.

TABLE 6.4
Microparticles as Biomarkers

Study	References
Circulating microparticles derived from endothelial cells as prognostic biomarkers for atherosclerosis	114
Endothelial microparticles as potential biomarkers for vascular dysfunction in antiphospholipid syndrome	115
Platelet-derived microparticles as biomarkers for thromboangiitis obliterans (Buerger's disease; a vascular disease associated with inflammation and thrombosis)	116
Evaluation of CD expression on microparticles as biomarkers for inflammation in acute coronary syndrome; use of a nitrocellulose membrane containing anti-CD antibody spots. The intracellular surface of the bound microparticle was labeled with a fluorescent anti-annexin antibody	117
Circadian variation in VCAM-1 expression on microparticles (flow cytometry) shows circadian variation with less variation in tissue factor; this variation may be important in the use of microparticles as biomarkers for cardiovascular risk	118
Platelet-derived microparticle as possible biomarker for recurrent spontaneous abortion	119
Platelet-derived microparticles and monocyte-derived microparticles as biomarkers for vascular complications in individuals with progressive systemic sclerosis (scleroderma)	120
Elevation of eight proteins in urinary microparticles in patients with bladder cancer as possible biomarkers	111
Microparticles derived from endothelial cells and leukocytes as biomarkers for hemodynamic severity of pulmonary hypertension	121, 122

Soluble forms of receptors have been of increasing interest over the past 20 years.[123,124] Early work focused on soluble receptors for cytokines and growth factors.[125] Soluble receptors do exist for a variety of other components such as IgA,[126] transferrin,[127] and urokinase-like plasminogen activator.[128] Soluble receptors are derived by the shedding of full-length membrane-bound receptors,[129–133] or from the expression of truncated forms of receptor.[134–136] Soluble receptors bind their respective ligands with equivalent or slightly reduced affinity, and were originally considered then to be antagonistic in that effective ligand concentration is thereby reduced.

Another possibility is the physiological function of a soluble receptor–ligand complex in the initiation of biological response.[137,138] Finally, in the case of shedding, the parent cell can lose the ability to be influenced by the ligand.[139] Soluble receptors that have a variety of potential functions are of interest as potential biomarkers (Table 6.5). A more specialized soluble receptor, receptor for advanced glycation end products deserves separate consideration in Table 6.6.

TABLE 6.5
Use of Soluble Receptors as Biomarkers

Study	Reference
Soluble IL-2 receptor is elevated in serum from subjects with systemic lupus erythematosus (SLE) and may be a useful biomarker for SLE; also, soluble IL-2 receptor is higher in a subpopulation with lupus nephritis	140
Serum soluble IL-2 receptor may be a biomarker for schizophrenia	141
Serum levels of TNF receptors are positively correlated with CRP levels in pediatric burn patients and may be biomarkers for the generalized inflammatory response in this patient population	142
Soluble IL-2 receptor is a predictive biomarker in diffuse large B-cell lymphoma	143
ST2 (a truncated, soluble form of an IL-1 receptor), which is a serum biomarker for acute heart failure	144
Soluble urokinase plasminogen activator receptor (suPAR) is a biomarker for thrombosis in paroxysmal nocturnal hemoglobinuria patients	145
Soluble TNF receptor in serum is a prognostic biomarker for cardiovascular disease in subject with rheumatoid arthritis	146
Soluble TNF receptor in serum is a predictive biomarker for phototherapy in subjects with psoriasis	147
Soluble CD163 is a prognostic biomarker in acute liver failure	148
Soluble IL-6 receptor in vitreous fluid is a biomarker for proliferative diabetic retinopathy	149

TABLE 6.6
Receptor for Advanced Glycation End Product (RAGE)[a] as Biomarker[b]

Study	Reference
Review of plasma sRAGE[b] as biomarker for inflammatory disease	150
sRAGE as a biomarker for coronary atherosclerosis (there is an inverse relationship between sRAGE and coronary atherosclerosis)	151
sRAGE and AGE are independent biomarkers for kidney function (glomerular filtration rate)	152
Endogenous secretory RAGE[c] is a biomarker for carotid atherosclerosis, while soluble RAGE[c] is not a biomarker	153
sRAGE is an independent prognostic biomarker for heart failure	154
sRAGE is positively correlated with AGEs and soluble vascular cell adhesion molecule-1 (sVCAM-1) in serum from patients with type 2 diabetes. sRAGE may be a biomarker for vascular injury in individuals with type 2 diabetes	155

(continued)

TABLE 6.6 (continued)

Receptor for Advanced Glycation End Product (RAGE)[a] as Biomarker[b]

Study	Reference
sRAGE is decreased in serum from subjects with Sjögren's syndrome and is suggested as a diagnostic biomarker for Sjögren's syndrome	156
The immunoreactivity of RAGE correlates with histologic differentiation in oral squamous cell carcinoma	157
Endogenous RAGE is plasma biomarker for cardiovascular disease in end-stage renal disease	158
sRAGE may be a prognostic biomarker in sepsis	159
Endogenous soluble RAGE is a biomarker for susceptibility to diabetic retinopathy demonstrated with a sensitive ELISA assay	160
sRAGE is a biomarker for coronary artery disease in nondiabetic men	161

[a] RAGE (receptor for advanced glycation end products) are scavenger receptors for advanced glycation end products (AGE) found in a variety of cells including monocytes, macrophages, endothelial cells, and astrocytes. Binding of AGE results in cellular activation and the formation of cytokines, growth factors, vascular adhesion products (see Browlee, M., Advanced protein glycosylation in diabetes and aging, *Annu. Rev. Med.* 46, 223–234, 1995; Thornally, P.J., Cell activation by glycated proteins. AGE receptors, receptor recognition factors and functional classification of AGEs, *Cell. Mol. Biol.* (Noisy-le-grand), 44, 1013–1023, 1998; Schmidt, A.M., Yan, S.D., Yan, S.F., and Stern, D.M., The biology of the receptor for advanced glycation end products and its ligands, *Biochim. Biophys. Acta* 1498, 99–111, 2000).

[b] RAGE as a biomarker is measured as soluble receptor for advanced glycation end products (sRAGE). sRAGE is a form of RAGE, which lacks the transmembrane domain derived from receptor ectodomain shedding (soluble RAGE) and/or splice variant secretion (endogenous secretory RAGE) (see Geroldi, D., Falcone, C., and Emanuele, E., Soluble receptor for advanced glycation end products: from disease marker to potential therapeutic target, *Curr. Med. Chem.* 13, 1971–1978, 2006; Santilli, F., Vazzana, N., Bucciarelli, L.G., and Davi, G., Soluble forms of RAGE in human diseases: clinical and therapeutic implications, *Curr. Med. Chem.* 16, 940–952, 2009).

REFERENCES

1. Szmitko, P.E., Fedak, P.W., Weisel, R.D. et al., Endothelial progenitor cells: New hope for a broken heart, *Circulation* 107, 3093–3100, 2003.
2. Hristov, M. and Weber, C., Endothelial progenitor cells: Characterization, pathophysiology, and possible clinical relevance, *J. Cell. Mol. Med.* 8, 498–508, 2004.
3. Blann, A.D., Woywodt, A., Bertolini, F. et al., Circulating endothelial cells. Biomarker of vascular disease, *Thromb. Haemost.* 93, 228–235, 2005.
4. Levenberg, S., Engineering blood vessels from stem cells: Recent advances and applications, *Curr. Opin. Biotechnol.* 16, 516–523, 2005.
5. Romagnani, P., Lasgni, L., and Romagnani, S., Peripheral blood as a source of stem cells for regenerative medicine, *Expert Opin. Biol. Ther.* 6, 193–202, 2006.
6. Edbruegger, U., Haubitz, M., and Woywodt, A., Circulating endothelial cells: Novel marker of endothelial damage, *Clin. Chim. Acta* 373, 17–26, 2006.
7. Miller-Kasprzak, E. and Jagodziński, P.P., Endothelial progenitor cells as a new agent contributing to vascular repair, *Arch. Immunol. Ther. Exp.* (Warsz.) 55, 247–259, 2007.

8. Balbarini, A., Barsotti, M.C., Di Stefano, R. et al., Circulating endothelial progenitor cells characterization, function and relationship with cardiovascular risk factors, *Curr. Pharm. Des.* 13, 1699–1713, 2007.
9. Marsbloom, G. and Janssens, S., Endothelial progenitor cells: New perspectives and applications in cardiovascular therapies, *Expert Rev. Cardiovasc. Ther.* 6, 687–701, 2008.
10. Avci-Adali, M., Paul, A., Ziemer, G., and Wendel, H.P., New strategies for in vivo tissue engineering by mimicry of homing factors for self-endothelialization of blood contacting materials, *Biomaterials* 29, 3936–3945, 2008.
11. Sekiguchi, H., Ii, M., and Losordo, D.W., The relative potency and safety of endothelial progenitor cells and unselected mononuclear cells for recovery from myocardial infarction and ischemia, *J. Cell. Physiol.* 219, 235–242, 2009.
12. Povsic, T.J. and Goldschmidt-Clermont, P.J., Endothelial progenitor cells: Markers of vascular reparative capacity, *Ther. Adv. Cardiovasc. Dis.* 2, 199–213, 2008.
13. Gaynor, E., Bouvier, C., and Spaet, T.H., Vascular lesions: Possible pathogenetic basis of the generalized Shwartman reaction, *Science* 170, 986–988, 1970.
14. Makin, A.J., Blann, A.D., Chung, N.A. et al., Assessment of endothelial damage in atherosclerotic vascular disease by quantification of circulating endothelial cells. Relationship with von Willebrand factor and tissue factor, *Eur. Heart J.* 25, 371–376, 2004.
15. Woywodt, A., Haubitz, M., Buchholz, S., and Hertenstein, B., Counting the cost: Markers of endothelial damage in hematopoietic stem cell transplantation, *Bone Marrow Transplant.* 34, 1015–1023, 2004.
16. Dignat-George, F. and Sampol, J., Circulating endothelial cells in vascular disorders: New insights into an old concept, *Eur. J. Haematol.* 65, 215–220, 2000.
17. Clancy, R.M., Circulating endothelial cells and vascular injury in systemic lupus erythematosus, *Curr. Rheumatol. Rep.* 2, 39–43, 2000.
18. Haubitz, M. and Woywodt, A., Circulating endothelial cells and vasculitis, *Intern. Med.* 43, 660–667, 2004.
19. Khan, S.S., Solomon, M.A., and McCoy, J.P. Jr., Detection of circulating endothelial cells and endothelial progenitor cells by flow cytometry, *Cytometry B: Clin. Cytom.* 64, 1–8, 2005.
20. Strijbos, M.H., Gratama, J.W., Kraan, J. et al., Circulating endothelial cells in oncology: Pitfalls and promises, *Br. J. Cancer* 98, 1731–1735, 2008.
21. Willett, C.G., Kozin, S.V., Duda, D.G. et al., Combined vascular endothelial growth factor-targeting therapy and radiotherapy for rectal cancer: Theory and clinical practice, *Semin. Oncol.* 33(5 Suppl 10), S35–S40, 2006.
22. Longo, R. and Garparini, G., Anti-VEGF therapy: The search for clinical biomarkers, *Expert Rev. Mol. Diagn.* 8, 301–314, 2008.
23. Zurita, A.J., Jonasch, E., Wu, H.K. et al., Circulating biomarkers for vascular endothelial growth factor inhibitors in renal cell carcinoma, *Cancer* 115(10 Suppl), 2346–2354, 2009.
24. Georgiou, H.D., Namdarian, B., Costello, A.J., and Hovens, C.M., Circulating endothelial cells as biomarkers of prostate cancer, *Nat. Clin. Pract. Urol.* 5, 445–454, 2008.
25. Kluz, J., Kopec, W., Jakobsche-Policht, U., and Adamiec, R., Circulating endothelial cells, endothelial apoptosis and soluble markers of endothelial dysfunction in patients with systemic lupus erythematosus-related vasculitis, *Int. Angiol.* 28, 192–201, 2009.
26. Dmadja, D.M., Gaussem, P., Mauge, L., Circulating endothelial cells. A new candidate biomarker of irreversible pulmonary hypertension secondary to congenital heart disease, *Circulation* 119, 374–381, 2009.
27. Strijbos, M.H., Landburg, P.P., Nur, E. et al., Circulating endothelial cells: A potential parameter for organ damage in sickle cell anemia?, *Blood Cells Mol. Dis.* 43, 63–67, 2009.
28. Strijbos, M.H., Verhoef, C., Gratama, J.W., and Sleijfer, S., On the origin of (CD105+) circulating endothelial cells, *Thromb. Haemost.* 102, 347–351, 2009.

29. Abdelmoneim, S.S., Talwalker, J., Sethi, S. et al., A prospective pilot study of circulating endothelial cells as a potential new biomarker in portal hypertension, *Liver Int.* 30, 191–197, 2010.

30. Allegra, A., Coppolino, G., Bolignano, D. et al., Endothelial progenitor cells: Pathogenetic role and therapeutic perspectives, *J. Nephrol.* 22, 463–475, 2009.

31. Furuya, M., Yonemitsu, Y., and Aoki, I., Angiogenesis: Complexity of tumor vasculature and microenvironment, *Curr. Pharm. Des.* 15, 1854–1867, 2009.

32. Yoder, M.D. and Ingram, D.A., The definition of EPCs and other bone marrow cells contributing to neoangiogenesis and tumor growth: Is there common ground for understanding the roles of numerous marrow-derived cells in the neoangiogenic process?, *Biochim. Biophys. Acta* 1796, 50–54, 2009.

33. Hibbert, B., Olsen, S., and O'Brien, E., Involvement of progenitor cells in vascular repair, *Trends Cardiovasc. Med.* 13, 322–326, 2003.

34. Zampetaki, A., Kirton, J.P., and Xu, Q., Vascular repair by endothelial progenitor cells, *Cardiovasc. Res.* 78, 413–421, 2008.

35. Fadini, G.P., Baesso, I., Albiero, M. et al., Technical notes on endothelial progenitor cells: Ways to escape from the knowledge plateau, *Atherosclerosis* 197, 496–503, 2008.

36. Matsumoto, T., Kuroda, R., Mifune, Y. et al., Circulating endothelial/skeletal progenitor cells for bone regeneration and healing, *Bone* 43, 434–439, 2008.

37. Besler, C., Doerries, C., Giannoti, G. et al., Pharmacological approaches to improve endothelial repair mechanisms, *Expert Rev. Cardiovasc. Ther.* 6, 1071–1082, 2008.

38. Van Craenenbroeck, E.M., Conraads, V.M., Van Bockstaele, D.R. et al., Quantification of circulating endothelial progenitor cells: A methodological comparison of six flow cytometric approaches, *J. Immunol. Methods* 332, 31–40, 2008.

39. Redondo, S., Hristov, M., Gordillo-Moscoso, A.A. et al., High-reproducible flow cytometric endothelial progenitor cell determination in human peripheral blood as CD34+/CD144+/CD3-lymphocyte sub-population, *J. Immunol. Methods* 335, 21–27, 2008.

40. Hristov, M., Schmitz, S., Schuhmann, C. et al., An optimized flow cytometry protocol for analysis of angiogenic monocytes and endothelia progenitor cells in peripheral blood, *Cytometry A* 75, 848–853, 2009.

41. Thomas, H.E., Redgrave, R., Cunnington, M.S. et al., Circulating endothelial progenitor cells exhibit diurnal variation, *Arterioscler. Thromb. Vasc. Biol.* 28, e21–e22, 2008.

42. Dome, P., Teleki, Z., Rihmer, Z. et al., Circulating endothelial progenitor cells and depression: A possible novel link between heart and soul, *Mol. Psychiatry* 14, 523–531, 2009.

43. Esposito, K., Ciotola, M., Mairorino, M.I. et al., Circulating CD34+ KDR+ endothelial progenitor cells correlate with erectile function and endothelial function in overweight men, *J. Sex. Med.* 6, 107–114, 2009.

44. Adams, V., Linke, A., Breuckmann, F. et al., Circulating progenitor cells decrease immediately after marathon race in advanced-age marathon runners, *Eur. J. Cardiovasc. Prev. Rehabil.* 15, 602–607, 2008.

45. Wu, H., Chen, H., and Hu, P.C., Circulating endothelial cells and endothelial progenitors as surrogate biomarkers in vascular dysfunction, *Clin. Lab.* 53, 285–295, 2007.

46. Rafat, N., Hanusch, C., Brinkkoetter, P.T. et al., Increased circulating endothelial progenitor cells in septic patients: Correlation with survival, *Crit. Care Med.* 35, 1677–1684, 2007.

47. Grundmann, F., Scheid, C., Braun, D. et al., Differential increase of CD34. KDR/CD34, CD133/CD34 and CD117CD34 positive cells in peripheral blood of patients with acute myocardial infarction, *Clin. Res. Cardiol.* 96, 621–627, 2007.

48. Werner, N., Wassmann, S., Ahlers, P. et al., Endothelial progenitor cells correlate with endothelial function in patients with coronary artery disease, *Basic Res. Cardiol.* 102, 565–571, 2007.

49. Yodoi, Y., Sasahara, M., Kameda, T. et al., Circulating hematopoietic stem cells in patients with neovascular age-related macular degeneration, *Invest. Ophthalmol. Vis. Sci.* 48, 5464–5472, 2007.

50. Kohagura, K., Ohya, Y., Miyagi, S. et al., rHuEPO does inversely correlated with the number of circulating CD34+ cells in maintenance hemodialysis patients, *Nephron Clin. Pract.* 108, c41–c46, 2008.

51. Chu, K., Jung, K.H., Lee, S.T. et al., Circulating endothelial progenitor cells as a new marker of endothelial dysfunction or repair in acute stroke, *Stroke* 39, 1441–1447, 2008.

52. Müller-Ehmsen, J., Braun, D., Schneider, T. et al., Decreased number of circulating progenitor cells in obesity: Beneficial effects of weight reduction, *Eur. Heart J.* 29, 1560–1568, 2008.

53. Diller, G.P., van Eijl, S., Okonko, D.O. et al., Circulating endothelial progenitor cells in patients with Eisenmenger syndrome and idiopathic pulmonary arterial hypertension, *Circulation* 117, 3020–3030, 2008.

54. Boilson, B.A., Kiernan, T.J., Harbuzariu, A. et al., Circulating CD34+ cell subsets in patients with coronary endothelial dysfunction, *Nat. Clin. Pract. Cardiovasc. Med.* 5, 489–496, 2008.

55. Nevskaya, T., Bykovskaia, S., Lyssuk, E. et al., Circulating endothelial progenitor cells in systemic sclerosis: Relation to impaired angiogenesis and cardiovascular manifestation, *Clin. Exp. Rheumatol.* 26, 421–429, 2008.

56. Stein, A., Montens, H.P., Steppich, B. et al., Circulating endothelial progenitor cells decrease in patients after endarterectomy, *J. Vasc. Surg.* 48, 1217–1222, 2008.

57. Brunner, M., Thurnher, D., Heiduschka, G. et al. Elevated levels of circulating endothelial progenitor cells in head and neck cancer patients, *J. Surg. Oncol.* 98, 545–550, 2008.

58. Maruyama, S., Taguchi, A., Iwashima, S. et al., Low circulating CD34+ cell count is associated with poor prognosis in chronic hemodialysis patients, *Kidney Int.* 74, 1603–1609, 2008.

59. Gössi, M., Mödder, U.I., Atkinson, E.J. et al., Osteocalcin expression by circulating endothelial progenitor cells in patients with coronary atherosclerosis, *J. Am. Coll. Cardiol.* 52, 1314–1325, 2008.

60. Delva, P., De Marchi, S., Prior, M. et al., Endothelial progenitor cells in patients with severe peripheral arterial disease, *Endothelium* 15, 246–253, 2008.

61. Van Craenenbroeck, E.M., Denollet, J., Paelinck, B.P. et al., Circulating CD34+/KDR+ endothelial progenitor cells are reduced in chronic heart failure patients as a function of Type D personality, *Clin. Sci.* 117, 165–172, 2009.

62. Makino, H., Okada, S., Nagumo, A. et al., Decreased circulating CD34+ cells are associated with progression of diabetic nephropathy, *Diabet. Med.* 26, 171–173, 2009.

63. Greenfield, J.P., Jin, D.K., and Young, L.M., Surrogate marker predict angiogenic potential and survival in patients with glioblastoma multiforme, *Neurosurgery* 64, 819–826, 2009.

64. Zhu, A.X., Sahani, D.V., Duda, D.G. et al., Efficacy, safety, and potential biomarkers of sunitinib monotherapy in advanced hepatocellular carcinoma, *J. Clin. Oncol.* 27, 3027–3035, 2009.

65. Manginas, A., Tsiavou, A., Sfyrakis, P. et al., Increased number of circulating progenitor cells after implantation of ventricular assist devices, *J. Heart Lung Transplant.* 28, 710–717, 2009.

66. Garolla, A., D'Incà, R., Checchin, D. et al., Reduced endothelial progenitor cell number and function in inflammatory bowel disease: A possible link to the pathogenesis, *Am. J. Gastroenterol.* 104, 2500–2507, 2009.

67. Taylor, M., Rössler, J., Georger, B. et al., High levels of circulating VEGFR2+ bone marrow-derived progenitor cells correlate with metastatic disease in patients with pediatric solid malignancies, *Clin. Cancer Res.* 15, 4561–4571, 2009.

68. Nowak, K., Rafat, N., Belle, S. et al., Circulating endothelial progenitor cells are increased in human lung cancer and correlated with stage of disease, *Eur. J. Cardiothorac. Surg.* 37, 758–763, 2010.

69. Fadini, G.P., de Kreutzenberg, S., Agostini, C. et al., Low CD34+ cell count and metabolic syndrome synergistically increase the risk of adverse outcomes, *Atherosclerosis* 207, 213–219, 2009.

70. Mostert, B., Sleijfer, S., Foekens, J.A., and Gratama, J.W., Circulating tumor cells (CTCs): Detection methods and their clinical relevance in breast cancer, *Cancer Treatment Rev.* 35, 463–474, 2009.

71. Dotan, E., Cohen, S.J., Alpaugh, K.R., and Meropol, N.J., Circulating tumor cells: Evolving evidence and future challenges, *Oncologist* 14, 1070–1082, 2009.

72. Pantel, K. and Riethdorf, S., Are circulating tumor cells predictive of overall survival?, *Nature Rev. Clin. Oncol.* 6, 190–191, 2009.

73. Ross, A.A., Cooper, B.W., Lazarus, H.M. et al., Detection and viability of tumor cells in peripheral blood stem cell collections from breast cancer patients using immunocytochemical and clonogenic assay techniques, *Blood* 82, 2605–2610, 1993.

74. Bidard, F.C., Mathiot, C., Delaloge, S. et al., Single circulating tumor cell detection and overall survival in nonmetastatic breast cancer, *Ann. Oncol.* 21, 729–733, 2010.

75. Allard, W.J., Matera, J., Miller, M.C. et al., Tumor cells circulate in the peripheral blood of all major carcinomas but not in healthy subjects or patients with nonmalignant diseases, *Clin. Cancer Res.* 10, 6897–6904, 2004.

76. Gates, J.D., Benavides, L.C., Stojadinovic, A. et al., Monitoring circulating tumor cells in cancer vaccine trials, *Hum. Vaccin.* 4, 389–392, 2008.

77. Sequist, L.V., Nagrath, S., Toner, M. et al., The CTC-chip: An exciting new tool to detect circulating tumor cells in lung cancer patients, *J. Thorac. Oncol.* 4, 281–283, 2009.

78. Xu, Y., Phillips, J.A., Yan, J. et al., Aptamer-based microfluidic device for enrichment, sorting, and detection of multiple cancer cells, *Anal. Chem.* 81, 7436–7442, 2009.

79. Dharmasiri, U., Balamurugan, S., Adams, A.A. et al., Highly efficient capture and enumeration of low abundance prostate cancer cells using prostate-specific membrane antigen aptamers immobilized to a polymeric microfluidic device, *Electrophoresis* 30, 3289–3300, 2009.

80. Segeant, G., Penninckx, F., and Topal, B., Quantitative TR-PCR detection of colorectal tumor cells in peripheral blood—A systematic review, *J. Surg. Rev.* 150, 144–152, 2008.

81. Tan, D.S.W., Thomas, G.V., Garrett, M.D. et al., Biomarker-driven early clinical trials in oncology. A paradigm shift in drug development, *Cancer J.* 15, 406–420, 2009.

82. de Bono, J.S., Attard, G., Adjei, A. et al., Potential applications for circulating tumor cells expressing the insulin-like growth factor-I receptor, *Clin. Cancer Res.* 13, 3611–3616, 2007.

83. Sawyers, C.L., The cancer biomarker problem, *Nature* 452, 548–552, 2008.

84. Apolo, A.B., Milowsky, M., and Bajorin, D.F., Clinical states model for biomarkers in bladder cancer, *Future Oncol.* 5, 977–992, 2009.

85. Tewes, M., Aktas, B., Welt, A. et al., Molecular profiling and predictive value of circulating tumor cells in patients with metastatic breast cancer: An option for monitoring response to breast cancer related therapies, *Breast Cancer Res. Treat.* 115, 581–590, 2009.

86. Olmos, D., Askenau, H.T., Ang, J.E. et al., Circulating tumor cell counts as intermediate end points in castration-resistant prostate cancer (CRPC): A single-centre experience, *Ann. Oncol.* 20, 27–33, 2009.

87. Koga, T., Tokunaga, E., Sumiyoshi, Y. et al., Detection of circulating gastric cells in peripheral blood using real time quantitative RT-PCR, *Hepatogastroenterology* 55, 1131–1135, 2008.

88. Okegawa, T., Nutahara, K., and Higashihara, E., Immunomagnetic quantification of circulating tumor cells as a prognostic factor of androgen deprivation responsiveness in patients with hormone naïve metastatic prostate cancer, *J. Urol.* 180, 1342–1347, 2008.

89. Yang, L., Lang, J.C., Balasubramanian, P. et al., Optimization of an enrichment process for circulating tumor cells from the blood of head and neck cancer patients through depletion of normal cells, *Biotechnol. Bioeng.* 102, 521–534, 2009.

90. Dawood, S., Broglio, K., Valero, V. et al., Circulating tumor cells in metastatic breast cancer: From prognostic stratification of modification of the staging system?, *Cancer* 113, 2422–2430, 2008.

91. Katsuno, H., Zacharakis, E., Aziz, O. et al., Does the presence of circulating tumor cells in the venous drainage of curative colorectal cancer resections determine prognosis? A meta-analysis, *Ann. Surg. Oncol.* 15, 3083–3091, 2008.

92. Taranova, A.G., Maldonado III D., Vachon, C.M. et al., Allergic pulmonary inflammation promotes the recruitment of circulating tumor cells to the lung, *Cancer Res.* 68, 8582–8589, 2008.

93. Findeisen, P., Röckel, M., Nees, M. et al., Systematic identification and validation of candidate genes for detection of circulating tumor cells in peripheral blood specimens of colorectal cancer patients, *Int. J. Oncol.* 33, 1001–1010, 2008.

94. Fan, T., Zhao, Q., Chen, J.J. et al., Clinical significance of circulating tumor cells determined by an invasion assay in peripheral blood of patients with ovarian cancer, *Gynecol. Oncol.* 112, 185–191, 2009.

95. Wu, C., Hao, H., Li, L. et al., Preliminary investigation of the clinical significance of detecting circulating tumor cells enriched from lung cancer patients, *J. Thorac. Oncol.* 4, 30–36, 2009.

96. Paris, P.L., Kobayashi, Y., Zhao, Q. et al., Functional phenotyping and genotyping of circulating tumor cells from patients with castration resistant prostate cancer, *Cancer Lett.* 277, 164–173, 2009.

97. Scher, H.I., Jia, X., de Bono, J.S. et al., Circulating tumour cells as prognostic markers in progressive, castration-resistant prostate cancer: A reanalysis of IMNC38 trial data, *Lancet Oncol.* 10, 233–239, 2009.

98. Weismann, P., Wesmanova, E., Masak, L. et al., The detection of circulating tumor cells expressing E6/E7 HR-HPV oncogenes in peripheral blood in cervical cancer patients after radical hysterectomy, *Neoplasma* 56, 230–238, 2009.

99. Nunes, R.A., Li, X., Kang, S.P. et al., Circulating tumor cells in HER-2 positive metastatic breast cancer patients treated with trastuzumab and chemotherapy, *Int. J. Biol. Markers* 24, 1–10, 2009.

100. Reckamp, K.L., Figlin, R.A., Burdick, M.D. et al., CXCR4 expression on circulating pan-cytokeratin positive cells is associated with survival in patients with advanced non-small cell lung cancer, *BMC Cancer* 9, 213, 2009.

101. Hou, J.M., Greystroke, A., Lancashire, L. et al., Evaluation of circulating tumor cells and serological cell death biomarker in small cell lung cancer patients undergoing chemotherapy, *Am. J. Pathol.* 175, 808–816, 2009.

102. Nakamura, S., Yagata, H., Ohno, S. et al., Multi-center study evaluating circulating tumor cells as a surrogate for response to treatment and overall survival in metastatic breast cancer, *Breast Cancer*, 2009. doi: 10.1007/s12282-009-0139-3.

103. Siuwerts, A.M., Kraan, J., Bolt, J. et al., Anti-epithelial cell adhesion molecule antibodies and the detection of circulating normal-like breast tumor cells, *J. Natl. Cancer Inst.* 101, 61–66, 2009.

104. Horstman, L.L. and Ahn, Y.X., Platelet microparticles: A wide-angle perspective, *Crit. Rev. Oncol. Hematol.* 30, 111–142, 1999.

105. Wischke, C. and Schwendeman, S.P., Principles on encapsulating hydrophobic drugs in PLA/PLGA microparticles, *Int. J. Pharm.* 364, 298–327, 2008.

106. Sehgal, P.K. and Srinivasan, A., Collagen-coated microparticles in drug delivery, *Expert Opin. Drug Deliv.* 6, 687–695, 2009.
107. Doeuvre, L., Plawinski, L., Toti, F., and Anglés-Cano, E., Cell-derived microparticles: A new challenge in neuroscience, *J. Neurochem.* 110, 457–468, 2009.
108. Chironi, G.N., Boulanger, C.M., Simon, A. et al., Endothelial microparticles in diseases, *Cell Tissue Res.* 335, 143–151, 2009.
109. Sabatier, F., Camoin-Jau, L., Anfosso, F. et al., Circulating endothelial cells, microparticles and progenitors: Key players towards the definition of vascular consequence, *J. Cell. Mol. Med.* 13, 454–471, 2009.
110. Shah, M.D., Bergeron, A.L., Dong, J.F., and López, J.A., Flow cytometric measurement of microparticles: Pitfalls and protocol modifications, *Platelets* 19, 365–372, 2008.
111. Smalley, D.M., Sheman, N.E., Nelson, K., and Theodorescu, D., Isolation and identification of potential urinary microparticle biomarkers of bladder cancer, *J. Proteome Res.* 7, 2088–2096, 2008.
112. Banfi, C., Brioschi, M., Wait, R. et al., Proteome of endothelial cell-derived procoagulant microparticles, *Proteomics* 5, 4443–4455, 2005.
113. Smalley, D.M. and Ley, K., Plasma-derived microparticles for biomarker discovery, *Clin. Lab.* 54, 67–79, 2008.
114. Boulanger, C.M., Amabile, N., and Tedgui, A., Circulating microparticles: A potential prognostic marker for artherosclerotic vascular disease, *Hypertension* 48, 180–186, 2006.
115. Pericleous, C., Giles, I., and Rahman, A., Are endothelial microparticles potential markers of vascular dysfunction in the antiphospholipid syndrome?, *Lupus* 18, 671–675, 2009.
116. Darnige, L., Helley, D., Fischer, A.M. et al., Platelet microparticle levels: A biomarker of thromboangiitis obliterans (Buerger's disease) exacerbation, *J. Cell. Mol. Biol.* 14, 449–451, 2010.
117. Lai, S., Brown, A., Nguyen, L. et al., Using antibody arrays to detect microparticles from acute coronary syndrome patients based on cluster of differentiation (CD) antigen expression, *Mol. Cell. Proteomics* 8, 799–804, 2009.
118. Madden, L.A., Vince, R.V., Sandström, M.E. et al., Microparticle-associated vascular adhesion molecule-1 and tissue factor follow a circadian variation in healthy human subjects, *Thromb. Haemost.* 99, 909–915, 2008.
119. Kaptan, K., Beyan, C., Ifran, A., and Pekel, A., Platelet-derived microparticle levels in women with recurrent spontaneous abortion, *Int. J. Gynaecol. Obstet.* 102, 271–274, 2008.
120. Nomura, S., Inami, N., Ozaki, Y. et al., Significance of microparticles in progressive systemic sclerosis with interstitial pneumonia, *Platelets* 19, 192–198, 2008.
121. Amabile, N., Heiss, C., Real, W.M. et al., Circulating endothelial microparticle levels predict hemodynamic severity of pulmonary hypertension, *Am. J. Respir. Crit. Care Med.* 177, 1268–1275, 2008.
122. Bakouboula, B., Morel, O., Faure, A. et al., Procoagulant membrane microparticles correlate with the severity of pulmonary arterial hypertension, *Am. J. Respir. Crit. Care Med.* 177, 536–543, 2008.
123. Sethi, K.K. and Näher, H., Elevated titers of cell-free interleukin-2 receptor in serum and cerebrospinal fluid specimens of patients with acquired immunodeficiency syndrome, *Immunol. Lett.* 13, 179–184, 1986.
124. Lotze, M.T., Custer, M.C., Sharrow, S.O. et al., In vivo administration of purified human interleukin-2 receptor positive cells and circulating soluble interleukin-2 receptors following interleukin-2 administration, *Cancer Res.* 47, 2188–2195, 1987.
125. Rose-John, S. and Heinrich, P.C., Soluble receptors for cytokines and growth factors: Generation and biological function, *Biochem. J.* 300, 281–290, 1994.
126. Monteiro, R.C., Pathogenic role of IgA receptors in IgA nephropathy, *Contrib. Nephrol.* 157, 64–69, 2007.

127. Thomas, C., Kirschbaum, A., Boehm, D., and Thomas, L., The diagnostic plot: A concept for identifying different states of iron deficiency and monitoring the response to epoetin therapy, *Med. Oncol.* 23, 23–36, 2006.

128. Montuori, N. and Ragno, P., Multiple activities of a multifaceted receptor: Roles of cleaved soluble uPAR, *Front. Biosci.* 14, 2494–2503, 2009.

129. Moldovan, I., Galon, J., Maridonneau-Parini, I. et al., Regulation of production of soluble Fc gamma receptors type III in normal and pathological conditions, *Immunol. Lett.* 68, 125–134, 1999.

130. Mayer, R.J., Flamberg, P.L., Katchur, S.R. et al., CD23 shedding: Requirements for substrate recognition and inhibition by dipeptide hydroxamic acids, *Inflamm. Res.* 51, 85–90, 2002.

131. Montuori, N., Visconte, V., Rossi, G., and Ragno, P., Soluble and cleaved forms of the urokinase-receptor: Degradation products or active molecules?, *Thromb. Haemost.* 93, 192–198, 2005.

132. Hikita, A. and Tanaka, S., Ectodomain shedding of receptor activator of NFκB ligand, *Adv. Exp. Med. Biol.* 602, 15–21, 2007.

133. Georges, S., Ruiz Velasco, C., Trichet, V. et al., Proteases and bone remodeling, *Cytokine Growth Factor Rev.* 20, 29–41, 2009.

134. Lainez, B., Fernandez-Real, J.M., Romero, X. et al., Identification and characterization of a novel spliced variant that encodes human soluble tumor necrosis factor receptor 2, *Int. Immunol.* 16, 169–177, 2004.

135. McLouglin, R.M., Hurst, S.M., Nowell, M.A. et al., Differential regulation of neutrophil-activating chemokines by IL-6 and its soluble receptor isoforms, *J. Immunol.* 172, 5676–5683, 2004.

136. Byström, J., Dyer, K.D., Ting-De Ravin, S.S. et al., Interleukin-5 does not influence differential transcription of transmembrane and soluble isoforms for IL-5Rα in vivo, *Eur. J. Haematol.* 77, 181–190, 2006.

137. Rose-John, S., Interleukin-6 biology is coordinated by membrane bound and soluble receptors, *Acta Biochim. Pol.* 50, 603–611, 2003.

138. Knüpfer, H. and Preiss, R., sIL-6R: More than an agonist?, *Immunol. Cell Biol.* 85, 87–89, 2008.

139. Berndt, M.C., Karunakaran, D., Gardiner, E.E., and Andrews, R.K., Programmed autologous cleavage of platelet receptors, *J. Thromb. Haemost.* 5(Suppl 1), 212–219, 2007.

140. El-Shafey, E.M., El-Naggar, G.F., El-Bendary, A.S. et al., Serum soluble interleukin-2 receptor alpha is systemic lupus erythematosus, *Iran J. Kidney Dis.* 2, 80–85, 2008.

141. Bresee, C. and Rapaport, M.H., Persistently increased serum soluble interleukin-2 receptors in continuously ill patients with schizophrenia, *Int. J. Neuropsychopharmacol.* 12, 861–865, 2009.

142. Sikera, J.P., Kuzański, W., and Andrzejewska, E., Soluble cytokine receptors sTNFR I and sTNFR II, receptor antagonist IL-1ra, and anti-inflammatory cytokines IL-10 and IL-13 in the pathogenesis of systemic inflammatory response syndrome in the course of burns in children, *Med. Sci. Monit.* 15, CR26–CR31, 2009.

143. Ennishi, D., Yokoyama, M., Terui, Y. et al., Soluble interleukin-2 receptor retains prognostic value in patients with diffuse large B-cell lymphoma receiving rituximab plus CHOP (RCHOP) therapy, *Ann. Oncol.* 20, 526–533, 2009.

144. Rehman, S.U., Mueller, T., and Januzzi, J.L. Jr., Characteristics of the novel interleukin family biomarker ST2 in patients with acute heart failure, *J. Am. Coll. Cardiol.* 52, 1458–1465, 2008.

145. Sloand, E.M., Pfannes, L., Scheinberg, P. et al., Increased soluble urokinase plasminogen activator receptor (suPAR) is associated with thrombosis and inhibition of plasmin generation in paroxysmal nocturnal hemoglobinuria (PNR) patients, *Exp. Hematol.* 36, 1616–1624, 2008.

146. Mattev, D.L., Glossop, J.R., Nixon, N.B., and Dawes, P.T., Circulating levels of tumor necrosis factor receptors are highly predictive of mortality in patients with rheumatoid arthritis, *Arthritis Rheum.* 56, 3940–3948, 2007.

147. Serwin, A.B. and Chodynicka, B., Soluble tumor necrosis factor-alpha receptor type I as a biomarker of response to phototherapy in patients with psoriasis, *Biomarkers* 12, 599–607, 2007.

148. Muller, H.J., Grønbaek, H., Schiodt, F.V. et al., Soluble CD163 from activated macrophages predicts mortality in acute liver failure, *J. Hepatol.* 47, 671–676, 2007.

149. Kawashima, M., Shoji, J., Nakajima, M. et al., Soluble IL-6 receptor in vitreous fluid of patients with proliferative diabetic retinopathy, *Jpn. J. Ophthalmol.* 5, 100–104, 2007.

150. Ramasamy, R., Yan, S.F., and Schmidt, A.M., RAGE: Therapeutic target and biomarker of the inflammatory response—The evidence mounts, *J. Leukoc. Biol.* 86, 505–512, 2009.

151. Lindsey, J.B., de Lemos, J.A., Cipollone, F. et al., Association between circulating soluble receptor for advanced glycation end products (sRAGE) and atherosclerosis: Observations from the Dallas Heart Study, *Diabetes Care* 32, 1218–1220, 2009.

152. Semba, R.D., Ferrucci, L., Fink, J.C. et al., Advanced glycation end products and their circulating receptors and level of kidney function in older community-dwelling women, *Am. J. Kidney Dis.* 53, 51–58, 2009.

153. Katakami, N., Matsuhisa, M., Kaneto, H. et al., Endogenous secretory RAGE but not soluble RAGE is associated with carotid atherosclerosis in type 1 diabetes patients, *Diab. Vasc. Dis. Res.* 5, 190–197, 2008.

154. Koyama, Y., Takeishi, Y., Niizeki, T. et al., Soluble receptor for advanced glycation end products (RAGE) is a prognostic factor for heart failure, *J. Card. Fail.* 14, 133–139, 2008.

155. Nakamura, K., Yamigishi, S., Adachi, H. et al., Serum levels of soluble form of receptor for advanced glycation end products (sRAGE) are positively associated with circulating AGEs and soluble form of VCAM-1 in patients with type 2 diabetes, *Microvasc. Res.* 76, 52–56, 2008.

156. Stewart, C., Cha, S., Caudle, R.M. et al., Decreased levels of soluble receptor for advanced glycation end products in patients with primary Sjögren's syndrome, *Rheumatol. Int.* 28, 771–776, 2008.

157. Landesberg, R., Woo, V., Huang, L. et al., The expression of the receptor for glycation endproducts (RAGE) in oral squamous cell carcinomas, *Oral Surg. Oral Med. Oral Pathol. Oral Radiol. Endod.* 105, 617–624, 2008.

158. Nishizawa, Y. and Koyama, H., Endogenous secretory receptor for advanced glycation end-products and cardiovascular disease in end-stage renal disease, *J. Ren. Nutr.* 18, 76–82, 2008.

159. Bopp, C., Hofer, S., Weitz, J. et al., sRAGE is elevated in septic patients and associated with patients outcome, *J. Surg. Res.* 147, 79–83, 2008.

160. Sakurai, S., Yamamoto, Y., Tamel, H. et al., Development of an ELISA for esRAGE and its application to type 1 diabetic patients, *Diabetes Res. Clin. Pract.* 73, 158–165, 2006.

161. Falcone, C., Emanuele, E., D'Angelo, A. et al., Plasma levels of soluble receptor for advanced glycation end products and coronary artery disease in nondiabetic men, *Arterioscler. Thromb. Vasc. Biol.* 25, 879–882, 2005.

7 Use of Microarray Technology in Biomarker Discovery and Development

Microarray infers a large number of probes or sample sites on a matrix. Within the context of this chapter, a microarray assay is a solid-phase assay where the analyte is (usually) transferred from the bulk solution onto a solid surface such as microplate or bead surface. This solid-phase step has the advantage of (1) providing specificity to the analytical procedure and (2) removing the substances that might interfere with the detection step. Solid-phase assays are related to solid-phase extraction,[1] which is a closely related procedure used in pharmaceutical analysis; both solid-phase extraction and solid-phase assays remove and concentrate the analyte from bulk solution.[2] Solid-phase assays are quite useful[3] but it has been argued that such assays may increase false-positive results when compared to fluid-phase assays.[4,5] Other results suggests that solid-phase assays may detect interactions not observed in solution phase.[6]

The term "microarray" originally referred to a small-scale or microscopic array of components but it is now more commonly used to describe a DNA microarray. However, in addition to DNA microarrays, there are protein, peptide, lectin, carbohydrate, and aptamer microarrays. These are all analytical microarrays and allow for the multiplex analysis of a single sample. Most microarrays have the capture molecule (probe) bound to the matrix surface. There are microarrays (reverse microarrays) where a sample is bound to the surface and many samples are simultaneously analyzed with a probe in solution phase.[7–9] Tissue microarrays are sample microarrays that allow for the processing of a large number of tissue samples under identical conditions. Immunohistochemistry has been a useful technology in identifying biomarkers in tissue microarrays.[10–15] Permuth-Wey and coworkers[16] observed that tissue microarrays enable the rapid analysis of biomarkers in archival tumor specimens. It is emphasized that samples should preferentially be taken from the periphery of tumor block where there is more complete exposure to tissue fixatives. Biomarker studies using tissue microarrays are shown in Table 7.1.

The use of tissue microarray technology has great importance for the study of archival samples.[10,25–28] The ability to use formalin-fixed tissue samples for gene expression studies has enabled the combination of tissue microarrays and cDNA analysis for the identification of biomarkers.[15,29–34] Some selected studies on the combination of immunohistochemistry and gene expression to identify biomarkers is shown in Table 7.2. The reader is referred to a recent study by Gry and coworkers,[45]

TABLE 7.1

The Use of Tissue Microarrays for the Identification of Biomarkers

Study	Reference
A 105 core tissue microarray from prostate cancer primary tumor samples (16 patients) was examined with immunohistochemical techniques for selected biomarkers. Elevated expression was observed for CD10, E-cadherin, and membranous β-catenin while decreased expression was observed for matrix metalloproteinase-9	17
Use of fluorescence in situ hybridization (FISH) to identify biomarkers in tissue microarrays	18
Tissue microarray samples obtained from lung adenocarcinoma showed decreased expression of CD63 consistent with RT-PCR results, which showed decreased CD63 mRNA levels. Decreased CD63 showed a weak association with poor survival	19
cDNA microarray analysis of lung cancer samples suggested that KIF4A (kinesin family member 4A) is a potential biomarker. Tissue microarray studies (immunohistochemistry) of samples from non-small-cell lung cancer subjects showed that KIF4A was a prognostic biomarker for non-small-cell lung cancer also associated with male gender and nonadenocarcinoma histology	20
A 256 core tissue microarray was prepared from tumor samples obtained from 47 prostate cancer patients. Immunohistochemistry was used to identify a number of biomarkers and correlate expression to Gleason score. Strong correlation was observed for bax, bin 1, FAS, p65, and p21	21
Sections from a tissue microarray prepared from diffuse large B-cell lymphoma (36 subjects) were evaluated with immunohistochemical techniques in eight different laboratories. The results were characterized by initial poor reproducibility. The standardization of technique improved the concordance between the laboratories	22
Tissue microarrays were prepared from 302 hepatocellular cancer subjects. The expression of osteopontin and CD44 were identified as prognostic biomarkers	23
Immunochemistry was used in bladder cancer tissue microarrays to identify LAMC2 (laminin V γ2) as a prognostic biomarker	24

which examined the correlation between RNA and protein expression in 23 human cell lines. RNA profiles using both cDNA and oligonucleotide platforms had a correlation coefficient of 0.52. A significant correlation between RNA and protein (immunohistochemistry) was obtained for a third of the proteins studied. These workers suggested that the correlation between RNA and immunohistochemistry provided support for the antibodies specificity.

Despite the use of the term microarray for tissue studies above and the protein studies described below, the term microarray as such is usually taken to refer to DNA microarrays, which are used to measure gene expression. Most approaches involve the preparation of tissue RNA (including blood and blood cells) and the subsequent preparation of cDNA by reverse transcription[46–48]; although the same technology is used for the assay of circulating nucleic acids as described in Chapter 5. The construction of multiple oligonucleotide arrays with specificity is comparatively easy because of knowledge derived from the genomic sequence.[49–52] DNA is less complex than protein; there are 4 monomer units instead of the 20 monomer units

TABLE 7.2
Some Selected Studies on the Combined Use of Immunohistochemistry and Gene Expression for the Identification of Biomarkers

Study	Reference
The correlation between mRNA and protein (immunohistochemical) for cluster designation (CD) was examined in prostate tissue. Correlation was said to be poor to moderate. Divergence seen most with gene shown mRNA expression but lack of immunohistochemical staining	35
Correlation of mRNA and immunohistochemistry in the evaluation of loricin and cartilage oligomeric matrix protein (COMP) in oral submucous fibrosis (OSF). Oral submucous fibrosis is a precancerous condition of the oral cavity resulting from the chewing of areca nuts. Increased expression of COMP may be a biomarker for the early diagnosis of OSF	36
Microarray analysis demonstrated increased expression of SOX genes in pediatric medulloblastoma and pediatric ependymoma; the differential expression was confirmed by immunohistochemical analysis. The expression of the SOX genes are potential diagnostic and prognostic biomarkers for pediatric brain tumors	37
Microarray analysis was used to study gene expression in glial tumors. Nine genes were selected for further study. Seven genes were over-expressed in low grade gliomas and under-expressed in glioblastoma while two genes were under-expressed in low grade gliomas and over-expressed in glioblastoma. This expressed pattern was confirmed by RT-PCR. Insulin growth factor binding protein-2 (IGFBP-2) and cell division cycle 20 homolog (CDC20) were over-expressed in glioblastoma and under-expressed in low grade gliomas. Approximately 90% of the glioblastomas were positive for IGFBP-2 by immunohistochemistry and only in one of the low-grade gliomas, while approximately 75% of the glioblastomas were positive for CDC-20 and none of the low-grade tumors were positive for CDC-20. It was concluded there is a close correlation between gene expression and immunohistochemical staining for IGFBP-2 and CDC-20 in glioblastoma and can serve as a diagnostic biomarker in biopsy samples	38
Lack of correlation between gene expression for cyclin proteins as measured by microarray or RT-PCR and immunohistochemical staining in colonic adenocarcinoma. Microarray and RT-PCR yielded comparable data	39
Suppression subtractive hybridization[a] and microarray is used to study differential expression of MMP and IL-8 in invasive oral carcinoma. The increased expression of these two genes was confirmed by RT-PCR and then identified by immunohistochemistry. The expression of MMP and IL-8 genes are potential prognostic biomarkers for invasive oral carcinoma	40
MMP-12 and ADAMDEC1 identified in pulmonary sarcoidosis by gene expression (oligonucleotide microarray analysis) and RT-PCR. Immunohistochemistry confirmed the increased protein expression. It is suggest that MMP-12 and ADAMDEC1 are biomarkers for pulmonary sarcoidosis	41
Insulin-like growth factor 2 (IGF2) and antigen Ki-67(Ki-67) as diagnostic biomarkers in adrenocortical tumors as identified with gene expression (microarray) and immunohistochemistry. IGF2 and Ki-67 expression differentiates adrenocortical carcinomas from adrenocortical adenomas as determined by microarray analysis	42

(*continued*)

TABLE 7.2 (continued)

Some Selected Studies on the Combined Use of Immunohistochemistry and Gene Expression for the Identification of Biomarkers

Study	Reference
Gene expression (oligonucleotide microarray) analysis of ovarian cancer tissue identified CDH1 expression (E-cadherin) as a prognostic biomarker. E-cadherin as prognostic biomarker (low expression) supported by subsequent immunohistochemical studies. Gene expression showed a relationship between high ZEB2 expression and low CDH1 expression by both microarray and RT-PCR. High ZEB2 expression and low CDH1 expression are correlated with each and associated with poor prognosis	43
Correlation between gene expression (FISH) and immunohistochemistry (polyclonal antibody) for HER-2/neu as a biomarker in esophageal squamous cell carcinoma. Increase in gene expression associated with poor prognosis	44

[a] See Diatchenko, L., Lukyanov, S., Lau, Y.F. et al., Suppression subtractive hybridization: A versatile method for identifying differentially expressed genes, *Methods Enzymol.* 303, 349–380, 1999.

for proteins. DNA microarrays are physically stronger than protein microarrays as it is difficult to irreversibly denature DNA permitting the facile regeneration of DNA microarrays.[53–57] DNA microarray technology is based on a technique referred to as Southern blotting based on the hybridization between DNA and RNA.[58–62] The terminology dates to the development of the Southern blot where the labeled RNA was the probes to label specific DNA sequences on the electrophoretogram.[59] DNA microarrays are composed of synthetic oligonucleotides or intact cDNA, which are used to analyze RNA probes prepared from the cDNA of samples.[63–65] Early hybridization experiments used radiolabeled probes; current technology used fluorescent dyes. The probes are synthesized *in situ* on the matrix with smaller oligonucleotides or "spotted" in the case of cDNA and longer oligonucleotides.[66–70] Zhang and coworkers have a recent work[71] on microarray quality control, which is quite useful. The "gold standard" for DNA probes is the cDNA, which presumably maximizes the specificity of binding. The efficacy of a synthetic oligonucleotide probe appears to be maximal at 712 base[72] while satisfactory results appear to be obtained with 100 mer probes. The binding of labeled cDNA to the oligonucleotide matrix is a time-dependent process[72] and Squires and coworkers have presented a discussion of the fluid engineering issues such as convection and diffusion, which must be considered in microarray and related analytical systems.[73] Commercial short-oligonucleotide microarrays are based on 35–30 mers.[74] More recent work suggests that short probes are more satisfactory than longer probes.[75] The stability of the probe-target duplex is influenced by mismatches, which can be position dependent.[76] Probe length is one of several issues that are important in the DNA microarray and summaries of selected studies are shown in Table 7.3. The comparison of results obtained with different DNA microarray platforms has proved to be a continuing issue[88,89]; although programs like minimal information about a microarray experiment (MIAME) are providing a basis for comparing data from different microarray platforms.[90–92]

TABLE 7.3
A Selection of Studies on the Experimental Variables Important for DNA Microarray Analysis[a]

Study	Reference
The rate and affinity of binding of Cy5-labeled DNA to immobilized DNA fragments increased as function of length of immobilized fragment up 712 bp where both rate of binding and affinity decreased at higher length (2057 bp). There was no difference in affinity between a 2D surface and a 3D surface	72
The effect of length of single-stranded DNA (8–48 bp) on the structure of the molecules bound to a gold surface either with or without a 5′-hexanethiol group. The thiol function increased oligonucleotide binding to the gold surface but the enhancement is decreased with increasing oligonucleotide length. Thiol-functionalized oligonucleotides shorter than 24 bp tended to organize in an extended configuration with surface coverage independent of oligonucleotide length. With oligonucleotide lengths longer than 24 bp, surface coverage decreased with increasing probe length. It is suggested that as oligonucleotide length increases, polymeric behavior becomes dominant	77
A review on the development of probes for DNA microarray with consideration of more restricted array platforms focused on specific targets[b]	78
Comparison of cDNA and shorter oligonucleotide platforms (25 bp) for the measurement of 2895 sequence-matched genes in 56 cell lines. There appeared to be poor correlation between the two platforms. GC content and sequence length were associated with the degree of correlation[b]	79
Longer probes provide better signal intensity than shorter probes. Accurate measure of gene expression is obtained with multiple probes for each gene and fewer probes are needed if longer probes are used. Probe length of 150 bp was shown to be optimal for the measurement of gene expression	80
Comparison of cDNA (PCR generated) and homologous oligonucleotide probes from *Shewanello oneidensis* MR-1 with respect to signal intensity, number of genes detected, specificity, sensitivity, and differential gene expression. A 70 bp probe provided intensity similar to that obtained with a full-length probe; a 70-mer had a detection sensitivity of 0.5 ng of genomic DNA while a 50-mer has a sensitivity of 100 ng of genomic DNA. Solvent conditions and temperature were also evaluated; good sensitivity was obtained at 45°C in the presence of 50% formamide	81
Comparison of Affymetrix gene expression arrays	82
Evaluation of various methods for labeling cDNA for microarray analysis	83
The solution-phase hybridization of 30 bp probes showed a strong correlation with binding to probes bound to 3D surface (the surface does not introduce steric effects). Also, specific binding takes a longer incubation time than nonspecific binding	84
Use of carbodiimide coupling of cyanine dyes for labeling cDNA obtained from messenger RNA by RT-PCR	85
Comparison of seven microarray labeling kits for the preparation of cDNA probes. This study suggests that it is useful to compare labeling technologies and essential to validate the method	86

(continued)

TABLE 7.3 (continued)
A Selection of Studies on the Experimental Variables Important
for DNA Microarray Analysis[a]

Study	Reference
Method of sample preparation designed to introduce quality control prior to array hybridization. Sample preparation is based on the use of mRNA separation using binding to oligo(T) magnetic beads	87
Use of sodium hydroxide for stripping oligonucleotide microarrays permitting the repeated use of the microarray	53

[a] Some of this material is dated by current microarray technology. However, the author is of the opinion that it is important to understand the various factors that have been responsible for today's technology. See Dufva, M., Introduction to microarray technology, *Methods Mol. Biol.* 529, 1–22, 2009; Bhattacharya, S. and Mariani, T.J., Array of hope: Expression profiling identifies disease biomarkers and mechanism, *Biochem. Soc. Trans.* 37, 855–862, 2009.

[b] Comparison across various microarray platforms continues to be an issue, which needs to be considered in evaluating data. See Archer, K.J., Dumur, C.I., Taylor, G.S. et al., Application of a correlation correction factor in a microarray cross-platform reproducibility study, *BMC Bioinformatics* 8, 447, 2007; Pedotti, P., 't Hoen, P.A.C., Vreugdenhil, E. et al., Can subtle changes in gene expression be consistently detected with different microarray platforms?, *BMC Genomics* 9, 124, 2008.

Nucleic acids can be bound to surfaces by several mechanisms[86,93] including the binding of biotin-labeled oligonucleotides to streptavidin-coated magnetic beads.[94–96] Another approach used the covalent attachment of 5'-aminohexyl oligonucleotides to polyethyleneimine-coated nylon beads.[97] Glass slides are cleaned with piranha solution (33% 30%H_2O_2–67% concentrated H_2SO_4).[98] The glass slides can be derivatized with silane derivatives to yield amino, carboxyl, or aldehyde derivatives (Figure 7.1)[99] for coupling to the probes. Binding to slides coated with poly-lysine followed by UV-crosslinking are also useful.[71,100–102] The point is that there are a variety of approaches for attaching cDNA or oligonucleotides to a matrix as well as several methods including photolithography, mechanical spotting, and ink-jet printing for application to the matrix.[103,104]

It is likely that gene expression biomarkers have great potential as predictive biomarkers[105–112] and, therefore, of great importance in personalized medicine.[113–120] However, much work remains to be done in the development and validation of gene expression platforms and the selection of appropriate target tissue(s). A promising example for the use of gene expression biomarkers is the study of the effect of intravenous immunoglobulin (IVIG; human polyvalent IgG) on a variety of diseases.[121,122] These studies enable the development of hypotheses regarding the mechanism action in the action of IVIG in immunomodulation. While the therapeutic effect of IgG in neutralizing and destroying bacterial and viral pathogens is well understood, the action in immunomodulation is, at best, poorly understood.[123,124] While some of the effects can be understood as neutralization of cytokines and chemokines[125,126] involved in inflammatory responses, the action of IVIG on cellular immunology is more complex.[127]

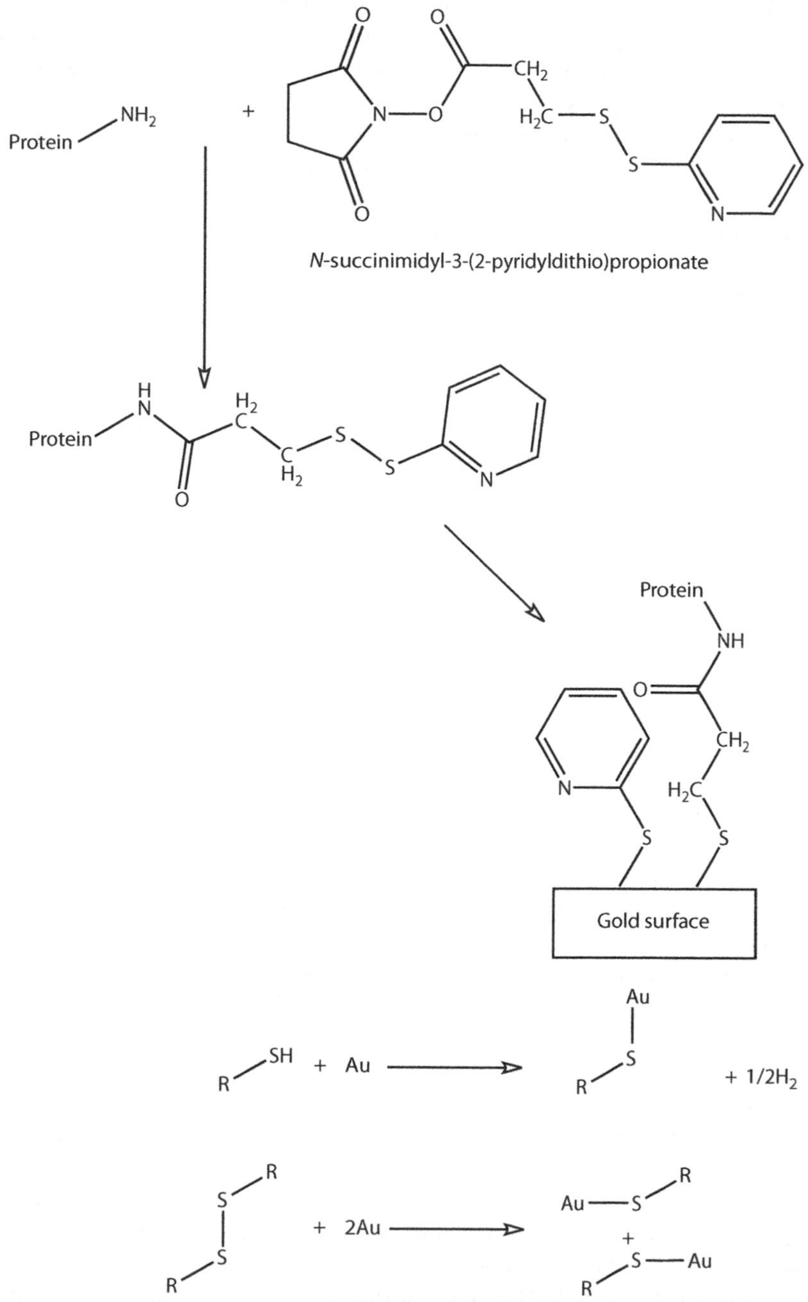

FIGURE 7.1 Binding of a protein to a gold surface. See Franzman, M.A. and Barrios, A.M., Spectroscopic evidence for the formation of goldfingers, *Inorg. Chem.* 47, 3928–3930, 2008. For the application of the pyridylthiopropionate, see Kohli, N., Hassler, B.L., Parthasarathy, L. et al., Tethered lipid bilayers on electrolessly deposited gold for bioelectronic applications, *Biomacromolecules* 7, 3327–3335, 2006.

As with the various biomarkers that are discussed elsewhere in this chapter, blood is the most useful sample for gene expression biomarkers. White blood cells including peripheral blood mononuclear cells and polymorphonuclear leukocytes are useful sources of gene expression biomarkers.[128–138] One particular concern in the use of white blood cells is the effect of pre-analytical variables on gene expression.[128,139] Siest and coworkers[119] emphasize the importance of controlling mRNA degradation with an appropriate systems such as PAXGene Blood RNA System or Tempus®.[140–145] Biological variation can also prove challenging in the understanding of changes in gene expression as, for example, in the effect of exercise on inflammatory biomarkers.[134,146] Tian and coworkers have a recent discussion of the processing of whole blood for DNA microarray studies.[147]

It is also possible to use DNA microarray analysis of whole peripheral blood[148–150] for the identification of gene expression as biomarkers. The major confounding factor in the use of whole blood is the presence of large amount of globin mRNA, but there are methods to reduce this problem.[151–153] Table 7.4 lists some selected studies on the identification of gene expression biomarkers in leukocytes and related blood cells while Table 7.5 lists some studies on gene expression in whole blood. It has been suggested that when it is possible to exclude the contribution of globin mRNA, the majority of gene expression in whole blood is due to the presence of leukocytes.[168] On balance, it would look as if meaningful data can be obtained from whole blood samples when care is taken to maintain DNA integrity and avoid complications from globin mRNA. This is quite useful as white blood cells are quite sensitive to solvent and storage conditions, which influence mRNA content.

Other examples of the use of gene expression as biomarkers were presented earlier in Table 2.20. DNA microarray technology is also useful for preclinical evaluation of drugs.[169–173]

The use of protein and peptide probes instead of cDNA and oligonucleotide probes for microarray has proved somewhat more challenging despite the long history of basic technologies such as enzyme-linked immunosorbent assays (ELISA). Specific base-pairing is well understood as the basis for oligonucleotide–polynucleotide interaction; similar specific phenomena are not available for protein–protein or protein–peptide interactions. While it is possible to use combinatorial chemistry or phage display to obtain "protein" microarrays similar to the DNA microarrays in binding site numbers, the rationale for site design is missing because absolute predictive rules for protein–protein interactions are still being developed. Furthermore, the primary structure complexity must be combined with co- and posttranslational modifications in the use of such rules. Antibodies are the most reasonable candidates for probes and the number of available antibodies has markedly increased in the last several years. Finally, with DNA microarrays, the samples are generated from isolated mRNA and can be labeled as part of the sample preparation process; a similar approach is not available for proteins. Despite these difficulties, there has been an increase in the use of protein microarrays resulting in part from a greater number of antibodies available as probes[174] and increase in solid-phase assay technology. Peptide probes can be used for serum antibody diagnostics.[175] In this approach, a peptide microarray platform is used to capture specific antibodies from serum, which are then detected by a fluorescent-labeled secondary antibody.

TABLE 7.4
Selected Studies on the Identification of Gene Expression Biomarkers Using DNA Microarray Technology in Leukocytes and Related Blood Cells

Study	Reference
Oligonucleotide (70 mer) microarrays used to study the effect of siRNA on gene expression in peripheral blood mononuclear cells	154
CD4 T cells, CD8 T cells, CD 19 B cells, CD14 monocytes and CD16 neutrophils were isolated from a single blood sample by positive selection (no consistent changes in gene expression with positive selection; poor results with negative selection reflecting purity issues) and studies by DNA microarray analysis (Affymetrix U133 and custom oligonucleotide probes [50 mer]). Storage of cells prior to separation results in large changes in expression patterns	133
Use of Codelink™ Human Whole Genome Microarray to study the effect of IFNβ on isolated lymphocytes obtained from patients with multiple sclerosis. Two different products of IFNβ were evaluated and there was no difference in global gene expression patterns; there was a change in accepted markers such as GTP cyclohydrolase and Myxococcus resistance protein for IFNβ action	155
DNA microarray technology (Affymetric GeneChip) used to study changes in gene expression induced by thymic stromal lymphopoietin in peripheral and cord blood mononuclear cells as well as isolated monocytes	156
DNA microarray used to study changes in gene expression (Affymetric HuGene FL) in HL60 granulocytoids and human polymorphonuclear leukocytes after exposure to *Candida albicans*	157
DNA microarray technology (Affymetrix) was used to make changes in gene expression during apoptosis following phagocytosis in human polymorphonuclear leukocytes. Over half of the differentially expressed genes (133/212) encoded proteins involved in some aspect of the inflammatory response	158
Leukocytes isolated from blood obtained from patients recovering from septic shock and used for gene expression studies with a DNA microarray based on genes involved in inflammation. CD74 gene expression increased and S100A8 and S100A12 showed reduced expression. The microarray results were confirmed by RT-qPCR. Plasma levels of S100A8 decreased and CD74 expression correlated with HLA-DR monocyte expression	159
Cytokine gene expression with microarray technology was evaluated in neutrophils isolated from presurgical peripheral blood and from the surgical wound site (hip drain neutrophils) in patients with total hip arthroplasty. Changes seen in gene expression were verified with RT-PCR. IL-1 receptor, IL-18 receptor, MIF, and macrophage inflammatory protein 3α (CCL20) were upregulated while IL-8 receptor β (IL8RB/CXCR2) was downregulated in surgical site neutrophils compared to presurgical peripheral blood neutrophils	160

Protein and other biological macromolecules bind to plastic (polystyrene, polypropylene, polyvinyl chloride) and other materials (glass, steel) in a "nonspecific" manner.[176–186] The adsorption of protein to plastic surfaces is frequently associated with conformational change/denaturation.[179,187–190] The binding of protein to plastic can be influenced by solvent conditions.[191–195] The binding of protein to microplate

TABLE 7.5
Selected Studies on the Identification of Gene Expression Biomarkers in Whole Blood

Study	Reference
Whole blood samples were used to examine factors blocking the action of IFNβ in multiple sclerosis patients. Whole blood was collected in EDTA for DNA microarray (Affymetrix) and in PAXgene® tubes for PCR amplification of specific genes. The induction of MxA (Myxococcus resistance protein) (PCR and microarray) is a biomarker for INFβ. MxA induction was absent in patients with antibodies against IFNβ; other IFNβ-regulated genes were also reduced in neutralizing antibody-positive patients. The lack of MxA expression on treatment with IFNβ in multiple sclerosis patients with neutralizing antibodies is an indication of a blocked response to IFNβ	161
Use of globin reduction to obtain a consistent increase in the quality of bioarray data. These studies used Illumina Beadchips[a] with globin reduction to perform genome-wide transcriptome analysis	147
Use of Affymetrix microarrays to study altered gene expression in psychotic patients	162
Use of Affymetric microarrays to study altered gene expression in colorectal cancer biopsy specimens and peripheral blood. It is suggested that the gene expression results in peripheral blood may be derived from circulating tumor cells or peripheral mononuclear cells	163
Gene microarray analysis (aligent oligonucleotide microarray) used to identify gene expression in bone marrow from gastric patients with and without metastasis. Membrane type 1 matrix metalloproteinase (MT1-MMP) as a candidate biomarker gene for metastasis. Real-time RT-PCR was used to study bone marrow and peripheral blood for MT1-MMP; MT1-MMP expression in peripheral blood was associated with peritoneal dissemination, lymphatic permeation, vascular permeation, and lymph node metastasis. MT1-MMP expression in bone marrow was also linked to distant metastasis an peritoneal dissemination	164
Comparison of methods for DNA microarray analysis of neonatal whole blood. Variables considered were anticoagulant and RNA extraction technical. Analysis used an oligonucleotide array (30 mer) for the human genome (55,000 probes; Codelink™ Human Whole Genome Bioarray[b])	165
DNA microarray (Affymetrix Human Genome U133A) used to study gene expression in whole blood samples patients with acute respiratory distress syndrome (ARDS); 126 genes were differentially expressed in ARDS patients compared to control subjects. Peptidase inhibitor 3 gene encoding pre-elafin was identified as a potential biomarker for ARDS	166
Method development for the use of whole blood for gene expression studies using DNA microarray technology	144
DNA Microarray (Affymetrix U133A) used to study gene expression in peripheral blood samples from patients with systemic lupus erythematosus as compared to a control population. Genes of interest were further studies with a custom focus oligonucleotide microarray and finally with RT-PCR. Response to INF therapy was also examined by gene expression profiling. It is possible to separate subjects into group with high and low INF signatures. It is also suggested that gene expression profiling might be a biomarker predictive of lupus flares	167

TABLE 7.5 (continued)

Selected Studies on the Identification of Gene Expression Biomarkers in Whole Blood

Study	Reference
Gene expression was studied with microarray (Affymetrix U133a Plus 2) analysis in whole blood samples (PAXgene®) obtained from patients with ischemic stroke. Most of the differentially expressed genes were shown in separate studies to be expressed by polymorphonuclear leukocytes. The use of aspirin also affected gene expression	168

[a] Kuhn, K., Baker, S.C., Chudin, E. et al., A novel, high-performance random array platform for quantitative gene expression profiling, *Genome Res.* 14, 2347–2356, 2004.

[b] Applied Microarrays, Tempe, AZ; http://appliedmicroarrays.com/index.php?option=com_content&view=article&id=15&Itemid=23; See Harris, L.W., Wayland, M., Lan, M. et al., The cerebral microvasculature in schizophrenia: A laser capture microdissection study, *PLoS One* 3, e3964, 2008.

wells can markedly influence cell-based assays.[196–200] Proteins can bind directly to gold via cysteine where a covalent bond is formed (Figure 7.1).[201] There has been considerable interest in the use of self-assembling monolayers on gold surfaces as used in surface plasmon resonance.[201–206] Self-assembled monolayers can also be formed on glass surfaces with the use of functionalized alkylsilane derivatives (Figure 7.2).[207]

Microplate/microarray surfaces and other surfaces such as beads may be modified to modulate binding.[208–214] Proteins with a special avidity for immunoglobulins such as protein A[215] and protein G[216,217] have been used to bind antibody matrices. Since protein A and protein G bind to the Fc domain of immunoglobulin, the bound antibody has the CDR region available for interaction with the analyte or target. Other approaches used protein engineering to attach affinity labels such as hexahistidine.[218,219]

While there are many possibilities, the majority of microarrays are prepared by coupling of molecules to matrices via aldehyde groups, amino groups, succinimide, maleimide groups, epoxy groups, and carboxyl groups (Figure 7.3).[220–223] Other approaches use the modification of protein as, for example, by the oxidation of carbohydrate with periodate to form aldehyde functions, which are in turn coupled to matrices with hydrazide or amino functions or by the inserting of a functional group either by chemical or genetic means. Reznik and colleagues[224] prepared a streptavidin mutant that could be covalently bound to a maleimide surface. Expression systems are available to prepare full-length proteins fused to a biotin-binding protein, which can then be used to bind to a streptavidin matrix.[225] Table 7.6 presents selected studies on the modification of proteins to provide linker functions for attachment to matrices. Table 7.7 presents selected studies on the coupling of antibodies (or related macromolecules) to planar or bead matrices. One of the major concerns in the interaction of any protein (or other macromolecule) to a matrix is conformational change (see above) so methods that are likely to preserve conformation are useful.[225]

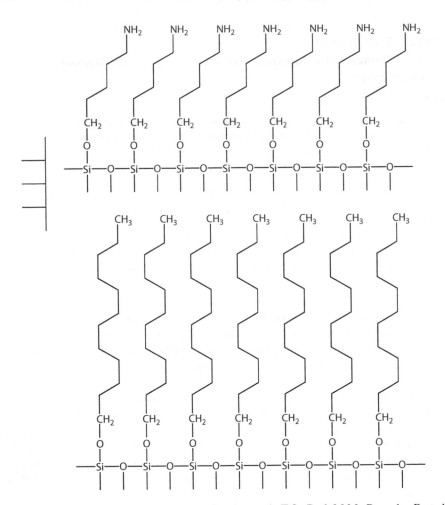

FIGURE 7.2   Self-assembled monolayers. See Acarturk, T.O., Peel, M.M., Petrosko, P. et al., Control of attachment, morphology, and proliferation of skeletal myoblasts on silanized glass, *J. Biomed. Mater. Res.* 44, 355–370, 1999; Tsai, P.-S., Yang, Y.-M., and Lee, Y.-L., Fabrication of hydrophobic surfaces by coupling of Langmuir-Blodgett deposition and a self-assembled monolayer, *Langmuir* 22, 5660–5665, 2006.

The great majority of protein-based microarrays have used antibodies bound to a solid-phase matrix[174] using the chemistry described above. It should be recognized that the technology is adapted from the various ELISA platforms.[251] ELISA were originally developed from radioimmunoassay (RIA) technology[252] by Engvall and coworkers.[253–255] It is useful at the point to consider the several different types of ELISA, which are in current use. The earliest form of an ELISA as developed by Engvall and Perlmann[253] used competition between labeled (alkaline phosphatase coupled to rabbit IgG with glutaraldehyde) and unlabeled IgG for binding to a solid-phase immunosorbent. This is referred to as a competitive direct ELISA; in a direct ELISA, the amount of antigen bound to the matrix would be measured by the amount

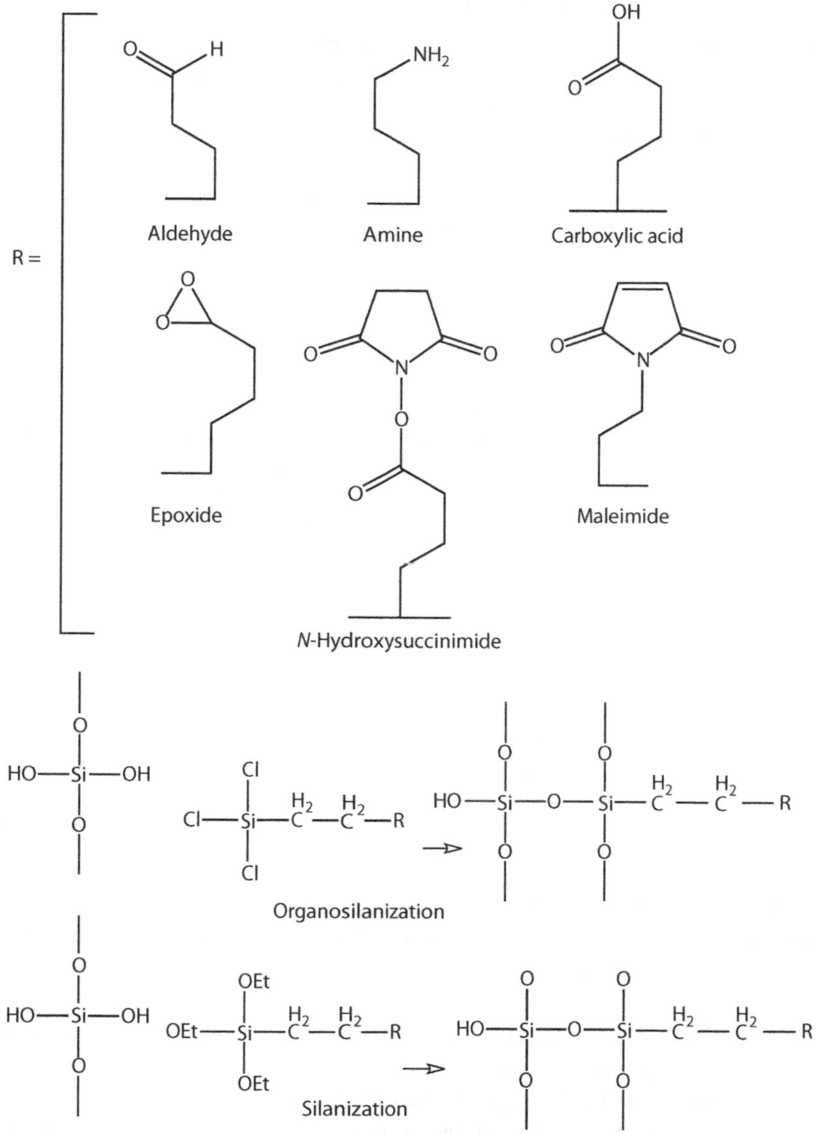

FIGURE 7.3 Microarray functional groups. The derivatives are prepared by coupling a trifunctional organic silane to a glass matrix. See Weetall, H.H., Preparation of immobilized proteins covalently coupled through silane coupling agents to inorganic supports, *Appl. Biochem. Biotechnol.* 41, 157–188, 1993; Matyska, M.T. and Pesek, J.J., Comparison of silanization/hydrosilation and organosilanization modification procedures on etched capillaries for electrokinetic chromatography, *J. Chromatogr. A* 1079, 366–371. 2005.

TABLE 7.6
Modification of Proteins for Attachment to Matrices

Functional Group	Comment	Reference
Thiol	3-Mercaptopropionic acid coupled to protein via lysine residues using carbodiimide	226
Aldehyde formed by periodate oxidation	Coupling of proteins to hydrazide or amino matrices	227
Biotin	Biotinylation using maleimide derivative at sulfhydryl group in antibody hinge region (after reduction with 2-mercaptoethanol) or at lysine residues with N-hydroxysuccinimide biotin derivatives	228
Metal-binding domain	A peptide of five glutamic and six histidine residues was either coupled or engineered into an antibody fragment[a]	229
Crosslinking reagent for coupling to various carbon-containing surfaces	Method for the immobilization of proteins and oligonucleotides to a variety of supports including polypropylene, polypropylene, nylon, and agarose. Linkage to the protein or nucleic acid occurs via amino or sulfhydryl function; coupling to polymer platform is via photochemical linkage	230
Thiol	Soluble scFv C-terminal free thiol	231
C-terminal or N-terminal hexalysine sequence	Coupling to maleic anhydride matrix	232
Aldehyde	Oxidation of carbohydrate with periodate to provide oriented antibody coupling to matrix for immunochromatography	233
Aldehyde	Kinetic model for oxidation of antibody carbohydrate by periodate	234

[a] Goel, A., Colcher, D., Koo, J.S. et al., Relative position of the hexahistidine tags effects binding properties of a tumor-associated single-chain Fv construct, *Biochim. Biophys. Acta* 1523, 13–20, 2000.

of labeled antibody bound. In an indirect ELISA, a secondary labeled antibody is used to detect the primary antibody, which recognizes the antibody bound to the antigen on the surface. In the original study,[255] the secondary antibody recognized IgG from rabbit antisera bound to antigen (bovine serum albumin or dinitrophenol) adsorbed in a polystyrene tube. While the early work was performed in polystyrene tubes, there was a rapid transition to microplates.[256–258] The reader is commended to the excellent article by Schuurs and VanWeemen[257] for a review of the basic technology base of ELISA systems. In a sandwich ELISA, a capture antibody is used to bind the analyte from bulk solution, which is recognized by a signal antibody. The signal was originally an enzyme permitting amplification; other approaches have used a fluorescent label, which then can be "read" by the same type of instrument as that used for DNA microarrays.[258] Additional sensitivity can be obtained by PCR

TABLE 7.7
Selected Studies on the Coupling of Antibodies and Related Proteins to Matrices

Antibody or Related Protein	Comment	Reference
Fab', F(ab'), IgG	SAM[a](thiols), random coupling via protein amino groups or specific attachment via protein sulfhydryl	235
Biotinylated antibody	Productive-oriented binding to streptavidin-coated plates	236
IgG, Fab'	Used random biotinylated (lysine) or specific (carbohydrate oxidation/coupling to resultant aldehyde to biotin hydrazide or free sulfhydryl in Fab'). Specific orientation provided greater system efficacy	237
N/A	Use of "his tag"–Protein A to immobilized IgG	238
N/A	Evaluation of commercial membranes for the manufacture of antibody microarrays	239
N/A	Purification of monospecific polyclonal antibodies for use in antibody microarrays	240
N/A	Autoantibody profiling microarray; comparison with multiplex beads	241
N/A	Phage versus phagemid libraries for the generation of human monoclonal antibodies	242
Antibodies to CD antigen	Measure leukocyte binding to microplate as index of CD expression	243
N/A	Internal control for antibody microarray	244
Antibodies to IL-1β, IL-1ra, IL-6, IL-8, MCP-1, TNFα	Piezoelectric application of antibody to conventional 96-well polystyrene microplate, ELISA-based assay, validation procedure	245
"Normal" proteins	Reverse-capture for the detection of autoantibodies	246
N/A	Comparison of bead and planar array technologies; rapid evaluation of antibody specificity	247
Candidate and control antigens	Profiling of autoantibodies in rheumatoid arthritis	248
N/A	Recombinant antibody-binding protein (hydrophobic domain fused with antibody-binding domain) for attachment of antibody to matrix	249
Lipopolysaccharide (LPS)	Use for analysis of anti-LPS antibodies	250

[a] SAM, self-assembled monolayer.

amplification of the signal antibody but this is not likely of value for protein microarrays but can be performed in a 96-well format.[259] ELISA systems and antibody microarray systems have mixed reviews with respect to comparability.[260,261]

Antibody microarrays do not have the flexibility of the ELISA platforms. Performance using "pins" on a matrix, either slide or bead, precludes the use of the generation of signal through enzyme action since wells do not limit the diffusion of product. It is possible to use denser microplates such as 384 well plates.

There are several approaches to the analysis of antibody microarrays. In one approach, there is direct labeling of the sample to provide a signal. Lin et al.[262] labeled the sample (lysate from cultured tumor cells) with biotin. An antibody microarray (cytokine antibodies) selects labeled cytokines from the sample and detection is accomplished with streptavidin–cyanine dye complexes. The use of two dyes, Cy3 (550 nm excitation, 570 nm emission) and Cy5 (650 nm excitation, 670 nm emission), allows the direct comparison of experimental and control samples. Miller et al.[263] employed Cy3 and Cy5 labeling of serum samples from prostate cancer patients and a control population to identify biomarkers using an antibody microarray either on a polyacrylamide-based hydrogel on glass or poly-L-lysine glass. The hydrogel matrix provided better data. Sandwich antibody microarrays are adapted from sandwich ELISA, where a secondary antibody is used to detect the antigen bound to the antibody microarray matrix.[251,264–266] Some examples of biomarker characterization using antibody microarrays are shown in Table 7.8.

Given the limitation of probes having the scope of DNA microarrays, there has been considerable interest in reverse protein microarrays where the sample is applied to the microplate and then probed with labeled antibodies. This approach is similar to the direct ELISA described above and it also possible to use a sandwich ELISA approach.[280] Reverse-phase microarrays also share technology with Western blotting as there is a correlation between the two analytical approaches.[281–283]

Reverse-phase technology is most often applied to blood or plasma but can be applied to whole cell lysates.[8,284] The use of fluorophore labels with different spectral characteristics as described above for Cy3 and Cy5 permits multiplexed analyses on reverse-phase microarrays.[285] Antibody specificity and technique are critical for the success of reverse-phase microarrays.[283] Table 7.9 presents some selected applications of reverse-phase microarray technology useful in biomarker research. The application of reverse-phase protein microarrays to the study of cellular responses to therapeutics may be the strongest use.[293–295]

The term "functional protein microarray" has been used to describe antibody microarrays, reverse-protein microarrays, and cDNA expression protein library microarrays.[296] The use of the term functional protein microarray is used to imply that the protein is bound in a manner that preserves biological function. Issues with respect to the binding of proteins to matrices in a manner that preserves biological function have been discussed above. The reader is directed to several recent reviews[297,298] that discuss the current status of functional protein microarrays.

DNA microarrays describe gene expression within a cell at a certain point in time. In basic science, this technology can provide great insight into the regulation of gene expression.[299] The comparison of data between various platforms is still a problem but is less of an issue than experimental design. The author sees considerable value in the use of DNA microarray studies in the preclinical evaluation of drugs and biopharmaceuticals. The author is less enthusiastic about protein microarrays although the major advances in antibody availability have great promise. The author is even less enthusiastic about reverse-protein microarrays. However, the author notes that

TABLE 7.8
A Selection of Studies on the Use of Microarrays for the Development of Biomarkers

Study	Reference
Protein array consisting of 329 proteins identified as antigenic in cancer patients by serological expression cloning (SEREX)[a] used to detect antibodies in sera from lung cancer patients. The proteins were immobilized in process,[b] which preserved conformational integrity. There was a 94% agreement between microarray binding and ELISA for antigens such as MAGEA4, TP53, SSX, and SOX2	225
Use of antibody microarray to measure CD expression on leukocyte surface.[c] The pattern of expression used to establish a CD expression signature on leukocytes, which could be used as biomarker	267
Use of allergen microarray (glass slides coated with a brush copolymer of N,N-dimethylacrylamide and N,N-acryloxysuccinimide immobilizes allergens in their native conformation) for the detection of specific IgE in serum	268
Use of phosphopeptide microarrays to screen for antibodies directed against posttranslational phosphorylation of proteins. A similar approach can be used for acetylated proteins	269
The use of phage display technology to identify ligands for patient serum antibodies in the development of biomarker B-cell epitope microarrays	270
Use of peptide microarrays to obtain antibody signature profiles in subjects with autoimmune disease. These antibody signature profiles can be biomarkers for the diagnosis of autoimmune disease	271
Allergen microarrays used to determine specific IgE patterns also allows the determination of specific IgG and IgM against the same allergens. The IgE pattern is a biomarker that can serve in a theranostic capacity with respect to immunotherapy	272
High-density protein microarrays were used to identify autoantibody patterns in colorectal cancer patients and normal subjects. Differential expression of 43 proteins was observed. The presence of specific autoantibodies for colorectal cancer and three antigens (PIMI, MAPKAPK3, and ACVR2B0) are possibly new biomarkers	273
A protein microarray was prepared using proteins from a pancreatic adenocarcinoma cell line. The proteins were separated by 2D-liquid-based technology (chromatofocusing following by reverse-phase HPLC) and spotted on nitrocellulose slides. The slides were probed with serum from pancreatic cancer patients and the patterns compared to those obtained with matched control subjects. Autoantibodies to phosphoglycerate kinase-1 and histone H4 were identified as possible serum biomarkers for pancreatic cancer	274
Preparation of peptide microarrays on bisphenol A polycarbonate using semicarbazone ligation with glyoxyl peptides. The peptide microarrays used to obtain antibody profiles in patient sera; the antibody profiles can serve as biomarkers[d]	275
Autoantibodies to tumor-associated antigens measure with protein microarrays to provide a profile or signature that can serve as biomarker	276

(continued)

TABLE 7.8 (continued)

A Selection of Studies on the Use of Microarrays for the Development of Biomarkers

Study	Reference
Use of tumor-associated MUC1 glycopeptides microarray to determine antibody specificity in sera in development of biomarkers for vaccine development[e]	277
Protein kinase Cζ as biomarker for risk of allograft lost in pediatric renal transplant patients as identified by protein microarray technology[f]	278
Use of a recombinant protein microarray[g] to define autoantibody profile as biomarker in colorectal cancer. Eighteen antigens were identified as being associated with cancer and four antigens associated without cancer, which may serve as biomarkers	279

[a] See Türeci, O., Sahin, U., and Pfreundschuh, M., Serological analysis of human tumor antigens: Molecular definition and implications, *Mol. Med. Today* 3, 342–349, 1997; Comtesse, N., Heckel, D., Maldener, E. et al., Probing the human natural autoantibody repertoire using an immunoscreening approach, *Clin. Exp. Immunol.* 121, 430–436, 2000.

[b] Full-length open reading frames for the target genes were cloned in-frame with a sequence encoding a C-terminal *E. coli* biotin carboxyl carrier protein fused to the myc epitope. The recombinant proteins that contained bound biotin were spotted on streptavidin-coated glass slides. See Blackburn, J.M. and Hart, D.J., Fabrication of protein function microarrays for systems-oriented proteomic analysis, *Methods Mol. Biol.* 310, 197–216, 2005.

[c] See Belov, L., de la Vega, O., dos Remedios, C.G. et al., Immunophenotyping of leukemias using a cluster of differentiation antibody microarray, *Cancer Res.* 61, 4483–4489, 2001.

[d] The concept of using autoantibodies as biomarkers is not new and the use of autoantibody profiling with microarray technology builds on established work in this area. See Neri, R., Tavoni, A., Cristofani, R. et al., Antinuclear antibody profile in Italian patients with connective tissue diseases, *Lupus* 1, 221–227, 1992; Molnár, K., Kovács, L., Kiss, M. et al., Antineutrophil cytoplasmic antibodies in patients with systemic lupus erythematosus, *Clin. Exp. Dermatol.* 27, 59–61, 2002.

[e] Kaiser, A., Gaidzik, N., Westerlind, U. et al., A synthetic vaccine consisting of a tumor-associated sialyl-T_N-MUC1 tandem-repeat glycopeptides and tetanus toxoid: Induction of a strong and highly selective immune response, *Angew. Chem. Int. Ed. Engl.* 48, 7551–7555, 2009.

[f] ProtoArray® Human Protein Microarray; Invitrogen, http://www.invitrogen.com. See Matton, D., Michaud, G., Merkel, J. et al., Biomarker discovery using protein microarray technology platforms: Antibody-antigen complex profiling, *Expert Rev. Proteomics* 2, 879–889, 2005.

[g] A 37,830 clone recombinant protein array representing 10,000 proteins was prepared for a human cDNA expression library. See Büssow, K., Nordhoff, E., Lübbert, C. et al., A human cDNA library for high-throughput protein expression screening, *Genomics* 65, 1–8, 2000; Walter, G., Büssow, K., Cahill, D. et al., Protein arrays for gene expression and molecular interaction screening, *Curr. Opin. Microbiol.* 3, 298–302, 2000; Weiner, H., Faupel, T., and Büssow, K., Protein arrays from cDNA expression libraries, *Methods Mol. Biol.* 264, 1–13, 2004.

TABLE 7.9
A Selection of Studies on the Use of Reverse Microarrays
for the Characterization of Biomarkers

Study	Reference
Reverse-phase protein microarray was used to study cell lysates from follicular lymphoma. There was specific interest in proteins involved in apoptosis	286
Reverse-phase protein microarrays were used to the phosphorylation of signaling components in Jurkat T cells stimulated through CD3 and CD28 and signal transducer and activator of transcription (STAT)[a] protein phosphorylation in IL-2 stimulated regulatory T-cells	287
Reverse-phase microarrays to measure tissue biomarkers after enrichment by immunoaffinity chromatography to remove hematopoietic cells. Purified cells are lysed and lysate is used as substrate for antibody probing in reverse-phase microarray	288
Reverse-phase microarrays can be used to measure biomarkers in lysates from formalin-fixed/paraffin-embedded tissues. This study compared Western blotting and reverse-phase microarray	289
Reverse-phase protein microarrays (cell lysates arrayed on nitrocellulose-coated slides and probed with antibody; one array/antibody) are used together with Western blotting and ELISA to identify potential theranostic biomarkers for the use gefitinib[b] in head and neck squamous cell carcinoma (HNSCC). Subsequent studies used tumor samples from patients. Gefitinib sensitivity is correlated with changes in p-AKT and p-STAT3 activation; these proteins are potential predictive biomarkers for the use of gefitinib in HNSCC	290
Use of reverse-phase protein microarray to identify biomarkers in colorectal cancer	291
Use of reverse-phase protein microarrays to validate clusterin as a biomarker. Various matrices were evaluated including Maxisorp®, nitrocellulose, and epoxy slides	292

[a] See Fu, X.Y., A direct signaling pathway through tyrosine kinase activation of SH2 domain-containing transcription factors, *J. Leukoc. Biol.* 57, 529–535, 1995.

[b] Genfitinib (Iressa®) is an inhibitor of EGFR protein tyrosine kinase. See Herbst, R.S. and Kies, M.S., Gefitinib: Current and future status in cancer therapy, *Clin. Adv. Hematol. Oncol.* 1, 466–472, 2003.

as a former football official, he was frequently wrong and there may still be films to prove it. I recommend a study of Keith Baggerly's work in this general area[300–303] lest you and your colleagues become candidates for investigation by a forensic bioinformatician.[304]

REFERENCES

1. Campíns-Falcó, P., Sevillano-Cabeza, A., Herráez-Hernández, R., and Molins-Legua, C., Solid-phase extraction and clean up procedures in pharmaceutical analysis, in *Encyclopedia of Analytical Chemistry*, R.A. Meyers (ed.), John Wiley & Sons, Chichester, U.K., pp. 7320–7336, 2000.
2. Lebert, J.M., Forsberg, E.M., and Brennan, J.D., Solid-phase assays for small molecule screening using sol-gel entrapped proteins, *Biochem. Cell Biol.* 86, 100–110, 2008.
3. Vynios, D.H., Solid phase assays: Theory and applications in clinical chemistry, biochemistry and biotechnology, *Pharmakeutike* 12, 33–41, 1999.

4. Liu, E. and Eisenbarth, G.S., Accepted clocks that tell time poorly: Fluid-phase versus standard ELISA autoantibody assays, *Clin. Immunol.* 125, 120–126, 2007.

5. Aubert, V., Venetz, J.P., Pantaleo, G., and Pascual, M., Low levels of human leukocyte antigen donor-specific antibodies detected by solid phase assay before transplantation are frequently clinical irrelevant, *Human Immunol.* 70, 580–583, 2009.

6. Rosenbluh, J., Kapelnikov, A., Shalev, D.E. et al., Positively charged peptides can interact with each other as revealed by solid phase binding assays, *Anal. Biochem.* 352, 157–168, 2006.

7. VanMeter, A., Signore, M., Pierobon, M. et al., Reverse-phase protein microarrays: Application to biomarker discovery and translational medicine, *Expert Rev. Mol. Diagn.* 7, 625–633, 2007.

8. Spurrier, B., Honkanen, P., Holway, A. et al., Protein and lysate array technologies in cancer research, *Biotechnol. Adv.* 26, 361–369, 2008.

9. Espina, V., Wulfkuhle, J., and Liotta, L.A., Application of laser microdissection and reverse-phase protein microarrays to the molecular profiling of cancer signal pathway networks in the tissue microenvironment, *Clin. Lab. Med.* 29, 1–13, 2009.

10. Bubendorf, L., Nocito, A., Moch, H., and Sauter, G., Tissue microarray (TMA) technology: Miniaturized pathology archives for high-throughput in situ studies, *J. Pathol.* 195, 72–79, 2001.

11. Mobasheri, A., Airley, R., Foster, C.S. et al., Post-genomic applications of tissue microarrays: Basic research, prognostic oncology, clinical genomics and drug discovery, *Histol. Histopathol.* 19, 325–335, 2004.

12. Bentzen, S.M., Buffa, F.M., and Wilson, G.D., Multiple biomarker tissue microarrays: Bioinformatics and practical approaches, *Cancer Metastasis Rev.* 27, 481–494, 2008.

13. Avninder, S., Yiaya, K., and Hewitt, S.M., Tissue microarray: A simple technology that has revolutionized research in pathology, *J. Postgrad. Med.* 54, 158–162, 2008.

14. Sullivan, C.A. and Chung, G.G., Biomarker validation: In situ analysis of protein expression using semiquantitative immunohistochemistry-based techniques, *Clin. Colorectal Cancer* 7, 172–177, 2008.

15. Camp, R.L., Neumeister, V., and Rimm, D.L., A decade of tissue microarrays: Progress in the discovery and validation of cancer biomarkers, *J. Clin. Oncol.* 26, 5630–5637, 2008.

16. Permuth-Wey, J., Boulware, D., Valkov, N. et al., Sampling strategies for tissue microarrays to evaluated biomarkers in ovarian cancer, *Cancer Epidemiol. Biomarkers Prev.* 18, 28–34, 2009.

17. Assikis, V.J., Do, K.A., Wien, S. et al., Clinical and biomarker correlates of androgen-independent, locally aggressive prostate cancer with limited metastatic potential, *Clin. Cancer Res.* 10, 6770–6778, 2004.

18. Bayani, J. and Squire, J.A., Application and interpretation of FISH in biomarker studies, *Cancer Lett.* 249, 97–109, 2007.

19. Kwon, M.S., Shin, S.H., Yim, S.H. et al., CD63 as a biomarker for predicting the clinical outcomes in adenocarcinoma of lung, *Lung Cancer* 57, 46–53, 2007.

20. Taniwaki, M., Takano, A., Ishikawa, N. et al., Activation of KIF4A as a prognostic biomarker and therapeutic target for lung cancer, *Clin. Cancer Res.* 13, 6624–6631, 2007.

21. McDonnell, T.J., Chari, N.S., Cho-Vega, J.H. et al., Biomarker expression patterns that correlate with high grade features in treatment naïve, organ-confined prostate cancer, *BMC Med. Genomics* 1, 1, 2008.

22. de Jong, D., Xie, W., Rosenwald, A. et al., Immunohistochemical prognostic markers in diffuse large B-cell lymphoma: Validation of tissue microarray as a perquisite for broad clinical applications (a study from the Lunenburg Lymphoma Biomarker Consortium), *J. Clin. Pathol.* 62, 128–138, 2009.

23. Yang, G.H., Fan, J., Xu, Y. et al., Osteopontin combined with CD44, a prognostic biomarker for patients with hepatocellular carcinoma undergoing curative resection, *Oncologist* 13, 1155–1165, 2008.

24. Smith, S.C., Nicholson, B., Nitz, M. et al., Profiling bladder cancer organ site-specific metastasis identifies LAMC2 as a novel biomarker of hematogenous dissemination, *Am. J. Pathol.* 174, 371–379, 2009.
25. Kallioniemi, O.P., Wagner, U., Kononen, J., and Sauter, G., Tissue microarray technology for high-throughput molecular profiling of cancer, *Hum. Mol. Genet.* 10, 657–662, 2001.
26. Braunschweig, T., Chung, J.Y., and Hewitt, S.M., Perspectives in tissue microarrays, *Comb. Chem. High Throughput Screen.* 7, 575–585, 2006.
27. Jubb, A.M., Pham, T.Q., Frantz, G.D. et al., Quantitative in situ hybridization of tissue microarrays, *Methods Mol. Biol.* 326, 255–264, 2006.
28. Brown, L.A. and Huntsman, D., Fluorescent in situ hybridization on tissue microarrays: Challenges and solutions, *J. Mol. Histol.* 38, 151–157, 2007.
29. Robertson, D., Savage, K., Reis-Filho, J.S., and Isacke, C.M., Multiple immunofluorescence labelling formalin-fixed paraffin-embedded (FFPE) tissue, *BMC Cell Biol.* 9, 13, 2008.
30. Bibikova, M., Yeakley, J.M., Wang-Rodriguez, J., and Fan, J.B., Quantitative expression profiling of RNA from formalin-fixed, paraffin-embedded tissues using randomly assembled bead arrays, *Methods Mol. Biol.* 439, 159–177, 2008.
31. Zhang, X., Chen, J., Radcliffe, T. et al., An array-based analysis of microRNA expression comparing matched frozen and formalin-fixed paraffin-embedded human tissue samples, *J. Mol. Diagn.* 10, 513–519, 2008.
32. Lassmann, S., Kreutz, C., Schoepflin, A. et al., A novel approach for reliable microarray analysis of microdissected tumor cells from formalin-fixed and paraffin-embedded colorectal cancer resection specimens, *J. Mol. Med.* 87, 211–224, 2009.
33. Koh, S.S., Opel, M.L., Wei, J.P. et al., Molecular classification of melanomas and nevi using gene expression microarray signatures and formalin-fixed and paraffin-embedded tissue, *Mol. Pathol.* 22, 538–546, 2009.
34. Schwers, S., Reifenberger, E., Gehrmann, M. et al., A high-sensitivity, medium-density, and target amplification-free planar waveguide microarray system for gene expression analysis of formalin-fixed and paraffin-embedded tissue, *Clin. Chem.* 55, 1995–2003, 2009.
35. Pascal, L.E., True, L.D., Campbell, D.S. et al., Correlation of mRNA and protein levels: Cell type-specific gene expression of cluster designation antigens in the prostate, *BMC Genomics* 9, 246, 2008.
36. Li, N., Jian, X., Yanjia, X. et al., Discovery of novel biomarkers in oral submucous fibrosis by microarray analysis, *Cancer Epidemiol. Biomarkers Prev.* 17, 2249–2259, 2009.
37. de Bont, J.M., Kros, J.M., Passier, M.M.C. et al., Differential expression and prognostic significance of SOX genes in pediatric medulloblastoma and ependymoma identified by microarray analysis, *Neuro-Oncology* 10, 648–660, 2008.
38. Macucci, G., Morandi, L., Magrini, E. et al., Gene expression profiling in glioblastoma and immunohistochemical evaluation of IGFBP-2 and CDC-20, *Virchows Arch.* 453, 599–609, 2008.
39. Jonsdottir, K., Stoerkson, R., Krog, A. et al., Correlation between mRNA detected by microarrays and qRT-PCR and protein detected by immunohistochemistry of cyclins in tumour tissue from colonic adenocarcinoma, *Open Pathol. J.* 2, 96–101, 2008.
40. Chiang, Y.Y., Tsai, M.-H., Lin, T.-Y. et al., Expression profile of metasis-related genes in invasive oral cancers, *Histol. Histopathol.* 23, 1213–1222, 2008.
41. Crouser, E.D., Culver, C.A., Knox, K.A. et al., Gene expression profiling identifies MMP-12 and ADAMDEC1 as potential pathogenic mediators of pulmonary sarcoidosis, *Am. J. Respir. Crit. Care Med.* 179, 928–938, 2009.
42. Soon, P.S.H., Gill, A.J., Benn, D.E. et al., Microarray gene expression and immunohistochemistry analysis of adrenocortical tumors identify IGF2 and Ki-67 as useful in differentiating carcinomas from adenomas, *Endocr. Relat. Cancer* 16, 573–583, 2009.

43. Yoshihari, K., Tajime, A., Dai, K. et al., Gene expression profiling of advanced serous ovarian cancer distinguishes novel subclasses and implicates ZEB2 n tumor progression and prognosis, *Cancer Sci.* 100, 1421–1428, 2009.

44. Sato-Kuwabara, Y., Neves, J.I., Fregnani, J.H. et al., Evaluation of gene amplification and protein expression of HER-2/neu in esophageal squamous cell carcinoma using Fluorescence in situ Hybridization (FISH) and immunohistochemistry, *BMC Cancer* 9, 6, 2009.

45. Gry, M., Rimini, R., Strömberg, S. et al., Correlations between RNA and protein expression profiles in 23 human cell lines, *BMC Genomics* 10, 365, 2009.

46. Ramsay, G., DNA chips: State-of-the-art, *Nat. Biotechnol.* 16, 40–44, 1998.

47. Mandruzzato, S., Technological platforms for microarray gene expression profiling, *Adv. Exp. Med. Biol.* 593, 12–18, 2007.

48. Sievertzon, M., Nilsson, P., and Lundeberg, J., Improving reliability and performance of DNA microarrays, *Expert Rev. Mol. Diagn.* 6, 481–492, 2006.

49. Lee, P.S. and Lee, K.H., Genomic analysis, *Curr. Opin. Biotechnol.* 11, 171–175, 2000.

50. Blohm, D.H. and Guiseppi-Elie, A., New developments in microarray technology, *Curr. Opin. Biotechnol.* 12, 41, 2001.

51. Olson, J.A., Application of microarray profiling to clinical trials in cancer, *Surgery* 136, 519–523, 2004.

52. Geschwind, D.H., DNA microarrays: Translation of the genome from laboratory to clinic, *Lancet Neurol.* 2, 275–282, 2003.

53. Hu, Z., Troester, M., and Perou, C.M., High reproducibility using sodium hydroxide-stripped long oligonucleotide DNA microarrays, *Biotechniques* 38, 121–124, 2005.

54. Jung, A., Stemmler, I., Brecht, A., and Gauglitz, G., Covalent strategy for immobilization of DNA-microspots suitable for microarrays with label-free and time-resolved optical detection of hybridization, *Fresen. J. Anal. Chem.* 371, 128–136, 2001.

55. Dolan, P.L., Wu. Y., Ista, L.K. et al., Robust and efficient synthetic method for forming DNA microarrays, *Nucleic Acids Res.* 29, E107, 2001.

56. Consolandi, C., Castiglioni, B., Bordoni, R. et al., Two efficient polymeric chemical platforms for oligonucleotide microarray preparation, *Nucleosides Nucleotides Nucleic Acids* 21, 561–580, 2002.

57. Hahnke, K., Jacobsen, M., Gruetzkau, A. et al., Striptease on glass: Validation of an improved stripping procedure for *in situ* microarrays, *J. Biotechnol.* 128, 1–13, 2007.

58. Gillespie, D. and Speigelman, S., A quantitative assay for DNA-RNA hybrids with DNA immobilized on a membrane, *J. Mol. Biol.* 12, 829–842, 1965.

59. Southern, E.M., Detection of specific sequences among DNA fragments separated by gel electrophoresis, *J. Mol. Biol.* 98, 503–517, 1975.

60. Thompson, J. and Gillespie, D., Molecular hybridization with RNA probes in concentrated solutions of guanidine thiocyanate, *Anal. Biochem.* 163, 281–291, 1987.

61. Southern, E.M., Case-Green, S.C., Elder, J.K. et al., Array of complementary oligonucleotides for analyzing the hybridization behavior of nucleic acids, *Nucleic Acids Res.* 22, 1368–1373, 1994.

62. Southern, E.M., DNA microarrays. History and overview, *Methods Mol. Biol.* 170, 1–15, 2001.

63. Schena, M. (ed.), *Microarray Analysis*, Wiley-Liss, Hoboken, NJ, 2001.

64. Simon, R.M., *Design and Analysis of DNA Microarray Investigations*, Springer, New York, 2003.

65. Bowtell, D. and Sambrook, J. (eds.), *DNA Microarrays: A Molecular Cloning Manual*, Cold Spring Harbor Laboratory Press, Cold Spring Harbor, NY, 2003.

66. Stec, J., Wang, J., Coombes, K. et al., Comparison of the predictive accuracy of DNA array-based multigene classifiers across cDNA arrays and Affymetrix GeneChips, *J. Mol. Diagn.* 7, 357–367, 2005.

67. Dalma-Weiszhausz, D.D., Warrington, J., Tanimoto, E.Y., and Miyada, C.G., The Affymetrix GeneChip® platform: An overview, *Methods Enzymol.* 410, 3–28, 2006.
68. Woblber, P.K., Collins, P.J., Lucas, A.B. et al., The Agilent *in-situ*-synthesized microarray platform, *Methods Enzymol.* 410, 28–57, 2006.
69. Hager, J., Making and using spotted DNA microarrays in an academic core laboratory, *Methods Enzymol.* 410, 125–168, 2006.
70. Tan, D.S., Lambros, M.B., Natrajan, R., and Reis-Filho, J.S., Getting it right: Designing microarray (and not 'microawry') comparative genomic hybridization studies for cancer research, *Lab. Invest.* 87, 737–754, 2007.
71. Zhang, W., Shmulevich, I., and Astola, J., *Microarray Quality Control*, Wiley-Liss, Hoboken, NJ, Chapter 2, 2004.
72. Stillman, B.A. and Tomkinson, J.L., Expression microarray hybridization kinetics depend on length of the immobilized DNA but are independent of immobilization substrate, *Anal. Biochem.* 295, 149–157, 2001.
73. Squires, T.M., Messinger, R.J., and Manalis, S.R., Making it stick: Convection, reaction and diffusion in surface-based biosensors, *Nat. Biotechnol.* 26, 417–425, 2008.
74. Shippy, R., Sendera, T.J., Lockner, R. et al., Performance evaluation of commercial short-oligonucleotide microarrays and the impact of noise in making cross-platform correlations, *BMC Genomics* 5, 61, 2004.
75. Suzuki, S., Ono, N., Furusawa, C. et al., Experimental optimization of probe length to increase the sequence specificity of high-density oligonucleotide microarrays, *BMC Genomics* 8, 373, 2007.
76. Rennie, C., Noyes, H.A., Kemp, S.J. et al., Strong position-dependent effects of sequence mismatches on signal ratios measured using long oligonucleotide microarrays, *BMC Genomics* 9, 317, 2008.
77. Steel, A.B., Levicky, R.L., Herne, T.M., and Tarlov, M.J., Immobilization of nucleic acids at solid surfaces: Effect of oligonucleotide length on layer assembly, *Biophys. J.* 79, 975–981, 2000.
78. Tomiuk, S. and Hofmann, K., Microarray probe selection strategies, *Brief Bioinform.* 2, 329–340, 2001.
79. Kuo, W.P., Jenssen, T.K., Butte, A.J. et al., Analysis of matched mRNA measurements from two different microarray technologies, *Bioinformatics* 18, 405–412, 2002.
80. Chou, C.C., Chen, C.H., Lee, T.T., and Peck, K., Optimization of probe length and the number of probes per gene for optimal microarray analysis of gene expression, *Nucleic Acids Res.* 32, e99, 2004.
81. He, Z., Wu, L., Fields, M.W., and Zhou, J., Use of microarrays with different probe sizes for monitoring gene expression, *Appl. Environ. Microbiol.* 71, 5154–5162, 2005.
82. Robinson, M.D. and Speed, T.P., A comparison of Affymetrix gene expression arrays, *BMC Bioinformatics* 8, 449, 2007.
83. Badiee, A., Eiken, H.G., Steen, V.M., and Løvlie, R., Evaluation of five different cDNA labeling methods for microarrays using spike controls, *BMC Biotechnol.* 3, 23, 2003.
84. Dorris, D.R., Nguyen, A., Gieser, L. et al., Oligonucleotide probe accessibility on a three-dimensional DNA microarray surface and the effect of hybridization time on the accuracy of expression ratios, *BMC Biotechnol.* 3, 6, 2003.
85. Kimura, N., Tamura, T., and Murakami, M., Evaluation of the performance of two carbodiimide-based cyanine dyes for detecting changes in mRNA expression with DNA microarrays, *Biotechniques* 38, 797–806, 2005.
86. Lynch, J.L., deSilva, C.J.S., Peeva, V.K., and Swanson, N.R., Comparison of commercial probe labeling kits for microarray: Towards quality assurance and consistency of reactions, *Anal. Biochem.* 355, 224–231, 2006.

87. Degenkolbe, T., Hannah, M.A., Freund, S. et al., A quality-controlled microarray method for gene expression profiling, *Anal. Biochem.* 346, 217–224, 2005.
88. Borozan, I., Chen, L., Paeper, B. et al., MAID: An effect size based model for microarray data integration across laboratories and platforms, *BMC Bioinformatics* 9, 305, 2008.
89. Shi, L., Perkins, R.G., Fang, H. et al., Reproducible and reliable microarray results through quality control: Good laboratory proficiency and appropriate data analysis practices are essential, *Curr. Opin. Biotechnol.* 19, 10–18, 2008.
90. Oliver, S., On the MIAME standards and central repositories of microarray data, *Comp. Funct. Genomics* 4, 1, 2003.
91. Tomlinson, C., Thimma, M., Alexandrakis, S. et al., MiMiR—An integrated platform for microarray data sharing, mining and analysis, *BMC Bioinformatics* 9, 379, 2008.
92. Zia, X.Q., McCelland, M., Porwollik, S. et al., WebArrayDB: Cross-platform microarray data analysis and public data repository, *Bioinformatics* 25, 2425–2429, 2009.
93. Kuhn, K., Baker, S.C., Chudin, E. et al., A novel, high-performance random array platform for quantitative gene expression profiling, *Genome Res.* 14, 2347–2356, 2004.
94. Schibler, U., Rifat, D., and Lavery, D.J., The isolation of differentially expressed mRNA sequences by selective amplification via biotin and restriction-mediated enrichment, *Methods* 24, 3–14, 2001.
95. Xu, Z., Jablonx, D.M., and Gruenert, D.C., Expression sequence tag-specific full-length cDNA cloning: Actin cDNAs, *Gene* 263, 265–272, 2001.
96. Liu, G. and Lin, Y., Electrochemical quantification of single-nucleotide polymorphisms using nanoparticle probes, *J. Am. Chem. Soc.* 129, 10394–10401, 2007.
97. Van Ness, J., Kalbfleisch, S., Petrie, C.R. et al., A versatile solid support system for oligonucleotide probe-based hybridization assays, *Nucleic Acids Res.* 19, 3345–3350, 1991.
98. Guo, W. and Ruckenstein, E., Crosslinked glass fiber affinity membrane chromatography and its application to fibronectin separation, *J. Chromatogr. B* 795, 61–72, 2003.
99. Zammatteo, N., Jeanmart, L., Hamels, S. et al., Comparison between different strategies of covalent attachment of DNA to glass surfaces to build DNA microarrays, *Anal. Biochem.* 280, 143–150, 2000.
100. DeRisi, J., Penland, L., Brown, P.O. et al., Use of a cDNA microarray to analyse gene expression patterns in human cancer, *Nat. Genet.* 14, 457–460, 1996.
101. Taylor, S., Smith, S., Windle, B., and Guiseppi-Elie, A., Impact of surface chemistry and blocking strategies on DNA microarrays, *Nucleic Acids Res.* 31, e87, 2003.
102. Hessner, M.J., Meyer, L., Tackes, J. et al., Immobilized probe and glass surface chemistry as variables in microarray fabrication, *BMC Genomics* 5, 53, 2004.
103. Selheyer, K. and Belbin, T.J., DNA microarrays: From structural genomics to functional genomics. The application of gene chips in dermatology and dermatopathology, *J. Am. Acad. Dermatol.* 51, 681–692, 2004.
104. Park, J.H., Lee, J.H., Paik, U. et al., Nanoscale patterns of oligonucleotides formed by electrohydrodynamic jet printing with applications in biosensing and nanomaterials assembly, *Nano Lett.* 8, 4210–4216, 2008.
105. Siena, S., Sartore-Bianchi, A., Di Nicolantonio, F. et al., Biomarkers predicting clinical outcome of epidermal growth factor receptor-targeted therapy in metastatic colorectal cancer, *J. Natl. Cancer Inst.* 101, 1308–1324, 2009.
106. West, H., Lilenbaum, R., Harpole, D. et al., Molecular analysis-based treatment strategies for the management of non-small cell lung cancer, *J. Thorac. Oncol.* 4(9 Suppl 2), S1029–S1039, 2009.
107. Harry, V.N., Gilbert, F.J., and Parkin, D.E., Predicting the response of advanced cervical and ovarian tumors to therapy, *Obstet. Gynecol. Surv.* 64, 548–560, 2009.

108. Walther, A., Johnstone, E., Swanton, C. et al., Genetic prognostic and predictive markers in colorectal cancer, *Nat. Rev. Cancer* 9, 489–499, 2009.
109. Kim, C., Taniyama, Y., and Paik, S., Gene expression-based prognostic and predictive markers in colorectal cancer: A primer for practicing pathologists, *Arch. Pathol. Lab.* 133, 855–859, 2009.
110. Oakman, C., Bessi, S., Zafarana, E. et al., Recent advances in systemic therapy: New diagnostics and biological predictors of outcome in early breast cancer, *Breast Cancer Res.* 11, 205, 2009.
111. Stimson, L. and La Thangue, N.B., Biomarkers for predicting clinical responses to HCAD inhibitors, *Cancer Lett.* 280, 177–183, 2009.
112. Pakkiri, P., Lakhani, S.R., and Smart, C.E., Current and future approach to the pathologist's assessment for targeted therapy in breast cancer, *Pathology* 41, 89–99, 2009.
113. Ozdemir, V., Williams-Jones, B., Cooper, D.M. et al., Mapping translational research in personalized therapeutics: Form molecular markers to health policy, *Pharmacogenomics* 8, 177–185, 2007.
114. van't Veer, L.J. and Bernards, R., Enabling personalized cancer medicine though analysis of gene-expression patterns, *Nature* 452, 564–570, 2008.
115. Shai, R.M., Reichardt, J.K., and Chen, T.C., Pharmacogenomics of brain cancer and personalized medicine in malignant gliomas, *Future Oncol.* 4, 525–534, 2008.
116. Dowsett, M. and Dunbier, A.K., Emerging biomarkers and new understanding of traditional markers in personalized therapy for breast cancer, *Clin. Cancer Res.* 14, 8019–8026, 2008.
117. Vosslamber, S., van Baarsen, L.G., and Verweij, C.L., Pharmacogenomics of IFN-β in multiple sclerosis: Towards a personalized medicine approach, *Pharmacogenomics* 10, 97–108, 2009.
118. Ross, J.S., Slodkowska, E.A., Symmans, W.F. et al., The HER-2 receptor and breast cancer: Ten years of targeted anti-HER-2 therapy and personalized medicine, *Oncologist* 14, 320–368, 2009.
119. Sawitzki, B., Pascher, A., Babel, N. et al., Can we use biomarkers and functional assays to implement personalized therapies in transplantation?, *Transplantation* 87, 1595–1601, 2009.
120. Smith, M.Q., Staley, C.A., Kooby, D.A. et al., Multiplexed fluorescence imaging of tumor biomarkers in gene expression and protein levels for personalize and predictive medicine, *Curr. Mol. Med.* 9, 1017–1023, 2009.
121. Sapan, C.V., Reisner, H.M., and Lundblad, R.L., Antibody therapy (IVIG): Evaluation of the use of genomics and proteomics for the study of immunomodulation therapeutics, *Vox Sang.* 92, 197–205, 2007.
122. Pigard, N., Elovaara, I., Kuusisto, H. et al., Therapeutic activities of intravenous immunoglobulins in multiple sclerosis involve modulation of chemokines expression, *J. Neuroimmunol.* 209, 114–120, 2009.
123. Sibéril, S., Elluru, S., Negi, V.S. et al., Intravenous immunoglobulin in autoimmune and inflammatory diseases: More than mere transfer of antibodies, *Transfus. Apher. Sci.* 37, 103–107, 2007.
124. Crow, A.R. and Lazarus, A.H., The mechanisms of action of intravenous immunoglobulin and polyclonal anti-d immunoglobulin in the amelioration of immune thrombocytopenic purpura: What do we really know?, *Transfus. Med. Rev.* 22, 103–116, 2008.
125. Winger, E.E., Reed, J.L., Ashoush, S. et al., Treatment with adalimumab (Humira) and intravenous immunoglobulin improves pregnancy rates in women undergoing IVF, *Am. J. Reprod. Immunol.* 61, 113–120, 2009.
126. Hirono, K., Kemmotsu, Y., Wittkowski, H. et al., Infliximab reduces the cytokine-mediated inflammation but does not suppress infiltration of the vessel wall in refractory Kawasaki disease, *Pediatr. Res.* 65, 696–701, 2009.
127. Willcocks, L.C., Smith, K.G., and Clatworthy, M.R., Low-affinity Fcγ receptors, autoimmunity and infection, *Expert Rev. Mol. Med.* 11, e24, 2009.

128. Woelk, C.H. and Burczynski, M.E., The clinical relevance of gene expression profiles in peripheral blood mononuclear cells, in *Oligonucleotide Array Sequence Analysis*, M.K. Movetti and L.J. Rizzo (eds.), Nova Scientific Publishers, New York, Chapter 2, pp. 37–82, 2008.

129. Aune, T.M., Maas, K., Moore, J.H., and Olsen, N.J., Gene expression profiles in human autoimmune disease, *Curr. Pharm. Des.* 9, 1905–1917, 2003.

130. Del Galdo, F., Artlett, C.M., and Jimenez, S.A., The role of allograft inflammatory factor 1 in systemic sclerosis, *Curr. Opin. Rheumatol.* 18, 588–593, 2006.

131. Visvikis-Siest, S. and Siest, G., The STANISLAS Cohort: A 10-year follow-up of supposed healthy families. Gene-environment interactions, reference values and evaluation of biomarkers in prevention of cardiovascular diseases, *Clin. Chem. Lab.* 46, 733–747, 2008.

132. Khatri, P. and Sarwal, M.M., Using gene arrays in diagnosis of rejection, *Curr. Opin. Organ Transplant.* 14, 34–39, 2009.

133. Lyons, P.A., Koukoulaki, M., Hatton, A. et al., Microarray analysis of human leucocyte subsets: The advantages of positive selection and rapid purification, *BMC Genomics* 8, 64, 2007.

134. Radom-Aizik, S., Zaldivar Jr. R., Leu, S.Y. et al., Effects of 30 min of aerobic exercise on gene expression in human neutrophils, *J. Appl. Physiol.* 104, 236–243, 2008.

135. Kobayashi, S.D., Sturdevant, D.E., and DeLeo, F.R., Genome-scale transcript analyses in human neutrophils, *Methods Mol. Biol.* 412, 441–453, 2007.

136. Geest, C.R., Buitenhuis, M., Groot Koerkamp, M.J. et al., Tight control of MEK-ERK activation is essential in regulating proliferation, survival, and cytokine production of CD34+-derived neutrophil progenitors, *Blood* 114, 3402–3412, 2009.

137. Smith, A.M., Rahman, F.Z., Hayee, B. et al., Disordered macrophage cytokine secretion underlies impaired acute inflammation and bacterial clearance in Crohn's disease, *J. Exp. Med.* 206, 1883–1897, 2009.

138. Daryadel, A., Yousefi, S., Troi, D. et al., RhoH/TTF negatively regulates leukotrienes production in neutrophils, *J. Immunol.* 182, 6527–6532, 2009.

139. Siest, G., Jeannesson, E., Marteau, J.B. et al., Drug metabolizing enzymes and transporters mRNA in peripheral blood mononuclear cells of healthy subjects: Biological variations and importance of pre-analytical steps, *Curr. Drug Metab.* 10, 410–419, 2009.

140. Rainen, L., Oelmueller, U., Jurgensen, S. et al., Stabilization of mRNA expression in whole blood samples, *Clin. Chem.* 48, 1883–1890, 2002.

141. Chai, V., Vassilakos, A., Lee, Y. et al., Optimization of the PAXgene blood RNA extraction system for gene expression analysis of clinical samples, *J. Clin. Lab. Anal.* 19, 182–188, 2005.

142. Yamamoto, T., Sekiyama, A., Sekiguchi, H. et al., Examination of stability of bone marrow blood RNA in the PAXgene tube, *Lab. Hematol.* 12, 143–147, 2006.

143. Kruhøffer, M., Dyrskjøt, L., Voss, T. et al., Isolation of microarray-grade total RNA, microRNA, and DNA from a single PAXgene blood RNA tube, *J. Mol. Diagn.* 9, 452–458, 2007.

144. Shou, J., Dotson, C., Qian, H.-C. et al., Optimized blood cell profiling method for genomic biomarker discovery using high-density microarray, *Biomarkers* 10, 310–320, 2005.

145. Matheson, L.A., Duong, T.T., Rosenberg, A.M., and Yeung, R.S.M., Assessment of sample collection and storage methods for multicenter immunologic research in children, *J. Immunol. Methods* 339, 82–89, 2008.

146. Radom-Aizik, S., Zaldivar Jr. F., Leu, S.Y., and Copper, D.M., A brief bout of exercise alters gene expression and distinct gene pathways in peripheral blood mononuclear cells of early- and late-pubertal females, *J. Appl. Physiol.* 107, 168–175, 2009.

147. Tian, Z., Palmer, N., Schmid, P. et al., A practical platform for blood biomarker study by using global gene expression profiling of peripheral whole blood, *PLoS One* 4, e5157, 2009.

148. Thach, D.C., Lin, B., Walter, E. et al., Assessment of two methods for handling blood in collection tubes with RNA stabilizing agent for surveillance of gene expression profiles with high density microarrays, *J. Immunol. Methods* 283, 269–279, 2003.

149. Pahl, A., Gene expression profiling using RNA extracted from whole blood: Technologies and clinical applications, *Expert Rev. Mol. Diagn.* 5, 43–52, 2005.

150. Debey-Pascher, S., Eggle, D., and Schultze, J.L., RNA stabilization of peripheral blood and profiling by bead chip analysis, *Methods Mol. Biol.* 496, 175–210, 2009.

151. Li, L., Ying, L., Naesens, M. et al., Interference of globin genes with biomarker discovery for allograft rejection in peripheral blood samples, *Physiol. Genomics* 32, 190–197, 2008.

152. Wright, C., Bergstrom, D., Dai, H. et al., Characterization of globin RNA interference in gene expression profiling of whole-blood samples, *Clin. Chem.* 54, 396–405, 2008.

153. Vartanian, K., Slottke, R., Johnstone, R. et al., Gene expression profiling of whole blood: Comparison of target preparation methods for accurate and reproducible microarray analysis, *BMC Genomics* 10, 2, 2009.

154. Cekaite, L., Furset, G., Hovig, G., and Sioud, M., Gene expression analysis in blood cells in response to unmodified and 2′-modified siRNAs reveals TLR-dependent and independent effects, *J. Mol. Biol.* 365, 90–108, 2007.

155. Prync, A.E., Yankilevich, P., Barrero, P.R. et al., Two recombinant human interferon-β 1a pharmaceutical preparations produce a similar transcriptional response determined using whole genome microarray analysis, *Int. J. Clin. Pharmacol. Ther.* 46, 64–71, 2008.

156. Urashima, M., Sakuma, M., Teramoto, S. et al., Gene expression profiles of peripheral and cord blood mononuclear cells altered by thymic stromal lymphopoietin, *Pediatr. Res.* 57, 563–569, 2005.

157. Mullick, A., Elias, H., Harakidas, P. et al., Gene expression in HL60 granulocytoids and human polymorphonuclear leukocytes exposed to *Candida albicans*, *Infect. Immun.* 72, 414–429, 2004.

158. Kobayashi, S.D., Voyich, J.M., Braughton, K.R., and DeLeo, F.R., Down-regulation of proinflammatory capacity during apoptosis in human polymorphonuclear leukocytes, *J. Immunol.* 170, 3357–3368, 2003.

159. Payen, D., Lukaszewicz, A.C., Belikova, I. et al., Gene profiling in human blood leukocytes during recovery from septic shock, *Intensive Care Med.* 34, 1371–1376, 2008.

160. Buvanedran, A., Mitchell, K., Kroin, J.S., and Iadarola, M.J., Cytokine gene expression after total hip arthroplasty: Surgical site versus circulating neutrophil response, *Anesth. Analg.* 109, 959–964, 2009.

161. Hesse, D., Sellebjerg, F., and Sorensen, P.S., Absence of MxA induction by interferon β in patients with MS reflects complete loss of bioactivity, *Neurology* 73, 372–377, 2009.

162. Kuzman, M.R., Medved, V., Terzic, J., and Krainc, D., Genome-wide expression analysis of peripheral blood identifies candidate biomarkers for schizophrenia, *Psychiatr. Res.* 43, 1073–1077, 2009.

163. Galamb, O., Sipos, E., and Solymosi, N., Diagnostic mRNA expression patterns of inflamed, benign, and malignant colorectal biopsy specimen and their correlation with peripheral blood results, *Cancer Epidemiol. Biomarkers Prev.* 17, 2835–2845, 2008.

164. Mimori, K., Fukagawa, T., Kosaka, Y. et al., A large-scale study of MT1-MMP as a marker for isolated tumor cells in peripheral blood and bone marrow in gastric cancer cases, *Ann. Surg. Oncol.* 15, 2934–2942, 2008.

165. Smith, C.L., Dickinson, P., Forster, T. et al., Quantitative assessment of human whole blood RNA as a potential biomarker for infectious disease, *Analyst* 132, 1200–1209, 2007.

166. Wang, Z., Beach, D., Su, L. et al., A genome-wide expression analysis in blood identifies pre-elafin as a biomarker in ARDS, *Am. J. Respir. Cell Mol. Biol.* 38, 724–732, 2008.

167. Nikpour, M., Dempsey, A.A., Urowitz, M.B. et al., Association of a gene expression profile from whole blood with disease activity in systemic lupus erythematosus, *Ann. Rheum. Dis.* 67, 1069–1075, 2008.

168. Tang, Y., Xu, H., Du, X. et al., Gene expression in blood changes rapidly in neutrophils and monocytes after ischemic stroke in humans: A microarray study, *J. Cereb. Blood Flow Metab.* 26, 1089–1102, 2006.

169. Lord, P.G., Nie, A., and McMillian, M., Application of genomics in preclinical drug safety evaluation, *Basic Clin. Pharmacol. Toxicol.* 98, 537–546, 2006.

170. Mendrick, D.L., Genomic and genetic biomarkers of toxicity, *Toxicology* 245, 175–181, 2008.

171. Iljin, K., Ketola, K., Vainio, P. et al., High-throughput cell-based screening of 4910 known drugs and drug-like small molecules identifies disulfiram as an inhibitor of prostate cancer cell growth, *Clin. Cancer Res.* 15, 6070–6078, 2009.

172. Python, F., Goebel, C., and Aeby, P., Comparative DNA microarray analysis of human monocyte derived dendritic cells and MUTZ-3 cells exposed to the moderate skin sensitizer cinnamaldehyde, *Toxicol. Appl. Pharmacol.* 239, 273–283, 2009.

173. Siegrist, F., Singer, T., and Certa, U., MicroRNA expression profiling by bead array technology in human tumor cell lines treated with interferon-α-2a, *Biol. Proc. Online* 11(1), 111–129, 2009. doi: 10.1007/s12575-009-9019-1.

174. Wingren, C. and Borrebaeck, C.A., Antibody-based microarrays, *Methods Mol. Biol.* 509, 57–84, 2009.

175. Andreson, H. and Bier, F.F., Peptide microarrays for serum antibody diagnostics, *Methods Mol. Biol.* 509, 123–134, 2009.

176. Gray, J.J., The interaction of proteins with solid surfaces, *Curr. Opin. Struct. Biol.* 14, 110–115, 2004.

177. Rosado, E., Caroll, H., Sánchez, O., and Peniche, C., Passive adsorption of human antirrabic immunoglobulin onto a polystyrene surface, *J. Biomater. Sci. Polym. Ed.* 16, 435–448, 2005.

178. Inouye, S., Nonspecific adsorption of proteins to microplates, *Appl. Microbiol.* 25, 279–283, 1973.

179. Butler, J.E., Ni, L., Nessler, R. et al., The physical and functional behavior of capture antibodies adsorbed on polystyrene, *J. Immunol. Methods* 150, 77–90, 1992.

180. Buteler, J.E., Solid supports in enzyme-linked immunosorbent assay and other solid-phase immunoassays, *Methods* 22, 4–23, 2000.

181. Nieto, A., Gays, A., Moreno, C. et al., Adsorption-desorption of antigen to polystyrene plates, *Ann. Inst. Pasteur Immunol.* 137C, 161–172, 1986.

182. Cantarero, L.A., Butler, J.E., and Osborne, J.W., The absorptive characteristics of proteins for polystyrene and their significance in solid-phase immunoassays, *Anal. Biochem.* 105, 375–382, 1980.

183. Pesce, A.J., Ford, D.J., Gaizutis, M., and Pollak, V.E., Binding of protein to polystyrene in solid-phase immunoassays, *Biochim. Biophys. Acta* 492, 399–407, 1977.

184. Kochanowaka, I.E., Rapak, A., and Szewczuk, A., Effect of pretreatment of wells in polystyrene plates on adsorption of some human serum proteins, *Arch. Immunol. Ther. Exp.* 42, 135–139, 1994.

185. Kenny, G.E. and Dunsmoor, C.L., Effectiveness of detergents in blocking nonspecific binding of IgG in the enzyme-linked immunosorbent assay (ELISA) depends upon the type of polystyrene used, *Isr. J. Med. Sci.* 23, 732–734, 1987.

186. Stevens, P.W., Hansberry, M.R., and Kelso, D.M., Assessment of adsorption and adhesion of proteins to polystyrene microwells by sequential enzyme-linked immunosorbent-assay analysis, *Anal. Biochem.* 225, 197–205, 1995.

187. Chang, H.Y. and Andrade, J.D., Immunochemical detection by specific antibody to thrombin of prothrombin; conformational changes upon adsorption to artificial surfaces, *J. Biomed. Mater. Res.* 19, 913–925, 1985.

188. Kilshaw, P.J., McEwan, E.J., Baker, K.C., and Cant, A.J., Studies on the specificity of antibodies to ovalbumin in normal human serum: Technical considerations in the use of ELISA methods, *Clin. Exp. Immunol.* 66, 481–489, 1986.

189. Hylton, D.M., Shalaby, S.W., and Latour Jr. R.A., Direct correlation between adsorption-induced changes in protein structure and platelet adhesion, *J. Biomed. Mater. Res. A.* 73, 349–358, 2005.

190. Kawamoto, N., Mori, H., Yui, N., and Terano, M., Mechanistic aspects of blood—Contacting properties of polypropylene surfaces—From the viewpoint of macromolecular entanglement and hydrophobic interaction via water molecules, *J. Biomater. Sci. Polym. Ed.* 9, 543–559, 1998.

191. Jørgensen, P.E., Eskildsen, L., and Nexø, E., Adsorption of EGF receptor ligands to test tubes—α factor with implications of studies on the potency of these peptides, *Scand. J. Clin. Invest.* 59, 191–197, 1999.

192. Sutjita, M., Hohmannm, A., Boey, M.L., and Bradley, J., Microplate ELISA for detection of antibodies to DNA in patients with systemic lupus erythematosus: Specificity and correlation with Farr radioimmunoassay, *J. Clin. Lab. Anal.* 3, 34–40, 1989.

193. Paczuski, R., Determination of von Willebrand factor activity with collagen-binding assay and diagnosis of von Willebrand disease: Effect of collagen source and coating conditions, *J. Lab. Clin. Med.* 140, 250–254, 2002.

194. Shrivastav, T.G., Bass, A., and Karlya, K.P., Substitution of carbonate buffer by water for IgG immobilization in enzyme linked immunosorbent assay, *J. Immunoassay Immunochem.* 24, 191–203, 2003.

195. Cavazzana, A., Ruffatti, A., Tonello, M. et al., An analysis of experimental conditions influencing the anti-β_2-glycoprotein I Elisa assay results, *Ann. N. Y. Acad. Sci.* 1109, 484–492, 2007.

196. Safer, D., Bolinger, L., and Leigh Jr. J.S., Undecagold clusters for site-specific labeling of biological macromolecules: Simplified preparation and model applications, *J. Inorg. Biochem.* 26, 77–91, 1986.

197. Sasaki, Y.C., Yasuda, K., Suzuki, Y. et al., Two-dimensional arrangement of a functional protein by cysteine-gold interaction: Enzyme activity and characterization of a protein monolayer on a gold substrate, *Biophys. J.* 72, 1842–1848, 1997.

198. Prisco, J., Leung, C., Xirouchaki, C. et al., Residue-specific immobilization of protein molecules by size-selected clusters, *J. R. Soc. Interface* 2, 169–175, 2005.

199. Lee, J.M., Park, H.K., and Jung, Y., Direct immobilization of protein G variants with various numbers of cysteine residues on a gold surface, *Anal. Chem.* 79, 2680–2687, 2007.

200. Andreescu, S. and Luck, L.A., Studies of the binding and signaling of surface-immobilized periplasmic glucose receptors on gold nanoparticles: A glucose biosensor application, *Anal. Biochem.* 375, 282–290, 2008.

201. Chaki, N.K. and Vijayamohanan, K., Self-assembled monolayers as a tunable platform for biosensor applications, *Biosens. Bioelectron.* 17, 1–12, 2002.

202. Sigal, G.B., Bamdad, C., Barberis, A. et al., A self-assembled monolayer for the binding and study of histidine-tagged proteins by surface plasmon resonance, *Anal. Chem.* 68, 490–497, 1996.

203. Disley, D.M., Morrill, P.R., Sproule, K., and Lowe, C.R., An optical biosensor for monitoring recombinant proteins in process, *Biosens. Bioelectron.* 14, 481–493, 1999.

204. Moon, J., Kang, T., Oh, S. et al., In situ sensing of metal ion adsorption to a thiolated surface using surface plasmon resonance spectroscopy, *J. Colloid Interface Sci.* 298, 543–549, 2006.

205. Ayela, C., Roquet, F., Valera, L. et al., Antibody-antigenic peptide interactions monitored by SPR and QCM-D. A model for SPR detection of IA-1 autoantibodies in human serum, *Biosens. Bioelectron.* 22, 3113–3119, 2007.

206. Jans, K., Bonroy, K., De Palma, R. et al., Stability of mixed PEO-thiol SAMs for biosensing applications, *Langmuir* 24, 3949–3954, 2008.

207. Mehne, J., Markovic, G., Pröll, F. et al., Characterization of morphology of self-assembled PEG monolayers: A comparison of mixed and pure coatings optimized for biosensor applications, *Anal. Bioanal. Chem.* 391, 1783–1791, 2008.

208. Suzuki, N., Quesenberry, M.S., Wang, J.K. et al., Efficient immobilization of proteins by modification of plate surface with polystyrene derivatives, *Anal. Biochem.* 247, 412–416, 1997.
209. Zouali, M. and Stollar, B.D., A rapid ELISA for measurement of antibodies to nucleic acid antigens using UV-treated polystyrene microplates, *J. Immunol. Methods* 90, 105–110, 1986.
210. Larsson, H., Johansson, S.G., Hult, A., and Göthe, S., Covalent binding of proteins to grafted plastic surfaces suitable for immunoassays, *J. Immunol. Methods* 98, 129–135, 1987.
211. Boudet, F., Thèze, J., and Zouali, M., UV-treated polystyrene microtitre plates for use in an ELISA to measure antibodies against synthetic peptides, *J. Immunol. Methods* 142, 73–82, 1991.
212. Yuan, S., Szakalas-Gratzl, G., Ziats, N.P. et al., Immobilization of high-affinity heparin oligosaccharides to radiofrequency plasma-modified polyethylene, *J. Biomed. Mater. Res.* 27, 811–819, 1993.
213. Dagenais, P., Desprez, B., Albert, J., and Escher, E., Direct covalent attachment of small peptide antigens to enzyme-linked immunosorbent assay plates using radiation and carbodiimide activation, *Anal. Biochem.* 222, 149–155, 1994.
214. Goldberg, J.S., Wagenknecht, D.R., and McIntrye, J.A., Alteration of the aPA ELISA by UV exposure of polystyrene microtiter plates, *J. Clin. Lab. Anal.* 10, 243–249, 1996.
215. Matson, R.S., Milton, R.C., Rampal, J.B. et al., Overprint immunoassay using protein A microarrays, *Methods Mol. Biol.* 382, 273–286, 2007.
216. Bae, Y.M., Oh, B.K., Lee, W. et al., Study on orientation of immunoglobulin G or protein G layer, *Biosens. Bioelectron.* 21, 103–110, 2005.
217. Tanaka, G., Funabashi, H., Mie, M., and Kobatake, E., Fabrication of an antibody microwell array with self-adhering antibody binding protein, *Anal. Biochem.* 350, 298–303, 2006.
218. Kato, K., Sato, H., and Iwata, H., Immobilization of histidine-tagged recombinant proteins onto micropatterned surfaces for cell-based functional assays, *Langmuir* 21, 7071–7075, 2005.
219. Kwon, K., Grose, C., Pieper, R. et al., High quality protein microarray using in situ protein purification, *BMC Biotechnol.* 9, 72, 2009.
220. Schaeferling, M. and Kambhampahi, D., Protein microarray surface chemistry and coupling scheme, in *Protein Microarray Technology*, D. Kambhampahi (ed.), Wiley-VCH, Weinheim, Germany, Chapter 2, pp. 11–38, 2004.
221. Seuryck-Servoss, S.L., White, A.M., Baird, C.L. et al., Evaluation of surface chemistries for antibody microarrays, *Anal. Biochem.* 371, 105–115, 2007.
222. Gauvreau, V., Chevallier, P., Vallières, K. et al., Engineering surfaces for bioconjugation: Developing strategies and quantifying the extent of reactions, *Bioconj. Chem.* 15, 1146–1156, 2004.
223. Mallik, R., Wa, C., and Hage, D.S., Development of sulfhydryl-reactive silica for protein immobilization in high-performance affinity chromatography, *Anal. Chem.* 79, 1411–1424, 2007.
224. Reznik, G.O., Vajda, S., Cantor, C.R., and Sano, T., A streptavidin mutant useful for direct immobilization on silica surfaces, *Bioconjug. Chem.* 12, 1000–1004, 2001.
225. Gnjatic, S., Wheeler, C., Ebner, M. et al., Seromic analysis of antibody responses in non-small cell lung cancer patients and healthy donors using conformational protein arrays, *J. Immunol. Methods* 341, 50–58, 2009.
226. Pyun, J.C., Kim, S.D., and Chung, J.W., New immobilization method for immunoaffinity biosensors by using thiolated proteins, *Anal. Biochem.* 347, 227–233, 2005.
227. O'Shannessy, D.J. and Quarles, R.H., Labeling of the oligosaccharide moieties of immunoglobulins, *J. Immunol. Methods* 99, 153–161, 1987.

228. Cho, H.-H., Pack, E.-H., Lee, H. et al., Site-directed biotinylation of antibodies for controlled immobilization on solid surfaces, *Anal. Biochem.* 365, 14–23, 2007.
229. Malecki, M., Hsu, A., Truong, L., and Sanchez, S., Molecular immunolabeling with recombinant single-chain variable fragment (scFv) antibodies designed with metal-binding domains, *Proc. Nat. Acad. Sci. USA* 99, 213–218, 2002.
230. Kumar, P., Agarwal, S.K., and Gupta, K.C., *N*-(3-trifluoroethanesulfonyloxypropyl) anthraquinone-2-carboxamide: A new heterobifunctional reagent for immobilization of biomolecules on a variety of polymer surfaces, *Bioconj. Chem.* 15, 7–11, 2004.
231. Albrecht, H., Burke, P.A., Natarajan, A. et al., Production of soluble scFvs with *C*-terminal-free thiol for site-specific conjugation or stable dimeric scFvs on demand, *Bioconj. Chem.* 15, 16–26, 2004.
232. Allard, L., Cheyne, V., Oriol, G. et al., Antigenicity of recombinant proteins after regioselective immobilization onto polyanhydride-based copolymers, *Bioconj. Chem.* 15, 458–466, 2004.
233. Vankova, R., Gaudinova, A., Sussenekova, H. et al., Comparison of oriented and random antibody immobilization in immunoaffinity chromatography of cytokines, *J. Chromatogr. A* 811, 77–84, 1998.
234. Hage, D.S., Wolfe, C.A., and Oates, M.R., Development of a kinetic model to describe the effective rate of antibody oxidation by periodate, *Bioconj. Chem.* 8, 914–920, 1997.
235. Bonray, K., Frederix, F., Reekmans, G. et al., Comparison of random and oriented immobilisation of antibody fragments on mixed self-assembled monolayers, *J. Immunol. Methods* 312, 167–181, 2006.
236. Davies, J., Roberts, C.J., Dawkes, A.C. et al., Use of scanning probe microscopy and surface plasmon resonance as analytical tools in the study of antibody-coated microtiter wells, *Langmuir* 10, 2654–2661, 1994.
237. Peluso, P., Wilson, D.S., Do, D. et al., Optimizing antibody immobilization strategies for the construction of protein microarrays, *Anal. Biochem.* 312, 113–124, 2003.
238. Johnson, C.P., Jensen, I.E., Prakasam, A. et al., Engineered protein A for the orientational control of immobilized proteins, *Bioconj. Chem.* 14, 974–978, 2003.
239. Huong, R.-P., Detection of multiple proteins in an antibody-based protein microarray system, *J. Immunol. Methods* 255, 1–13, 2001.
240. Agaton, C., Falk, R., Guthenberg, I.H. et al., Selective antibody-based proteomics efforts, *J. Chromatogr. A* 1043, 33–40, 2004.
241. Hueber, W., Utz, P.J., Steinman, L., and Robinson, W.H., Autoantibody profiling for the study and treatment of autoimmune disease, *Arthritis Res.* 4, 290–295, 2002.
242. O'Connell, D.O., Becerrill, B., Ray-Burman, A. et al., Phage versus plasmid libraries for the generation of human monoclonal antibodies, *J. Mol. Biol.* 321, 49–56, 2002.
243. Lai, S., Lui, R., Nguyen, L. et al., Increases in leukocyte cluster of differentiation antigen expression during cardiopulmonary bypass in patients undergoing heart transplantation, *Proteomics* 4, 1916–1926, 2004.
244. Olie, E.W., Sreekumar, A., Warner, R.L. et al., Development of an internally controlled antibody microarray, *Mol. Cell. Proteomics* 4, 1664–1672, 2005.
245. Urbanowska, T., Managialaio, S., Zickler, C. et al., Protein microarray platform for multiple analysis of biomarkers in human sera, *J. Immunol. Methods* 316, 1–7, 2006.
246. Qin, S., Qin, W., Ehrlich, J.R. et al., Development of a "reverse capture" autoantibody microarray for studies of antigen profiling, *Proteomics* 6, 3199–3209, 2006.
247. Schwenk, J.M., Lindberg, J., Sundberg, M. et al., Determination of binding specificities in highly multiplexed bead-based assays for antibody proteomics, *Mol. Cell. Proteomics* 6, 125–132, 2007.
248. Hueber, W., Kidd, B.A., Tomooka, B.H. et al., Antigen microarray profiling of autoantibodies in rheumatoid arthritis, *Arthritis Rheum.* 53, 2645–2655, 2005.

249. Sugihara, T., Seong, G.H., Kobatake, E., and Aizawa, M., Genetically synthesized anti-body binding proteins self-assembled on hydrophobic matrix, *Bioconj. Chem.* 11, 789–794, 2000.

250. Thirunmalapura, N.R., Morton, R.J., Ramachandran, A., and Malayer, J.R., Lipopoly-saccharide microarrays for the analysis of antibodies, *J. Immunol. Methods* 298, 73–81, 2005.

251. Nielsen, U.B. and Geierstanger, B.H., Multiplexed sandwich assays in microarray format, *J. Immunol. Methods* 290, 107–120, 2004.

252. Berson, S.A. and Yalow, R.S., General principles or radioimmunoassay, *Clin. Chim. Acta* 22, 51–69, 1968.

253. Engvall, E. and Perlmann, P., Enzyme-linked immunosorbent assay (ELISA) quantita-tive assay of immunoglobulin G, *Immunochemistry* 8, 871–874, 1971.

254. Engvall, E., Jonssan, K., and Perlmann, P., Enzyme-linked immunosorbent assay II. Quantitative assay of protein antigen, immunoglobulin G, by means of enzyme-labelled antigen and antibody-coated tubes, *Biochim. Biophys. Acta* 251, 427–434, 1977.

255. Engvall, E. and Perlmann, P., Enzyme-linked immunosorbent assay, ELISA III. Quantitation of specific antibodies by enzyme-labeled anti-immunoglobulin in antigen coated tubes, *J. Immunol.* 109, 129–135, 1972.

256. Ruitenberg, E.J., Brosi, B.J.M., and Steerenberg, P.A., Direct measurement of micro-plates and its applications to enzyme-linked immunosorbent assay, *J. Clin. Microbiol.* 3, 541–542, 1976.

257. Schuurs, A. and VanWeemen, B.K., Enzyme-immunoassay, *Clin. Chim. Acta* 81, 1–40, 1977.

258. Ruitenberg, E.J., Sekhuis, V.M., and Brosi, B.J.M., Some characterization of a new multiple-channel photometer for through-the-plate reading of microplates to be used in enzyme-linked immunosorbent-assay, *J. Clin. Microbiol.* 11, 132–134, 1980.

259. St-Louse, M., PCR-ELISA for high-throughput blood group genotyping, *Methods Mol. Biol.* 496, 3–13, 2009.

260. Pang, S., Smith, J., Onley, D. et al., A comparability study of the emerging protein array platforms with established ELISA procedures, *J. Immunol. Methods* 302, 1–12, 2005.

261. Lebrum, S.J. and VanRenterghern, B., Performance characteristics of colorimetric pro-tein microarrays compared to ELISA, *Assay Drug Dev. Technol.* 4, 197–202, 2006.

262. Lin, Y., Huang, R., Chen, L.P. et al., Profiling of cytokine expression by biotin-labeled-based protein arrays, *Proteomics* 3, 1750–1757, 2003.

263. Miller, J.C., Zhou, H., Kwekel, J. et al., Antibody microarray profiling of human prostate cancer sera: Antibody screening and identification of potential biomarkers, *Proteomics* 3, 56–63, 2003.

264. Nielsen, U.B., Cardone, M.H., Sinskey, A.J. et al., Profiling receptor tyrosine kinase activation by using Ab microarrays, *Proc. Natl. Acad. Sci. USA* 100, 9330–9335, 2003.

265. Haab, B.B. and Zhou, H., Multiplexed protein analysis using spotted antibody microarrays, *Methods Mol. Biol.* 264, 33–45, 2004.

266. Haab, B.B., Applications of antibody array platforms, *Curr. Opin. Biotechnol.* 17, 415–421, 2006.

267. Barber, N., Gez, S., Belov, L. et al., Profiling CD antigens on leukemias with an antibody microarray, *FEBS Lett.* 583, 1785–1791, 2009.

268. Cretich, M., Di Carlo, G., Giudici, C. et al., Detection of allergen specific immuno-globulins by microarrays coupled to microfluidics, *Proteomics* 9, 2098–2107, 2009.

269. Zerweck, J., Masch, A., and Schutkowski, M., Peptide microarrays for profiling of mod-ification state-specific antibodies, *Methods Mol. Biol.* 524, 169–180, 2009.

270. Cekaite, L., Hovig, E., and Sioud, M., Monitoring B cell response to immunoselected phage-displayed peptides by microarrays, *Methods Mol. Biol.* 524, 273–285, 2009.

271. Lorenz, P., Kreutzer, M., Zerweck, J. et al., Probing the epitope signatures of IgG antibodies in human serum from patients with autoimmune disease, *Methods Mol. Biol.* 524, 247–258, 2009.

272. Ferrer, M., Sanz, M.L., Sastre, J. et al., Molecular diagnosis in allergology: Application of the microarray technique, *J. Investig. Allergol. Clin. Immunol.* 19(Suppl 1), 19–24, 2009.

273. Babel, I., Baderas, R., Diaz-Uriarte, R. et al., Identification of tumor-associated autoantigens for the diagnosis of colorectal cancer in serum using high density protein microarrays, *Mol. Cell. Proteomics* 8, 2382–2395, 2009.

274. Patwa, T.H., Poisson, L.M., Kim, H.Y. et al., The identification of phosphoglycerate kinase-1 and histone H4 autoantibodies in pancreatic cancer patient serum using a natural protein microarrays, *Electrophoresis* 30, 2215–2226, 2009.

275. Souplet, V., Roux, C., and Melnyk, O., Peptide microarrays of bisphenol A polycarbonate, *Methods Mol. Biol.* 570, 287–297, 2009.

276. Desmetz, C., Maudelonde, T., Mangé, A., and Solassol, J., Identifying autoantibody signatures in cancer: A promising challenge, *Expert Rev. Proteomics* 6, 377–386, 2009.

277. Westerlind, U., Schröder, H., Hobel, A. et al., Tumor-associated MUC1 tandem-repeat glycopeptides microarrays to evaluate serum- and monoclonal-antibody specificity, *Angew. Chem. Int. Ed. Engl.* 48, 8263–8267, 2009.

278. Sutherland, S.M., Li, L., Sigdel, T.K. et al., Protein microarrays identify antibodies to protein kinase Cζ that are associated with a greater risk of allograft loss in pediatric renal transplant recipients, *Kidney Int.* 76, 1277–1283, 2009.

279. Kijanka, G.S., Hector, S., Kay, E.W. et al., Human IgG antibody profiles differentiate between symptomatic patients with and without colorectal cancer, *Gut* 59, 69–78, 2010.

280. Järås, K., Ressine, A., Nillson, E. et al., Reverse-phase versus sandwich antibody microarray, technical comparison from a clinical perspective, *Anal. Chem.* 79, 5817–5825, 2007.

281. Tibes, R., Qiu, Y., Lu., Y. et al., Reverse-phase protein array: Validation of a novel proteomic technology and utility for analysis of primary leukemia specimens and hematopoietic stem cells, *Mol. Cancer* 5, 2512–2521, 2006.

282. Spurrier, B., Ramalingam, S., and Nishizuka, S., Reverse-phase protein lysate microarrays for cell signaling analysis, *Nat. Protoc.* 3, 1796–1808, 2008.

283. Ambroz, K.L.H., Zhang, Y., Schutz-Geschwander, A., and Olive, D.M., Blocking and detection chemistries affect antibody performance on reverse phase protein arrays, *Proteomics* 8, 2379–2383, 2008.

284. Gromov, P., Celis, J.E., Gromova, I. et al., A single lysis solution for the analysis of tissue samples by different proteomic technologies, *Mol. Oncol.* 2, 368–379, 2008.

285. Haab, B.B., Antibody arrays in cancer research, *Mol. Cell. Proteomics* 4, 377–383, 2005.

286. Zha, H., Raffeld, M., Charboneau, L. et al., Similarities of prosurvival signals I Bcl-2-positive and Bcl-2-negative follicular lymphomas identified by reverse phase protein microarray, *Lab. Invest.* 84, 235–244, 2004.

287. Chan, S.M., Ermann, J., Su., L. et al., Protein microarrays for multiplex analysis of signal transduction pathways, *Nat. Med.* 10, 1390–1396, 2004.

288. Romeo, M.J., Wunderlich, J., Ngo, L. et al., Measuring tissue-based biomarkers by immunochromatography coupled with reverse-phase lysate microarray, *Clin. Cancer Res.* 12, 2463–2467, 2006.

289. Becker, K.F., Schott, C., Hipp, S. et al., Quantitative protein analysis from formalin-fixed tissues: Implications for translational clinical research and nanoscale molecular diagnosis, *J. Pathol.* 211, 370–378, 2007.

290. Pernas, F.G., Allen, C.T., Winters, M.E. et al., Proteomic signatures of epidermal growth factor receptor and survival signal pathways correspond to gefitinib sensitivity in head and neck cancer, *Clin. Cancer Res.* 15, 2361–2372, 2009.

291. Pierbon, M., Calvert, V., Belluco, C. et al., Multiplexed cell signaling analysis of metastatic and nonmetastatic colorectal cancer reveals COX2-EGFR signaling activation as a potential prognostic pathway biomarker, *Clin. Colorectal Cancer* 8, 110–117, 2009.

292. Aguilar-Macecha, A., Cantin, C., O'Connor-McCourt, M. et al., Development of reverse phase protein microarrays for the validation of clusterin, a mid-abundant blood biomarker, *Proteome Sci.* 7, 15, 2009.

293. Espina, V., Wulfkuhle, J.D., Calvert, V.S. et al., Reverse phase protein microarrays for monitoring biological responses, *Methods Mol. Biol.* 383, 321–336, 2207.

294. Hu, J., He, X., Baggerly, K.A. et al., Non-parametric quantification of protein lysate arrays, *Bioinformatics* 23, 1986–1994, 2007.

295. Boyd, Z.S., Wu, Q.J., O'Brien, C. et al., Proteomic analysis of breast cancer molecular subtypes and biomarkers of response in targeted kinase inhibitors using reverse-phase protein microarrays, *Mol. Cancer Ther.* 7, 3695–3706, 2008.

296. Wilson, D.S. and Nock, S., Functional protein microarrays, *Curr. Opin. Chem. Biol.* 6, 81–85, 2002.

297. Mattoon, D.R. and Schweitzer, B., Antibody specificity profiling on functional protein microarrays, *Methods Mol. Biol.* 524, 213–223, 2009.

298. Sboner, A., Karpikov, A., Chen, G. et al., Robust-linear-model normalization to reduce technical variability in functional protein microarrays, *J. Proteome Res.* 8, 5451–5464, 2009.

299. Allison, D.B., Page, G.P., Beasely, T.M., and Edwards, J.W. (eds.), *DNA Microarrays and Related Genomic Techniques. Design, Analysis, and Interpretation of Experiments,* Chapman & Hall/CRC press, Boca Raton, FL, 2006.

300. Hess, K.R., Zhang, W., Baggerly, K.A. et al., Microarrays: Handling the deluge of data and extracting reliable information, *Trends Biotechnol.* 19, 463–468, 2001.

301. Wu, C., Zhao, H., Baggerly, K. et al., Short oligonucleotide probes containing G-stacks display abnormal binding affinity on Affymetrix microarrays, *Bioinformatics* 23, 2566–2572, 2007.

302. Coomes, K.R., Wang, J., and Baggerly, K.A., Microarrays: Retracing steps, *Nat. Med.* 13, 1276–1277, 2007.

303. Baggerly, K.A., Coombes, K.R., and Neeley, E.S., Run batch effects potentially compromise the usefulness of genomic signatures for ovarian cancer, *J. Clin. Oncol.* 26, 1186–1187, 2008.

304. Savage, L., Forensic bioinformatician aims to solve mysteries of biomarker studies, *J. Natl. Cancer Inst.* 100, 983–987, 2008.

8 Development of Assays for Biomarkers

The goal of biomarker research is the development of assays, which can be used for the more effective screening and diagnosis of disease. It is also possible that these assays will be used as bioanalytical assays for clinical trials or assays in the laboratory/preclinical development of a biopharmaceutical product. For the purpose of the current discussion, it is assumed that the validity of the biomarker has been established; that is, a relationship (diagnostic, prognostic, theranostic, predictive, etc.) between the pathology and the biomarker has been established. The validation of the biomarker in this sense is not the same as establishing the validity of the assays as discussed below. If the plan is to have an assay that will go "commercial," it is critical to use Design Control in the development of the assay.[1-7] "Design controls are an interrelated set of practices and procedures that are incorporated into the design and development process, that is, a system of checks and balances. Design controls make systematic assessment of the design an integral part of development and as a result, deficiencies in design input requirements and discrepancies between the proposed designs and requirements are made evident and corrected earlier in the development process. Design controls increase the likelihood that the design transferred to production will translate into a device that is appropriate for its intended use."[8] In other words: Design Control is an iterative process that requires one to know what one want in a final product (specifications; see below for discussion of assay attributes). Furthermore, the application of Design Control needs to start early in the development process, as an inadequate design and development plan is a major design deficiency cited by the FDA in 483 citations.[9] There are similarities between Design Control and Quality-by-Design, which is a process for biopharmaceutical product development.[10]

An analyte-specific reagent[11-16] is defined as "antibodies, both polyclonal and monoclonal, specific receptor proteins, ligands, nucleic acid sequences, and similar reagents, which, through specific binding or chemical reactions with substances in a specimen, are intended for use in a diagnostic application for identification and quantification of an individual chemical substance or ligand in biological specimens." Many of the assays used in biomarker assay are considered to be analyte-specific reagents.

I want to emphasize that clinical laboratory assays, which would include assays developed from the discovery of biomarkers, are considered to be medical devices like an artificial knee or surgical dressing.[17,18] The great majority of medical devices including diagnostic assays are used in the United States (41%) and Europe (30%)[17] but it is likely that this will change in the coming years. In general, regulatory approval is required for reimbursement by third parties. The complexity of the regulatory approval process depends on the criticality of the device[19] as does the requirement for cGMP for that product. However, the more planning and effort that goes into the development of a biomarker as a diagnostic product, the better the quality of the product.

There are a number of assays that are used in the development, manufacture, and use of biopharmaceuticals, which are categorized by function rather than composition. A listing of various assays is presented in Table 8.1. It is most likely that an assay designed for a biomarker will be a bioanalytical assay. It should be emphasized that designation is not exclusive, a bioanalytical assay can be a regulatory assay as well as a stability-indicating assay; this is true of the activated partial thromboplastin time used for hemophilia. Much of the work on biomarkers is focused toward the broad area of diagnostics; however, development of biopharmaceutical products requires the use of biomarkers in clinical trials.[28-44] Some aspects of the use of biomarkers in the preclinical development of biopharmaceuticals are also discussed in Chapter 2.

It would be remiss not to discuss "fit-for-purpose" assays. The *Oxford English Dictionary* defines "fit-for-purpose" as "suitable for the intended use; fully capable of performing the required task."[45] Many of the early citations for fit-for-purpose have to do with education and equipment.[46,47] To be perfectly honest, the author fails to see the designation of fit-for-purpose as value added in describing an assay; a validated assay for a biomarker implies fit-for-purpose. There are, however, several publications that embrace the concept of fit-for-purpose.[48-53] It is likely that fit-for-purpose is a bit like the older concept of applicability-to-system.

The next issue is the reference standard; assay results are usually expressed data relative to those obtained with a reference standard (or equivalent) and in comparison to reference interval. A reference standard is usually a material approved by WHO or related national organization such as NIBSC or USP[54-64] but other reference standards are proposed in the literature.[65-69] In an increasing number of pathologies, reference standards are not available or applicable[70-72]; in addition, as technology improves and changes, the question of when a new test becomes a reference standard is being considered.[73-78] There is also the issue of emerging therapeutics where new reference standards are developed.[79-81] High priority should be given to the development of a reference standard in parallel with the development of an assay. Novel approaches to the determination of reference standard concentration include isotope dilution mass spectrometry after enzymatic digestion.[82] This study established reference standards for ricin and influenza hemagglutinin.

The commutability of an assay standard is desirable[83-85] as this property allows the comparison of data obtained with different assay platforms.[86] A working standard may be prepared from the parent reference standard; the process of preparation and storage of a working standard must be contained with the standard operating procedure (SOP) (see Appendix 8.1) for the assay.

The assay matrix (see Table 8.2) is important if an assay will be either multifunctional (e.g., bioanalytical and stability indicating) or used on different biological fluids (e.g., blood and urine). The effect of serum is observed in immunological assays[87-91] while tissue fluorescence is a problem in fluorescence-based assay.[92] Transferability between assay matrices is an example of commutability of the assay

TABLE 8.1
Various Types of Assays Used in Biopharmaceutical Research

Assay	Use	References
Regulatory	An analytical procedure used to evaluate a defined characteristic of the drug substance or drug product. This is a legally recognized procedure for the purpose of measuring a compendial item. A regulatory analytical procedure usually uses an International Reference Standard. Regulatory analytical procedures are in the manufacturing of cGMP products and may include assays for identification, quantitative tests for the active pharmaceutical ingredient, and impurities	20, 21
Alternative analytical procedure	A validated assay that can be used in the absence of a regulatory assay (usually used when a regulatory analytical procedure is not available)	22
Stability-indicating assays	An assay used to demonstrate the stability of a biopharmaceutical. These assays are used to demonstrate the stability of a biopharmaceutical during development	23
Bioanalytical assay	An assay that measures a biomarker (analyte) within a biological matrix such as blood or urine	24–27
Chemical assay	An assay that is considered more chemical than biological in nature. Such assays include assays for electrolytes and metabolites like glucose. Colorimetric assays for protein and carbohydrate are chemical assays. Classical drugs (as opposed to biopharmaceuticals) are usually (but not always[a]) measured with chemical assays	27
Microbiological or ligand binding assay	These assays are biological assays and this category would also include cell-based assays other than microbiological assays. Examples include immunological assays such as ELISA assays and multiplex assays. In general, biological assays are more complex (and expensive) than chemical assays[b]	27

[a] The development of high-throughput immunological assays has proved useful for the assay of drugs. See Kraemer, T. and Paul, L.D., Bioanalytical procedures for determination of drugs of abuse in blood, *Anal. Bioanal. Chem.* 388, 1415–1435, 2007.

[b] The characterization of a biopharmaceutical product can include both chemical and biological assays. ICH Harmonized Tripartite Guideline; Specifications: Test Procedures and Acceptable Criteria for Biotechnological/Biological Products, Q6B, http://www.ich.org/LOB/media/MEDIA432.pdf; Bristow, A.F., When are bioassays necessary and when not?; http://www.usp.org/pdf/EN/meetings/bioassayWorkshop/session3a.pdf 2008

TABLE 8.2

Definition of Assay Matrices

Matrix	Description
Biological matrix	A discrete material of biological origin that can be sampled and processed in a reproducible manner. Examples of a biological matrix include tissue, blood, urine, and saliva
Experimental matrix	A discrete material generally of manufacturing origin that can be sampled and processed in a reproducible manner. Examples include chromatographic effluent fractions, conditioned culture media, and intact cells or cell mass
Matrix effect	The direct or indirect alteration or interference in assay response due to components in the assay matrix

and assay reference standard. Table 8.3 contains some useful definitions for assay components and various critical attributes of analytical procedures are contained in Table 8.4. The stability of the components[93–95] of an approved assay is a regulatory consideration during assay development. Very little of the above is concerned with the science of biomarker assays or the technical aspects of the development of an assay for a biomarker. However, if care is not taken to carefully prepare a development plan and assure documentation of the plan and progress, a lot of time and money could be wasted.

Assays for biomarkers are either activity assays such as enzyme activity or involve a cellular response such as one detecting the presence of neutralizing antibodies to erythropoietin,[96] or a binding assay relying on the specificity of a probe such as an aptamer or antibody; example of binding assays are enzyme-linked immunosorbent assay (ELISA) and DNA microplates. These assays are described as solid-phase assays, indicating that the reaction occurs on a surface.[97–105] Microarray (high-density) technology is discussed in Chapter 7 while solid-phase immunoassays are discussed below. There are a large number of enzymes that are currently used as biomarkers such as alkaline phosphatase, aspartate transaminase (AST/SGOT), lactic dehydrogenase, and alanine aminotransferase (ALT/SGPT). Other enzymes such as kallikrein (see Chapter 3) are under development. The largest problem with the development of biomarker enzymes is specificity in the assay system. Most of the current enzyme biomarkers such as AST/SGOT are markers of cell damage while new enzyme biomarkers are likely to be enzymes involved in regulation such as thrombin on kallikrein. The reactions need to be carefully described with respect to time, temperature, and solvent conditions (buffer salt selection can be critical). The reader is directed to several recent articles on the validation of enzyme assays.[106–113] Cell-based assays present issues of reproducibility and precision (high coefficient of variation). Response lines that are parallel with standards[114–116] are critical for successful interpretation. Parallel lines obtained for test materials and standards provides support for analytic identity and lack of matrix effects.[117] The largest problem

TABLE 8.3
Definition of Assay Components and Other Terms[a]

Component	Definition
Reagent	Any substance used in a reaction for the purpose of detecting, measuring, examining, or analyzing other substances; a component of a chemical reaction, which is used to measure an analyte[b]
Solvent	A liquid that may be a pure substance such as water, acetonitrile or glacial acetic acid, or a solution such as a buffer or dilute salt solution. It is noted that in the formal sense, a buffer solution or salt solution is a solute contained in a solvent[b]
Spiking	The addition of a known amount of the analyte to standard, sample, or blank; typically for the purpose of confirming the performance or an analytical procedure or for calibration of an instrument
Reference standard	A validated source of the analyte, which may be used as a primary standard for the determination of said analyte; a reference standard is usually provided by a national regulatory agency or delegate
Calibration standard	A biological matrix to which a known amount of analyte has been added or "spiked." Calibration standards are used to construct calibration curves from which the concentration of the desired analyte in study samples are determined
Internal standard	Test compounds added to both calibration standards and samples at a known and constant concentration to facilitate the quantitation of the target analyte(s). Internal standards are used in mass spectrometry studies of biomarkers[d]
Sample	A generic term encompassing controls, blanks, unknowns, and processed samples; also referred to as test materials, test items, or control materials
Blank	A sample of the sample matrix (see Table 8.2) to which no analytes have been added. The blank is used for instrument calibration and for the determination of the specificity of the analytical method. There are various types of blanks depending on use: • *Optical blank*: Used for spectrophotometric/fluorometric procedures where instrument calibration is necessary. • *Instrument blank*: Maybe the same as the optical blank; for example the use of pure water in a spectrophotometer to correct for solvent absorbance. • *Solvent blank*: A sample of the experimental matrix. • *Experimental blank*: A sample where the analyte has not been added.

[a] See also Stewart, K.K. and Ebel, R.E., *Chemical Measurements in Biological Systems*, John Wiley & Sons, New York, 2000.

[b] Analyte is a generic term referring to the substance being measured in an analytical reaction; a biomarker is an analyte.

[c] See Sharp, D.W.A., *Dictionary of Chemistry*, 2nd edn., Penguin Books, London, U.K., 1990.

[d] Brönstrup, M., Absolute quantification strategies in proteomics based on mass spectrometry, *Expert Rev. Proteomics* 1, 503–512, 2004; Albrethsen, J., Reproducibility in protein profiling by MALDI-TOF mass spectrometry, *Clin. Chem.* 53, 852–858, 2007; Li, H., Rose, M.J., Tran, L. et al., Development of a method for the sensitive and quantitative determination of hepcidin in human serum using LC-MS/MS, *J. Pharmacol. Toxicol. Methods* 59, 171–180, 2009.

TABLE 8.4
Critical Attributes of Assays

Attribute	Description
Range	That span of analyte concentration where it is possible to obtain an acceptable degree[a] of linearity, accuracy, and precision
Specificity[b]	The degree to which the procedure measures the desired analyte in the presence of degradation products,[c] matrix components, and other interfering substances
Sensitivity[b]	The degree to which as assay can measure or detect the analyte. Sensitivity can refer to the detection limit (limit of detection, LOD), which is the lowest concentration at which the analyte may be detected or to the quantitation limit, which is the lowest level at which the concentration of the analyte may be determined with acceptable accuracy and precision (lower limit of quantitation, LLOQ)
Accuracy	The degree of closeness of the determined value to the nominal or known true value under prescribed conditions. This is sometimes termed *trueness*
Method	A comprehensive description of all procedures used in a sample analysis
Precision	The closeness of agreement (*degree of scatter*) between a series of measurements obtained from multiple sampling of the same homogeneous sample under the prescribed conditions. Precision does not infer accuracy
Reproducibility	It also represents precision of the method under the same operating conditions. Reproducibility may be intra-laboratory, intraoperator, inter-operator, or inter-laboratory
Robustness	The capacity of the assay to remain unaffected by small but deliberate variations in method parameters, which provides an indication of the reliability of the assay under normal operating conditions

[a] For example, for the assay of a drug substance, this should be a minimum of 70%–120% of the test concentration.

[b] These definitions for specificity and sensitivity differ from their statistical definitions. In statistics, specificity is the proportion of negative tests to the total number of negative tests; specificity is the proportion of positive tests as part of the total number of positive tests. See Sharples, L.D., Statistical approaches to rational biomarker selection, in *Biomarkers in Disease. An Evidence-Based Approach*, Trull, A.K. et al. (eds.), Cambridge University Press, Cambridge, U.K., Chapter 3, pp. 24–31, 2002.

[c] An example is provided for the heterogeneity of the cardiac troponins (Gaze, D.C. and Collinson, P.O., Multiple molecular forms of circulating cardiac troponin: Analytical and clinical significance, *Ann. Clin. Biochem.* 45, 349–355, 2008).

with cell-based assays is the high degree of variance (coefficient of variation)[118] and the interaction(s) with matrix.[119–122]

Given that the final assay must satisfy a spectrum of attributes (Table 8.4), where does one start? While the information may be imperfect, there should be information on the necessary sensitivity derived from the biomarker discovery studies. Geng and coworkers[123] described a scheme for the validation of assays for immunogenicity of therapeutics. Initial work would best focus on specificity, linearity, sensitivity, and accuracy as these factors are amenable to experimental

TABLE 8.5
Immunoassays-/Receptor-Based Assay

A. Electrophoresis
  1. Serum electrophoresis
  2. Immunoelectrophoresis
B. Radial immunodiffusion
  1. Double immunodiffusion
C. Precipitation methods
  1. Direct agglutination
    • Latex
    • Dextran
D. Nephelometry
E. Immunoprecipitation methods
  1. Direct agglutination
    • Latex
    • Dextran
F. Immunoblotting
  1. Dot blots
  2. Western blots
G. Enzyme immunoassays
  1. Radioimmunoassays
  2. Enzyme-linked enzyme immunoassays

optimization. Attributes such as precision and ruggedness are controlled by system development. Precision and accuracy must not be confused; one can be precisely wrong! As described in Chapters 2 and 3, most of the current protein/peptide biomarker research and development uses solid-phase assay technology, while nucleic acid biomarkers use techniques of molecular diagnostics. The remainder of this chapter contains a very limited consideration of solid-phase assay technologies for the assay of biomarkers.

A variety of assays have been developed for the measurement of protein–protein interactions in immunology and receptor binding (Table 8.5). The early techniques of nephelometry and precipitation evolved into the development of what are today referred to as solid-phase assays. Solid-phase assays include microplate-based ELISA assays where a capture antibody (usually polyclonal) is absorbed into the surface[124] and analogous assays where a receptor is bound to a microplate.[125,126] There is considerable similarity in the study of ligand binding to receptors and antigen–antibody interaction[127] as both use biospecific interactions (characterized by biological specificity) such as an antigen–antibody reaction or ligand binding to a receptor. Solid-phase assays can also be used to measure the interaction of a protein with a ligand or for a specific protein–protein interaction. For example, thrombin bound to a matrix was used to study the binding of DNA aptamers[128] while ecotin was bound to matrix to study the interaction with thrombin.[129] Binding assays based on chemical specificity include the use of aptamers.[130] DNA microarray assays, which measure gene expression, are examples of biospecific assays and are discussed in Chapter 7.

It is most likely that an assay developed for a biomarker will be based on a biospecific interaction, which is most commonly an interaction between an antigen and an antibody. The first issue is the production of an antibody to the biomarker. It is advised to produce both monoclonal and polyclonal antibodies; in the absence of production of polyclonal antibodies, it is useful to then produce several monoclonal antibodies with differing epitopic specificity. It is desirable to have the option of maximum avidity in the capture step and greater avidity is obtained with several different antibodies rather than a polyclonal antibody or a mixture of monoclonal antibodies.[131–133] In addition, the use of several monoclonal antibodies to differing epitopes appear to provide more accurate assays.[134,135] Robertson and coworkers[134] demonstrated that a combination of monoclonal antibodies provided data for the assay of the inhibin α subunit matching that obtained with a polyclonal antibody. McGuinness and Mantis[135] demonstrated that a combination of two monoclonal antibodies functioned synergistically to neutralize ricin. Another approach is to increase the density of capture antibody by using site-oriented binding of recombinant antibody fragments,[136] which also has the advantage of avoiding interference with heterophilic antibody.

There are a variety of approaches to the use of solid-phase immunoassay technology including the classical sandwich ELISA, the competitive ELISA, and the label-free assay.[137–139] The ELISA can be performed in several formats. In the direct ELISA, the antigen is adsorbed to the microplate and detected by an antibody that is linked to an enzyme such as horseradish peroxidase, which is then measured. In the indirect or sandwich ELISA, the antigen is "captured" by an antibody (the primary antibody) bound to the microplate: detection then occurs via the use of a secondary antibody that is linked to an enzyme such as horseradish peroxidase. In the competitive ELISA, labeled and unlabeled antigens (samples) compete for binding to a primary antibody bound to the microplate. In the label-free system, the binding of analyte and probe occurs on a surface resulting in an increase in size resulting in the change in the diffraction of light (a quantum optical phenomena referred to as surface plasmon resonance).[140,141]

The critical part of the assay is the antibody, which must have sufficient specificity and sensitivity (avidity); an equally critical consideration is the assurance of a continuous supply of antibody. Thus, while, a classical polyvalent antibody is an attractive consideration for a capture antibody, the combination of several monoclonal antibodies with differing epitopic specificity would accomplish the same purpose. It is also possible that a single monoclonal antibody could serve as the capture antibody. In addition, the production of monoclonal antibodies provides the technological basis for recombinant antibodies and antibody fragments, which could be more useful in label-free system and to improve capture density.

The author's bias is to start with the development of a sandwich ELISA. However, the author does acknowledge that, if sufficient antigen was available (which is unlikely), a competitive ELISA could be developed more rapidly and would lead more directly to a label-free assay. The author also acknowledges that there are a number of more sophisticated assay systems using microarrays, sensors, and microfluidic systems.[141–145] However, furthermore, it is the author's sense that the development of a classical sandwich ELISA yields considerable information useful in the

TABLE 8.6
Cell-Based Immunoassays

Flow cytometry (FACS analysis)
Cytotoxicity assays
Chemotaxis
Adherence
Phagocytosis
Bacteriocidal
Direct cell binding
Cellular response (i.e., platelet aggregation and release)

Design Control development process, which is applicable to a spectrum of final end diagnostic device products. There are also a number of cell-based immunoassays (Table 8.6), which would benefit from such studies.

While technology has advanced to 384- and 1536-well plates, the majority of work uses 96-well microplates, which can be read in a variety of microplate readers. Most microplates are constructed from plastic although quartz microplates that provide superior performance in the UV-spectral region are available.[146] Microplates are not identical and care needs to be taken in selection and use (Table 8.7). Table 8.7 also contains a variety of studies on the use of microplates for ELISA assays. In addition, the specificity of capture antibody binding to the matrix can be improved through the use of protein A.[161–164]

Technology is moving away from the 96-well microplate, and while higher order plates are useful for combinatorial chemistry,[165,166] immunoassay and related technologies are moving toward microarray and microfluidic systems. An intermediate approach is the use of bead technology for multiplexed assays.[167–171] The use of multiplexed assays would appear to be future for the use of biomarkers.[168,172–174] The most popular system has been developed by Luminex.[172,175] The system is based on a population of small (5 nm) beads, which have encoded fluorophores that can be sorted by flow cytometry. Antibody coupled to the bead gives the specificity to assay and the signal is read in a flow cytometer.[173] A dye attached to the reporter antibody increases the multiplex number. Current technology allows 100- and 200-plexing, with 500-plexing becoming available. Sensitivity and specificity in the multiplexed bead system is similar to that obtained with ELISA assays.[175–179] The strength of the approach is demonstrated by the studies of Kellar and associated on the measurement of cytokines in serum and cell culture,[176] which has been validated and extended by many investigators.[180–183] While the classical ELISA assays are robust and quite useful for single analytes, the necessity for multiplex analytical platforms for biomarkers make the fluorescent bead/flow cytometry technology quite appealing.[184] The other advantage of flow cytometry is its existence as a mature analytical technology in the clinical laboratory.

The above technologies use "labels" for reporting the interaction. These labels are usually enzymes such as horseradish peroxidase[185] or fluorophores.[186] There are methods for evaluating interactions in the absence of label; the most popular is

TABLE 8.7
Microplate/Material Issues Important in Solid-Phase Assays

Study	References
This shows that the release of IL-1ra (IL-1 receptor antagonist) peripheral blood mononuclear cells cultured in 10% human serum differed depended on the source (manufacturer) of the microplate. IL-1ra release from PBMC is stimulated by solid-phase IgG and it was presumed that the above results reflect differences in the adsorption of IgG from serum; pretreatment of the plates with human serum albumin decreased release—if IgG was included with human serum albumin, stimulation was restored. Differences in human serum albumin binding between plates could also be observed	147
The nature of the polymer matrix is important as the same composition with different stereochemistry results in binding differences	148
Materials leaching from plastic that affect assays	149
This study compared the adsorption of albumin and IgG to several different microplates. There were two classes of microplates; one group adsorbed albumin poorly and the other group adsorbed albumin well; IgG adsorbed well on both classes of plates. In antigen-capture assays, normal serum components blocked attachment of antigen-specific IgG, but this competition could be lessened to a degree by the use of strongly binding polystyrene plates. There may be problems in the adsorption of antigen-specific IgG from crude mixtures onto polystyrene plates	150
Effect of detergents on binding of protein to microplates	151, 152
Estimation of protein binding to polystyrene matrices	153
Covalent attachment of protein to microplates—Prior adsorption of polyvinylbenzyl lactonoylamide to the microplate surface. The oxidation of lactonoylamide with periodate to generate aldehyde function, which reacts with the protein amino groups followed by reductive amination	154
Materials leached from plastics	149
Modification of polystyrene with sulfuric/nitric acid followed reduction to form amino groups for coupling to proteins	155, 156
Effect of pH and buffer ion on adsorption of IgG to polystyrene plates	157
Effect of Tween 20 on the binding of glycolipids to microplates	158
Species-specific inhibition of an ELISA system using murine monoclonal antibodies	159
Specific immobilization of intact antibody or F(ab)' fragments increases binding capacity	160

surface plasmon resonance.[144,187,188] Surface plasmon resonance arises at the surface of a metallic film as a result of changes in the refractive index of the medium on the other side (opposite side from the reflected light) of the film.[189–193] A capture material, which may be an antibody,[194] carbohydrate,[195] aptamer,[196] or other potential binding ligand for the analyte/biomarker, is on the opposite side. As the desired analyte/biomarker binds, the thickness of the layer increases, changing the refractive index. This increase in thickness is a measure of the concentration of analyte/biomarker; it is also possible to measure binding kinetics. There are several studies[137,197–199] that compare SPR and ELISA technologies, which suggest that there is good correlation between the two assay systems with the SPR having higher sensitivity. Surface plasmon resonance does have the advantage over the sandwich ELISA of requiring only a single antibody or binding agent.[200] The method of attachment of the antibody–antibody fragment to the sensor surface is an important consideration (Chapter 7).[201–207]

The issue of serum versus plasma was discussed in Chapter 4. I wish to raise one more striking example of an issue of serum versus plasma combined with a known source of biomarker, which is activated during the process of the conversion of blood to serum. All of this is combined with isoforms, latent forms, and complexes. In addition, there are various ELISA and bioassays. Transforming growth factor-β (TGF-β) is a cytokine, which was originally purified from platelets,[208] and has been proposed as a biomarker for several disease states.[209–211] Grainger and colleagues[212] reviewed various studies (ELISA and bioassay) on the determination of TGF-β concentration in blood and reported a range from <0.1 to 25 ± 21 ng/mL. It is suggested that the preparation of plasma is the most likely source of variation. The presence of a latent form of TGF-β[213] provides additional complication, which can be addressed by acid activation prior to assay.[214] The influence of different immunoassays for the determination of TGF-β in urine has been reported by Tsakas and Goumenos.[215] A consideration of recent studies where TGF-β is used as biomarker provides a concentration of 0.19 ng/mL in plasma for both controls and patients,[216] 10.41 ng/mL in preeclamptic women compared to 7.01 ng/mL for normotensive women,[217] and 6.9 ng/mL in patients anticoagulated with enoxaparin versus 8.4 ng/mL in patients anticoagulated with unfractionated heparin.[218]

I want to again emphasize the importance of keeping excellent records from the very start of research work to the actual process of assay development. Such records make the process of using Design Control much more effective and decrease development time. The use of Design Control enables a rigorous development process directed at satisfying customer requirements. It is important to remember that a good product requires good science but great science does not necessarily make a great product. It is recognized that the development of assays for biomarkers may be more challenging than usual product development and requirement changes in managerial style.[219] Focus on customer requirements is essential; there is no need to make an assay more sensitive than necessary even if possible from a technical perspective. It is also important to keep a product within the technology competence of customers.[220]

I have tried to provide an objective overview of biomarkers. While there has been great excitement about biomarkers, there has only been limited transition from biomarker discovery to clinical laboratory application.[221–223] Transition from biomarker to clinical analyte is also challenged by the success of current analytes.[224] The greatest value of biomarkers would be in the early diagnosis of cancer but as the data in Chapter 3 shows, this is challenging. It should be emphasized that biomarker is as much a concept as it is a laboratory analyte.[225–228] The discovery and development of biomarkers is likely to add at least as much, if not more, value from the increased understanding of the pathology as from the development of a new analyte.

APPENDIX 8.1: MODEL STANDARD OPERATING PROCEDURE FOR AN ASSAY

Title for procedure
Document number
Date of origination
Revision number (if applicable)
Date of latest revision
Supercedes revision (if applicable)
Approval signatures

1. Purpose
   1.1 Application (if for specific process or analyte)
   1.2 Scientific basis for procedure
   1.3 Definitions
   1.4 References for procedure
2. Equipment
3. Solutions and reagents
4. Reference standards
5. Method of procedure
6. Data management

REFERENCES

1. Design Control (FDA/CRDH), http://www.fda.gov/crdh/comp/designd.html
2. Stoeger, K.J., Implementing the new quality system requirements: Design controls, *Biomed. Instrum. Technol.* 31, 119–127, 1997.
3. Lasky, F.D. and Boser, R.B., Designing in quality through design control: A manufacturer's perspective, *Clin. Chem.* 43, 866–872, 1997.
4. Powers, D.M. and Greenberg, N., Development and use of analytical quality specifications in the *in vitro* diagnostics medical device industry, *Scand. J. Clin. Lab. Invest.* 59, 539–543, 1999.
5. Gouget, B., Barclay, J., and Rakotoambinina, B., Impact of emerging technologies and regulations on the role of POCT, *Clin. Chim. Acta* 307, 235–240, 2001.
6. Von Versen, R., Mönig, H.J., Salai, M., and Bettin, D., Quality issues in tissue banking: Quality management systems—A review, *Cell Tissue Bank.* 1, 181–192, 2000.
7. Panteghini, M., The importance of analytical quality specifications for biomarker assays currently used in acute cardiac care, *Acute Card. Care* 8, 133–138, 2006.
8. FDA Design Control Guidance for Medical Device Manufacturers; http://www.fda.gov/MedicalDevices/DeviceRegulationandGuidance/GuidanceDocuments/ucm070627.htm
9. Justiniano, J.M. and Gopalaswamy, V., *Practical Design Control Implementation for Medical Devices*, Interpharm/CRC, Boca Raton, FL, 2003.
10. Rathore, A.S., Roadmap for implementation of quality by design (QbD) for biotechnology products, *Trends Biotechnol.* 27, 546–553, 2009.
11. 21 CFR 864.4020; Analyte Specific Reagents.
12. FDA/CRDH/CBER Guidance for Industry and FDA Staff Commercially Distributed Analyte Specific Reagents (ASRs): Frequently asked questions, http://www.fda.gov/downloads/MedicalDevices/DeviceRegulationandGuidance/GuidanceDocuments/ucm071269.pdf

13. Goodrich, J.S. and Miller, M.B., Comparison of culture and 2 real-time polymerase chain reaction assays to detect group B Streptococcus during antipartum screening, *Diagn. Microbiol. Infect. Dis.* 59, 17–22, 2007.
14. Sábato, M.F., Shiffman, M.L., and Langley, M.R., Comparison of performance characteristics of three real-time reverse transcription-PCR test systems for detection and quantification of hepatitis C virus, *J. Clin. Microbiol.* 45, 2529–2536, 2007.
15. Tang, W., Elmore, S.H., Fan, H. et al., Cytomegalovirus DNA measurement in blood and plasma using Roche LightCycler CMV quantification reagents, *Diagn. Mol. Pathol.* 17, 166–173, 2008.
16. Selvaraju, S.B., Wurst, M., Horvat, R.T., and Selvarangan, R., Evaluation of three analyte-specific reagents for detection and typing of herpes simplex virus in cerebrospinal fluid, *Diagn. Microbiol. Infect. Dis.* 63, 286–291, 2009.
17. Altenstetter, C., *Medical Devices*, Transaction Publishers, New Brunswick, NJ, 2008.
18. King, P.H. and Fries, R.C., *Design of Biomedical Devices and Systems*, 2nd edn., CRC Press, Boca Raton, FL, 2009.
19. An overview of the medical device approval process, FDA/CRDH; http://www.fda.gov/MedicalDevices/DeviceRegulationandGuidance/Overview/default.htm
20. Validation of Analytical Procedures: Text and Methodology Q2R1, International Conference on Harmonization; http://www.ich.org/LOB/media/MEDIA417.pdf
21. FDA Guidance for Industry Analytical Procedures and Methods Validation; http://www.fda.gov/downloads/Drugs/GuidanceComplianceRegulatoryInformation/Guidances/UCM070489.pdf
22. Hauck, W.W., DeStefano, A.J., Cecil, T.L. et al., Acceptable, equivalent, or better: Approaches for alternatives to official compendial procedures, *Pharmacopeial Forum* 35, 772–778, 2009.
23. Bakshi, M. and Singh, S., Development of validated stability-indicating assay methods—Critical review, *J. Pharm. Biopharm. Anal.* 28, 1011–1040, 2002.
24. Smolec, J., DeSilva, B., Smith, W. et al., Bioanalytical method validation for macromolecules in support of pharmacokinetic studies, *Pharm. Res.* 22, 1425–1431, 2005.
25. Nowatzke, W. and Woolf, E., Best practices during bioanalytical method validation for the characterization of assay reagents and the evaluation of analyte stability in assay standards, quality controls, and study samples, *AAPS J.* 9, E117–E122, 2007.
26. Kelley, M. and DeSilva, B., Key elements of bioanalytical method validation for macromolecules, *AAPS J.* 9, E156–E163, 2007.
27. FDA Guidance for Industry Bioanalytical Assay Validation: http://www.fda.gov/downloads/Drugs/GuidanceComplianceRegulatoryInformation/Guidances/UCM070107.pdf
28. Downing, D.J. and Biomarkers Definition Working Group, Biomarkers and surrogate endpoints: Preferred definitions and conceptual framework, *Clin. Pharmacol. Therapeut.* 69, 89–95, 2001.
29. Floyd, E. and McShane, T.M., Development and use of biomarkers in oncology drug development, *Toxicol. Pathol.* 32(Suppl 1), 106–115, 2004.
30. Pien, H.H., Fischman, A.J., Thrall, J.H., and Sorensen, A.G., Using imaging biomarkers to accelerate drug development and clinical trials, *Drug Discov. Day* 10, 259–266, 2005.
31. Zhang, H., Chung, D., Yang, Y.-C. et al., Identification of new biomarkers for clinical trials of Hsp90 inhibitors, *Mol. Cancer Therapeut.* 5, 1256–1264, 2006.
32. Mohs, R.D., Kawas, C., and Carrillo, M.C., Optimal design of clinical trials for drugs designed to slow the course of Alzheimer's disease, *Alzheimer Dementia* 2, 131–139, 2006.
33. Olofsson, M.H., Ueno, T., Pan, X. et al., Cytokeratin-18 is a useful serum biomarker for early determination of response of breast carcinomas to chemotherapy, *Clin. Cancer Res.* 13, 3198–3206, 2007.
34. Steinerman, J.R. and Honig, L.S., Laboratory biomarkers in Alzheimer's disease, *Curr. Neurol. Neurosci. Rep.* 7, 381–387, 2007.

35. Fraser, G.A.M. and Meyer, R.M., Biomarkers and the design of clinical trials in cancer, *Biomarkers Med.* 1, 387–397, 2007.
36. Dalgleish, A.G., Practical aspects in the use of biomarkers for the development of cancer vaccines, *Curr. Cancer Therapy Rev.* 4, 161–165, 2008.
37. Sessa, C., Guibal, A., Del Conte, G. et al., Biomarkers of angiogenesis for the development of antiangiogenic therapies in oncology: Tools or decorations?, *Nat. Clin. Pract. Oncol.* 5, 378–391, 2008.
38. Krause, M. and Baumann, M., Clinical biomarkers of kinase activity: Examples from EGFR inhibition trials, *Cancer Metastasis Rev.* 27, 387–402, 2008.
39. Liu, X., Palma, J., Kinders, R. et al., An enzyme-linked immunosorbent poly(ADP-ribose) polymerase biomarker assay for clinical trials of PARP inhibitors, *Anal. Biochem.* 381, 240–247, 2008.
40. Carden, C.P., Banerji, U., Kaye, S.B. et al., From darkness to light with biomarkers in early clinical trials of cancer drugs, *Clin. Pharmacol. Therapeut.* 85, 131–133, 2009.
41. Adjei, A.A., Christian, M., and Ivy, P., Novel designs and end points for phase II clinical trials, *Clin. Cancer Res.* 15, 1866–1872, 2009.
42. Mutsaers, A.J., Francia, G., Man, S. et al., Dose-dependent increases in circulating TGF-α and other EGFR ligands act as pharmacodynamic markers for optimal biological dosing of cetuximab and are tumor independent, *Clin. Cancer Res.* 15, 2397–2405, 2009.
43. Cazaubiel, M. and Bard, J.-M., Use of biomarkers for the optimization of clinical trials in nutrition, *Agro Food Industry Hi-Tech* 19, 22–24, 2008.
44. Muller, P.Y. and Dieterle, F., Tissue-specific, non-invasive biomarkers: Translation from preclinical safety assessment to clinical safety monitoring, *Expert Opin. Drug Metab. Toxicol.* 5, 1023–1038, 2009.
45. *Oxford English Dictionary*, Oxford University Press, Oxford, U.K.; http://www.oup.com/online/us/oed/
46. Norton, D., Equipment fit for purpose, *Nurs. Times* 74(Suppl), 73–76, 1978.
47. Rushforth, H. and Ireland, L., Fit for whose purpose? The contextual forces underpinning the provision of nurse education in the UK, *Nurse Educ. Today* 17, 437–441, 1997.
48. Wright, P., Edwards, S., Diallo, A., and Jacobson, R., Development of a framework of international certification by the OIE of diagnostic tests validated as fit for purpose, *Dev. Biol.* 128, 27–35, 2007.
49. Wagner, J.A., Williams, S.A., and Webster, C.J., Biomarkers and surrogate end points for fit-for-purpose development and regulatory evaluation of new drugs, *Clin. Pharmacol. Ther.* 81, 104–107, 2007.
50. Resch-Genger, U., Hoffman, K., and Hoffman, A., Standardization of fluorescence measurements: Criteria for the choice of suitable standards and approaches to fit-for-purpose calibration tools, *Ann. N. Y. Acad. Sci.* 1130, 35–43, 2008.
51. Wagner, J.A., Strategic approach to fit-for-purpose biomarkers in drug development, *Annu. Rev. Pharmacol. Toxicol.* 48, 631–651, 2008.
52. Mander, A., Chowdhury, F., Low, L., and Ottensmeier, C.H., Fit for purpose? A case study: Validation of immunological endpoint assays for the detection of cellular and humoral responses to anti-tumor DNA fusion vaccines, *Cancer Immunol. Immunother.* 58, 789–800, 2009.
53. Backen, A.C., Cummings, J., Mitchell, C. et al., 'Fit-for-purpose' validation of Searchlight multiplex ELISAs of angiogenesis for clinical trial use, *J. Immunol. Methods* 342, 106–114, 2009.
54. Kanis, J.A., McCloskey, E.V., Johansson, H. et al., A reference standard for the description of osteoporosis, *Bone* 42, 467–475, 2007.
55. Saldanha, J., Validation and standardisation of nucleic acid amplification technology (NAT) assays for the detection of viral contamination of blood and blood products, *J. Clin. Virol.* 20, 7–13, 2001.

56. Ward, A.M., Catto, J.W., and Hamdy, F.C., Prostate specific antigen: Biology, biochemistry and available commercial assays, *Ann. Clin. Biochem.* 38, 633–651, 2001.

57. Barrowcliffe, T.W., Raut, S., Sands, D., and Hubbard, A.R., Coagulation and chromogenic assays of Factor VII activity: General aspects, standardization, and recommendations, *Semin. Thromb. Hemost.* 28, 247–256, 2002.

58. Theakston, R.D., Warrell, D.A., and Griffiths, E., Report of a WHO workshop on the standardization and control antivenoms, *Toxicon* 41, 541–557, 2003.

59. Jódar, L., Griffiths, E., and Feavers, I., Scientific challenges for the quality control and production of group C meningococcal conjugate vaccine, *Vaccine* 22, 1047–1053, 2004.

60. Williams, R.L., 2000–2005 Reference Standards Committee of the USP Council of Experts and Its Advisory Panel; USP Staff and Consultant. Official USP Standards: Metrology concepts, overview, and scientific issues and opportunities, *J. Pharm. Biomed. Anal.* 40, 3–15, 2006.

61. Meager, A., Measurement of cytokines by bioassays: Theory and application, *Methods* 38, 237–252, 2006.

62. Jones, R.G., Corbel, M.J., and Sesardic, D., A review of WHO International Standards of botulinum antitoxins, *Biologicals* 34, 223–226, 2006.

63. World Health Organization. WHO Expert Committee on Specifications for Pharmaceutical Preparations. Forty-first report. World Health Organ. Tech. Rep. Ser. (943), 1–156, 2007.

64. WHO Expert Committee on Biological Standardization (1992), 42nd report, http://www.who.org

65. Glas, A.S., Roos, D., Deutekom, M. et al., Tumor markers in the diagnosis of primary bladder cancer. A systematic review, *J. Urol.* 169, 1975–1982, 2003.

66. Clemmons, D.R., Quantitative measurement of IGF-1 and its use in diagnosing and monitoring treatment of growth hormone secretion, *Endocr. Dev.* 9, 55–65, 2005.

67. Goatman, K.A., A reference standard for the measurement of macular oedema, *Br. J. Opthalmol.* 90, 1197–1202, 2006.

68. Pinzani, M., Vizzutti, F., Arena, U., and Marra, F., Technology insight: Noninvasive assessment of liver fibrosis by biochemical scores and elastography, *Nat. Clin. Pract. Gastroenterol. Hepatol.* 5, 95–106, 2008.

69. Guha, I.N. and Rosenberg, W.M., Noninvasive assessment of liver fibrosis: Serum markers, imaging, and other modalities, *Clin. Liver Dis.* 12, 883–900, 2008.

70. Reitsma, J.B., Rutjes, A.W., Khan, K.S. et al., A review of solutions for diagnostic accuracy studies with an imperfect or missing reference standard, *J. Clin. Epidemiol.* 62, 797–806, 2009.

71. Kondratovich, M.V., Comparing two medical tests when results of reference standard are unavailable for those negative via both tests, *J. Biopharm. Stat.* 18, 145–166, 2008.

72. Lee, J.W. and Hall, M., Method validation of protein biomarkers in support of drug development or clinical diagnosis/prognosis, *J. Chromatogr. B Analyt. Technol. Biomed. Life Sci.* 877, 1259–1271, 2009.

73. Glasziou, P., Irwig, L., and Deeks, J.J., When should a new test become the current reference standard?, *Ann. Int. Med.* 149, 816–822, 2008.

74. Sheehan, K.M., Calvert, V.S., Kay, E.W. et al., Use of reverse phase protein microarrays and reference standard development for molecular network analysis of metastatic ovarian cancer, *Mol. Cell. Proteomics* 4, 346–355, 2005.

75. Gorreta, F., Barzaghi, D., VanMeter, A.J. et al., Development of a new reference standard for microarray experiments, *Biotechniques* 36, 1002–1009, 2004.

76. Henderson, T.J., Quantitative NMR spectroscopy using coaxial inserts containing a reference standard: Purity determinations for military nerve agents, *Anal. Chem.* 74, 191–198, 2002.

77. Masters, J.R., Thomson, J.A., Daly-Burns, B. et al., Short tandem repeat profiling provides an international reference standard for human cell lines, *Proc. Natl. Acad. Sci. USA* 98, 8012–8017, 2001.

78. Greener, M., Reference standard for gene therapy closer, *Mol. Med. Today* 6, 454, 2000.

79. Niimi, S., Oshizawa, T., Naotsuka, M. et al., Establishment of a standard assay method for human thromomodulin and determination of the activity of the Japanese reference standard, *Biologicals* 30, 69–76, 2002.

80. Ho, H.S. and Cheng, C.W., Bipolar transurethral resection of prostate: A new reference standard?, *Curr. Opin. Urol.* 18, 50–55, 2008.

81. Leeflang, M.M., Debets-Ossenkopp, Y.J., Visser, C.E. et al., Galactomannan detection of invasive aspirillosis in immunocompromised patients, *Cochrane Database Syst. Rev.* 8, CD007394, 2008.

82. Norrgran, J., Williams, T.I., Woolfitt, A.R. et al., Optimization of digestion parameters for protein quantification, *Analyt. Biochem.* 393, 48–55, 2009.

83. Miller, W.G., Myers, G.L., and Rej, R., Why commutability matters, *Clin. Chem.* 52, 553–554, 2006.

84. Vesper, H.W., Miller, W.G., and Myers, G.L., Reference materials and commutability, *Clin. Biochem. Rev.* 28, 139–147, 2007.

85. Kimberly, M.M., Caudill, S.P., Vesper, H.W. et al., Standardization of high-sensitivity immunoassay for measurement of C-reactive protein II. Two approaches for assessing commutability of a reference material, *Clin. Chem.* 55, 342–350, 2009.

86. Canalias, F., García, E., and Sánchez, M., Metrological traceability of values for α-amylase catalytic concentration assigned to a commutable calibrator materials, *Clin. Chim. Acta* 411, 7–12, 2010.

87. Gilfillan, C.P. and Robertson, D.M., Development and validation of a radioimmunoassay for follistatin in human serum, *Clin. Endocrinol.* 41, 453–461, 1994.

88. Liang, M., Klakamp, S.L., Funelas, C. et al., Detection of high- and low-affinity antibodies against a human monoclonal antibody using various technology platforms, *Assay Drug Dev. Technol.* 5, 655–662, 2007.

89. Wood, W.G., "Matrix effects" in immunoassays, *Scand. J. Clin. Lab. Invest.* 205, 105–112, 1991.

90. Mitchell, J.S. and Lowe, T.E., Matrix effects on an antigen immobilized format for competitive enzyme immunoassay of salivary testosterone, *J. Immunol. Methods* 349, 61–66, 2009.

91. Fichorova, R.N., Richardson-Harman, N., Alfano, M. et al., Biological and technical variables affecting immunoassay recovery of cytokines from human serum and stimulated vaginal fluid: A multicenter study, *Anal. Chem.* 80, 4741–4751, 2008.

92. Gagne, A., Banks, P., and Hurt, S.D., Use of fluorescence polarization detection for the measurement of fluopeptidetm binding to G protein-coupled receptors, *J. Recept. Signal. Transduct. Res.* 22, 333–343, 2002.

93. European Standard EN13640:2002; Stability testing of in vitro diagnostic reagents. European Committee on Standardization, http://www.cen.eu

94. Lisinger, T.P., Homogeneity and stability of reference materials, *Accred. Qual. Assur.* 6, 20–25, 2001.

95. van der Veen, A.M.H., Trends in the certification of reference materials, *Accred. Qual. Assur.* 9, 232–236, 2004.

96. Wei, X., Swanson, S.J., and Gupta, S., Development and validation of a cell-based bioassay for the detection of neutralizing antibodies against recombinant human erythropoietin in clinical studies, *J. Immunol. Methods* 293, 115–126, 2004.

97. Normalsell, D.E., Quantitation of serum immunoglobulins, *Crit. Rev. Clin. Lab. Sci.* 17, 103–170, 1982.

98. Hemmilä, I., Fluoroimmunoassays and immunofluorometric assays, *Clin. Chem.* 31, 359–370, 1985.
99. Gosling, J.P., A decade of development in immunoassay methodology, *Clin. Chem.* 36, 1408–1427, 1990.
100. Porstmann, T. and Kiessig, S.T., Enzyme immunoassay techniques. An overview, *J. Immunol. Methods* 150, 5–21, 1992.
101. Meldal, M., Properties of solid supports, *Methods Enzymol.* 289, 83–104, 1997.
102. Mould, A.P., Solid phase assays for studying ECM protein-protein interactions, *Methods Mol. Biol.* 139, 295–299, 2000.
103. Kay, C., Lorthioir, O.E., Parr, N.J. et al., Solid-phase reaction monitoring—Chemical derivatization and off-bead analysis, *Biotechnol. Bioeng.* 71, 110–118, 2000–2001.
104. Kusnezow, W. and Hoheisel, J.D., Solid supports for microarray immunoassays, *J. Mol. Recognit.* 16, 165–176, 2003.
105. Ziouti, N., Triantaphyllidou, I.E., Assouti, M. et al., Solid phase assays in glycoconjugate research: Applications to the analysis of proteoglycans, glycosaminoglycans and metalloproteinases, *J. Pharm. Biomed. Anal.* 34, 771–789, 2004.
106. Tuomainen, P., Reenila, I., and Mannisto, P.T., Validation of assay of catechol-*O*-methyltransferase activity in human erythrocytes, *J. Pharm. Biomed. Anal.* 14, 515–523, 1996.
107. Vermeirssen, V., Van Camp, J., and Verstraete, W., Optimisation and validation of an angiotensin-converting enzyme inhibition assay for the screening of bioactive peptides, *J. Biochem. Biophys. Methods* 51, 75–87, 2002.
108. Deng, G.J., Gu, R.F., Marmor, S. et al., Development of an LC-MS based enzyme activity assay for MurC: Application to evaluation of inhibitors and kinetic analysis, *J. Pharm. Biomed. Anal.* 35, 817–828, 2004.
109. Perdicakis, B., Montgomery, H.J., Guilllemette, J.G., and Jervis, E., Validation and characterization of uninhibited enzyme kinetics performed in multiwell plates, *Analyt. Biochem.* 332, 122–136, 2004.
110. Perdicakis, B., Montgomery, H.J., Guillemeet, J.G., and Jervis, E., Analysis of slow-binding enzyme inhibitors at elevated enzyme concentrations, *Analyt. Biochem.* 337, 211–223, 2005.
111. Zhou, J.Y. and Prognon, P., Raw-material enzymatic activity determination: A specific case for validation and comparison of analytical methods—The example of superoxide dismutase (SOD), *J. Pharm. Biomed. Anal.* 40, 1143–1148, 2006.
112. Khalil, P.N., Erb, N., Khalil, M.N. et al., Validation and application of a high-performance liquid chromatographic-based assay for determination of the inosine-5'-monophosphate dehydrogenase activity in erythrocytes, *J. Chromatogr. B.* 842, 1–7, 2006.
113. Montavon, P., Kuki, K.R., and Bortlik, K., A simple method to measure effective catalase activities: Optimization, validation, and application in green coffee, *Analyt. Biochem.* 360, 207–215, 2007.
114. Hammerling, U., Kroon, R., Wilhelmsen, T., and Sjödin, L., In vitro bioassay for human erythropoietin based on proliferative stimulation of an erythroid cell line and analysis of carbohydrate-dependent microheterogeneity, *J. Pharm. Biomed. Anal.* 14, 1455–1469, 1996.
115. Ogawa, Y., Fawaz, F., Reyes, C. et al., Development of parallel line analysis criteria for recombinant adenovirus potency assay and definition of a unit of potency, *PDA J. Pharm. Sci. Technol.* 61, 183–193, 2007.
116. Zimmermann, H., Gerhard, D., Dingermann, T., and Hothorn, L.A., Statistical aspects of design and validation of microtitre-plate-based linear and non-linear parallel in vitro bioassays, *Biotechnology*, in press, 2009.
117. Calabozo, B., Duffort, O., Carpizo, J.A. et al., Monoclonal antibodies against the major allergen of *Plantago lanceolata*, Pla l 1: Affinity chromatography purification of the allergen and development of an ELISA method for Pla l 1 measurement, *Allergy* 56, 429–435, 2001.

118. Ren, S. and Frymier, P.D., Reducing bioassay variability by identifying sources of variation and controlling key parameters in assay protocol, *Chemosphere* 57, 81–90, 2004.

119. Allen, L.T., Tosetto, M., Miller, I.S. et al., Surface-induced changes in protein adsorption and implications for cellular phenotypic responses to surface interaction, *Biomaterials* 27, 3096–3108, 2006.

120. Meade, A.D., Lyng, F.M., Knief, P., and Byrne, H.J., Growth substrate induced functional changes elucidated by FTIR and Raman spectroscopy in in-vitro cultured human keratinocytes, *Anal. Bioanal. Chem.* 387, 1717–1728, 2007.

121. Lee, J., Cuddihy, M.J., and Kotov, N.A., Three-dimensional cell culture matrices: State of the art, *Tissue Eng. Part B Rev.* 14, 86, 2008.

122. Dainiak, M.B., Savina, I.N., Musolino, I. et al., Biomimetic macroporous hydrogel scaffolds in a high-throughput screening format for cell-based assays, *Biotechnol. Prog.* 24, 1373–1383, 2008.

123. Geng, D., Shankar, G., Schantz, A. et al., Validation of immunoassays used to assess immunogenicity to therapeutic monoclonal antibodies, *J. Pharm. Biomed. Anal.* 39, 364–375, 2005.

124. Crowther, J.R., *ELISA: Theory and Practice*, Humana Press, Totowa, NJ, 1995.

125. Smisterová, J., Ensing, K., and De Zeeuw, R.A., Methodological aspects of quantitative receptor assays, *J. Pharm. Biopharm. Anal.* 12, 723–745, 1994.

126. Wilton, R., Yousef, M.A., Saxena, P. et al., Expression and purification of recombinant human receptor for advanced glycation endproducts in *Escherichia coli, Protein Expr. Purif.* 47, 25–35, 2006.

127. Englebienne, P., *Immune and Receptor Assays in Theory and Practice*, CRC Press, Boca Raton, FL, 2000.

128. Paborsky, L.R., McCurdy, S.N., Griffith, L.C. et al., The single-stranded DNA aptamersbinding sites of human thrombin, *J. Biol. Chem.* 268, 20808–20811, 1993.

129. Castro, H.C., Monteiro, R.Q., Assafim, M. et al., Ecotin modulates thrombin activity through exosite-2 interactions, *Int. J. Biochem. Cell. Biol.* 38, 1893–1900, 2006.

130. McCauley, T.G., Hamaguchi, N., and Stanton, M., Aptamer-based biosensor arrays for detection and quantification of biological macromolecules, *Anal. Biochem.* 319, 244–250, 2003.

131. Lin, C.T., Chen, L.H., and Chan, T.S., A comparative study of polyclonal and monoclonal antibodies for immunocytochemical localization of cytosolic aspartate aminotransferase in rat liver, *J. Histochem. Cytochem.* 31, 920–926, 1983.

132. Marks, J.D., Deciphering antibody properties that lead to potent Botulinum neurotoxin neutralization, *Mov. Disord.* 19(Suppl 8), S101–S108, 2004.

133. Wiberg, F.C., Rasmussen, S.K., Frandsen, T.P. et al., Production of target-specific recombinant human polyclonal antibodies in mammalian cells, *Biotechnol. Bioeng.* 94, 396–405, 2006.

134. Robertson, D.M., Stephenson, T., Cahir, N. et al., Development of an inhibin α subunit ELISA with broad specificity, *Mol. Cell. Endocrinol.* 180, 79–86, 2001.

135. McGuinness, C.R. and Mantis, N.J., Characterization of a novel, high-affinity monoclonal immunoglobulin G antibody against the ricin B subunit, *Infect. Immun.* 74, 3463–3470, 2006.

136. Brockmann, E.-C., Vehniänen, M., and Pettersson, K., Use of high-capacity surface with oriented recombinant antibody fragments in a 5-min immunoassay for thyroid-stimulating hormone, *Anal. Biochem.* doi:10.1016/j.ab.2009.10.002.

137. Cho, H.S. and Park, N.Y., Serodiagnostic comparison between two methods, ELISA and surface plasmon resonance for the detection of antibodies of classical swine fever, *J. Vet. Med. Sci.* 68, 1327–1329, 2006.

138. Cho, H.S. and Kim, T.J., Comparison of surface plasmon resonance imaging and enzyme-linked immunosorbent assay for the detection of antibodies against iridovirus in rock bream (*Oplegnathus fasciatus*), *J. Vet. Diagn. Invest.* 19, 414–416, 2007.
139. Vaisocherová, H., Faca, V.M., Taylor, A.D. et al., Comparative study of SPR and ELISA methods based on analysis of CD166/ALCAM levels in cancer and control human sera, *Biosens. Bioelectron.* 24, 2143–2148, 2009.
140. Arima, Y., Teramura, Y., Takiguchi, H. et al., Surface plasmon resonance and surface plasmon field-enhanced fluorescence spectroscopy for sensitive detection of tumor markers, *Methods Mol. Biol.* 503, 3–20, 2009.
141. Jain, K.K., Applications of nanobiotechnology in clinical diagnostics, *Clin. Chem.* 53, 2002–2009, 2007.
142. Ziober, B.L., Mauk, M.B., Falls, E.M. et al., Lab-on-a-chip for oral cancer screening and diagnosis, *Head Neck* 30, 111–121, 2008.
143. Kerschgens, J., Egner-Kuhn, T., and Mermod, N., Protein-binding microarrays: Probing disease markers at the interface of proteomics and genomics, *Trends Mol. Med.* 15, 352–358, 2009.
144. Sadik, O.A., Aluoch, A.O., and Zhou, A., Status of biomolecular recognition using electrochemical techniques, *Biosens. Bioelectron.* 24, 2749–2765, 2009.
145. Tothill, I.E., Biosensors for cancer markers diagnosis, *Semin. Cell Dev. Biol.* 20, 55–62, 2009.
146. Biotek; http://www.biotek.com/resources/articles/absorbance-measurement-ultraviolet-spectrum.html
147. Clinchy, B., Youssefi, M.R., and Hakansson, L., Differences in adsorption of serum proteins and production of IL-1ra by human monocytes incubated in different tissue culture plates, *J. Immunol. Methods* 282, 53–61, 2003.
148. Matsuno, H., Nagasaka, Y., Kurita, K., and Serizawa, T., Superior activities of enzymes physically immobilized on structurally regular poly(methyl methacrylate) surfaces, *Chem. Mater.* 19, 2174–2179, 2007.
149. McDonald, G.R., Hudson, A.L., Dunn, S.M.J. et al., Bioactive contaminants leach from disposable laboratory plastic wear, *Science* 322, 917, 2008.
150. Kenny, G.E. and Dunsmoor, C.L., Principles, problems and strategies in the use of antigenic mixtures for the enzyme-linked immunosorbent assay, *J. Clin. Microbiol.* 17, 655–656, 1983.
151. Kenny, G.E. and Dunsmoor, C.L., Effectiveness of detergents in blocking nonspecific binding of IgG in the enzyme-linked immunosorbent assay (ELISA) depends upon the type of polystyrene used, *Isr. J. Med. Sci.* 23, 732–734, 1987.
152. Stevens, P.W., Hansberry, M.R., and Kelso, D.M., Assessment of adsorption and adhesion of proteins to polystyrene microwells by sequential enzyme-linked-immunosorbent-assay analysis, *Anal. Biochem.* 225, 197–205, 1995.
153. Stevens, P.W. and Kelso, D.M., Estimation of the protein-binding capacity of microplate well using sequential ELISAs, *J. Immunol. Methods* 178, 59–70, 1995.
154. Suzuki, N., Quesenberry, M.S., Wang, J.K. et al., Efficient immobilization of proteins by modification of plate surface with polystyrene derivatives, *Analyt. Biochem.* 247, 412–416, 1997.
155. Page, J.D., Derango, R., and Huang, A.E., Chemical modification of polystyrene's surface and its effect on immobilized antibodies, *Colloids Surf. A: Physicochem. Eng. Aspects* 132, 193–201, 1998.
156. Derango, R. and Page, J., The quantitation of coupled bead antibody by enzyme-linked immunosorbent assay, *J. Immunoassay* 17, 145–153, 1996.
157. Cuvelier, A., Bourguignon, J., Muir, J.F. et al., Substitution of carbonate by acetate buffer for IgG coating in sandwich ELISA, *J. Immunoassay* 17, 371–382, 1996.

158. Julián, E., Cama, M., Martínez, P. et al., An ELISA for five glycolipids from the cell wall of *Mycobacterium tuberculosis*: Tween 20 interference in the assay, *J. Immunol. Methods* 251, 21–30, 2001.

159. DeForge, L.E., Shih, D.H., Kennedy, D. et al., Species-dependent serum interference in a sandwich ELISA for Apo2L/TRAIL, *J. Immunol. Methods* 320, 58–69, 2007.

160. Peluso, P., Wilson, D.S., and Do, D., Optimizing antibody immobilization strategies for the construction of protein microarrays, *Analyt. Biochem.* 312, 113–124, 2003.

161. Ngai, P.K., Ackermann, F., Wendt, H. et al., Protein A antibody-capture ELISA (PACE): An ELISA format to avoid denaturation of surface-adsorbed antigens, *J. Immunol. Methods* 158, 267–276, 1993.

162. Widjojoatmodjo, M.N., Fluit, A.C., Torensma, R. et al., Comparison of immunomagnetic beads coated with protein A, protein G, or goal anti-mouse immunoglobulins. Applications in enzyme immunoassays and immunomagnetic separations, *J. Immunol. Methods* 165, 11–19, 1993.

163. Dahlbom, I., Adardh, D., and Hansson, T., Protein A and protein G ELISA for the detection of IgG autoantibodies against tissue transglutaminase in childhood celiac disease, *Clin. Chim. Acta* 395, 72–76, 2008.

164. Yuan, Y., He, H., and Lee, L.J., Protein A-based antibody immobilization onto polymeric microdevices for enhanced sensitivity of enzyme-linked immunosorbent assay, *Biotechnol. Bioeng.* 102, 891–901, 2009.

165. Wunder, F., Kalthof, B., Müller, T., and Hüser, J., Functional cell-based assay in microliter volumes for ultra-high throughput screening, *Comb. Chem. High Throughput Screen.* 11, 495–504, 2008.

166. Pfeifer, M.J. and Scheel, G., Long-term storage of compound solutions for high-throughput screening by using a novel 1536-well microplate, *J. Biomol. Screen.* 14, 492–498, 2009.

167. Swartzman, E.E., Miraglia, S.J., Mellentin-Michelotti, J. et al., A homogeneous and multiplexed immunoassay for high-throughput screening using fluorometric microvolume assay technology, *Anal. Biochem.* 271, 143–151, 1999.

168. Vignali, D.A., Multiplex particle-based flow cytometric assays, *J. Immunol. Methods* 243, 243–255, 2000.

169. Morgan, E., Varro, R., Sepulvada, H. et al., Cytometric bead array: A multiplexed assay platform with applications in various areas of biology, *Clin. Immunol.* 110, 252–266, 2004.

170. Schwenk, M.J., Lindberg, J., Sundberg, M. et al., Determination of binding specificities in highly multiplexed bead-based assays for antibody proteomics, *Mol. Cell. Proteomics* 6, 125–132, 2007.

171. Djoba-Siawaya, J.F., Roberts, R., Babb, C. et al., An evaluation of commercial fluorescent bead-based luminex cytokine assays, *PLoS One* 3, e2535, 2008.

172. Kettman, J.R., Davies, T., Chandler, D. et al., Classification and properties of 64 multiplexed microsphere sets, *Cytometry* 33, 234–243, 1998.

173. Lalvani, A., Meroni, P.L., Millington, K.A. et al., Recent advances in diagnostic technology: Applications in autoimmune and infectious disease, *Clin. Exp. Rheumatol.* 26 (1 Suppl 48), S62–S66, 2008.

174. Krishhan, V.V., Khan, I.H., and Luciw, P.A., Multiplexed microbead immunoassays by flow cytometry for molecular profiling: Basic concepts and proteomics applications, *Crit. Rev. Biotechnol.* 29, 29–43, 2009.

175. Luminex Corporation, http://www.luminexcorp.com/products/index.html

176. Kellar, K.L., Kalwar, R.R., Dubois, K.A. et al., Multiplexed fluorescent bead-based immunoassays for quantitation of human cytokines in serum and culture supernatants, *Cytometry* 45, 27–36, 2001.

177. Dasso, J., Lee, J., Bach, H., and Mage, R.G., A comparison of ELISA and flow micro sphere-based assays for quantification of immunoglobulins, *J. Immunol. Methods* 263, 23–33, 2002.

178. Vedrine, C., Caraion, C., Lambert, C., and Genin, C., Cytometric bead assay of cytokines in sepsis: A clinical evaluation, *Cytometry B: Clin. Cytom.* 60B, 14–22, 2004.

179. DuPont, N.C., Wang, K.H., Wadhwa, P.D. et al., Validation and comparison of luminex cytokine analysis kits with ELISA: Determinations of a panel of nine cytokines in clinical sample culture supernatants, *J. Reprod. Immunol.* 66, 175–191, 2005.

180. LaFrance, M.W., Kehinde, L.E., and Fullard, R.J., Multiple cytokine analysis in human tears: An optimized procedure for cytometric bead-based assay, *Curr. Eye Res.* 33, 525–544, 2008.

181. Wong, H.L., Pfeiffer, R.M., Fears, T.R. et al., Reproducibility and correlations of multiplex cytokine levels in asymptomatic persons, *Cancer Epidemiol. Biomarkers Prev.* 16, 3450–3456, 2008.

182. Gu, Y., Zeleniuch-Jacquotte, A., Linkov, F. et al., Reproducibility of serum cytokines and growth factors, *Cytokine* 45, 44–49, 2009.

183. Merchant, T.E., Li, C.H., Xiong, X., and Gaber, M.W., Cytokine and growth factor responses after radiotherapy for localized ependymona, *Int. J. Radiat. Oncol. Biol. Phys.* 74, 159–167, 2009.

184. de Jager, W. and Rijkers, G.T., Solid-phase and bead-based cytokine immunoassay: A comparison, *Methods* 38, 294–303, 2006.

185. Thermo Scientific (Pierce); http://www.piercenet.com/Products/Browse.cfm?fldID=01030102

186. Molecular Probes (Invitrogen); http://www.invitrogen.com/site/us/en/home/References/Molecular-Probes-The-Handbook.html

187. Ahmed, F.E., Mining the oncoproteome and studying molecular interactions for biomarker development by 2DE, ChIP and SPR technologies, *Expert Rev. Proteomics* 5, 469–496, 2008.

188. Richens, J.L., Urbanowicz, R.A., Lunt, E.A. et al., Systems biology coupled with label-free high-throughput detection as a novel approach for diagnosis of chronic obstructive pulmonary disease, *Respir. Res.* 10, 29, 2009.

189. Malmqvist, M., Surface plasmon resonance for detection and measurement of antibody-antigen affinity and kinetics, *Curr. Opin. Immunol.* 5, 282–286, 1993.

190. Kricka, L.J., Selected strategies for improving sensitivity and reliability of immunoassays, *Clin. Chem.* 40, 347–357, 1994.

191. Schuck, P., Use of surface plasmon resonance to probe the equilibrium and dynamic aspects of interactions between biological macromolecules, *Annu. Rev. Biophys. Biomol. Struct.* 26, 541–566, 1997.

192. Phillips, K.S. and Cheng, Q., Recent advances in surface plasmon resonance based techniques for bioanalysis, *Anal. Bioanal. Chem.* 387, 1831–1840, 2007.

193. Paul, S., Vadgama, P., and Ray, A.K., Surface plasmon resonance imaging for biosensing, *IET Nanobiotechnol.* 3, 71–80, 2009.

194. Ladd, J., Taylor, A.D., Piliarik, M. et al., Label-free detection of cancer biomarker candidates using surface plasmon resonance imaging, *Anal. Bioanal. Chem.* 393, 1157–1163, 2009.

195. de Boer, A.R., Hokke, C.H., Deelder, A.M., and Wuhrer, M., Serum antibody screening by surface plasmon resonance using a natural glycan microarray, *Glycoconj. J.* 25, 75–84, 2008.

196. Lee, S.J., Youn, B.S., Park, J.W. et al., ssDNA aptamers-based surface plasmon resonance biosensor for the detection of retinol binding protein 4 for the early diagnosis of type 2 diabetes, *Anal. Chem.* 80, 2867–2873, 2008.

197. Saenko, E., Kannicht, C., Loster, K. et al., Development and applications of surface plasmon resonance-based von Willebrand factor-collagen binding assay, *Anal. Biochem.* 302, 252–262, 2002.

198. Shelver, W.L. and Smith, D.J., Determination of ractopamine in cattle and sheep urine samples using an optical biosensor analysis: Comparative study with HPLC and ELISA, *J. Agric. Food Chem.* 51, 3715–3721, 2003.

199. Akerstedt, M., Björk, L., Persson Waller, K., and Sternesjö, A., Biosensor assay for determination of haptoglobin in bovine milk, *J. Dairy Res.* 73, 299–305, 2006.

200. Wang, Y., Zhu, X., Wu, M. et al., Simultaneous and label-free determination of wild-type and mutant p53 at a single surface plasmon resonance chip preimmobilized with consensus DNA and monoclonal antibody, *Anal. Chem.* 81, 8441–8446, 2009.

201. Townsend, S., Finlay, W.J., Hearty, S., and O'Kennedy, R., Optimizing recombinant antibody function in SPR immunosensing. The influence of antibody structural format and chip surface chemistry on assay sensitivity, *Biosens. Bioelectron.* 22, 268–274, 2006.

202. Lindquist, G., Edström, A., Müller Hillgren, R.M., and Hansson, A., Comparison of methods for immobilization to carboxymethyl dextran sensor surfaces by analysis of the specific activity of monoclonal antibodies, *J. Mol. Recognit.* 8, 125–131, 1995.

203. Catimel, B., Nerrie, M., Lee, F.T. et al., Kinetic analysis of the interaction between the monoclonal antibody A33 and its colonic epithelial antigen by the use of an optical biosensor. A comparison of immobilisation strategies, *J. Chromatogr. A* 776, 15–30, 1997.

204. Frederix, F., Bonroy, K., Reekmans, G. et al., Reduced nonspecific adsorption on covalently immobilized protein surfaces using poly(ethylene oxide) containing blocking agents, *J. Biochem. Biophys. Methods* 58, 67–74, 2004.

205. Kang, J.H., Choi, H.J., Hwang, S.Y. et al., Improving immunobinding using oriented immobilization of an oxidized antibody, *J. Chromatogr. A* 1161, 9–14, 2007.

206. Jung, Y., Lee, J.M., Kim, J.W. et al., Photoactivatable antibody binding protein: Site-selective and covalent coupling of antibody, *Anal. Chem.* 81, 936–942, 2009.

207. Kausaite-Minkstimiene, A., Ramanaviciene, A., and Ramanavicius, A., Surface plasmon resonance biosensor for direct detection of antibodies against human growth hormone, *Analyst* 134, 2051–2057, 2009.

208. Childs, C.B., Proper, J.S., Tucker, R.F. et al., Serum contains a platelet-derived transforming growth factor, *Proc. Natl. Acad. Sci. USA* 79, 5312–5316, 1982.

209. Flisiak, I., Zaniewski, P., and Chodynicka, B., Plasma TGF-β1, TIMP-1, MMP-1 and IL-18 as a combined biomarker of psoriasis activity, *Biomarkers* 13, 549–556, 2008.

210. Suthanthiran, M., Gerber, L.M., Schwartz, J.E. et al., Circulating transforming growth factor-β1 levels and the risk for kidney disease in African Americans, *Kidney Int.* 76, 72–80, 2009.

211. Ybarra, J., Pou, J.M., Romeo, J.H. et al., Transforming growth factor β1 as a biomarker of diabetic peripheral neuropathy: Cross-sectional study, *J. Diabetes Complications,* in press, 2009.

212. Grainger, D.J., Mosedale, D.E., and Metcalfe, J.C., TGF-β in blood: A complex problem, *Cytokine Growth Factor Rev.* 11, 133–145, 2000.

213. Lawrence, D.A., Latent-TGF-β: An overview, *Mol. Cell. Biochem.* 219, 163–170, 2001.

214. Phillipis, A.O., Steadman, R., and Donovan, K.D., A new antibody capture enzyme linked immunoassay specific for transforming growth factor β, *Int. J. Biochem. Cell Biol.* 27, 207–213, 1995.

215. Tsakas, S. and Goumenos, D.S., Accurate measurement and clinical significance of urinary transforming growth factor-β1, *Am. J. Nephrol.* 26, 186–193, 2006.

216. Duranyildiz, D., Camlica, H., Soydine, H.O. et al., Serum levels of angiogenic factors in early breast cancer remain close to normal, *Breast* 18, 26–29, 2009.

217. Peracoli, M.T.S., Terezinha, M., Menegon, F.T.F. et al., Platelet aggregation and TGF-β1 plasma levels in pregnant women with preeclampsia, *J. Reprod. Immunol.* 79, 79–84, 2008.

218. Maumnik, B., Borawski, J., Pawlak, K., and Mysliwiec, M., Enoxaparin but not unfractionated heparin causes a dose-dependent increase in plasma TGF-β1 during haemodialysis: A cross-over study, *Nephrol. Dial. Transplant.* 22, 1690–1696, 2007.

219. Hatchuel, A., Le Masson, P., and Weil, B., The development of science-based products: Managing by design spaces, *Creativity Innov. Manag.* 14, 345–354, 2005.
220. Calantone, R.J., Chan, K., and Cui, A.S., Decomposing product innovativeness and its effects on new product success, *J. Prod. Innov. Manag.* 23, 408–421, 2006.
221. Anon, Lost in validation, *Nat. Biotechnol.* 24, 1398, 2006.
222. Rifsi, N., Gillette, M.A., and Carr, S.A., Protein biomarker discovery and validation: The long and uncertain path to clinical utility, *Nat. Biotechnol.* 24, 971–983, 2006.
223. Kiernan, U.A., Biomarker rediscovery in diagnostics, *Expert Opin. Med. Diagn.* 2, 1391–1400, 2008.
224. O Collinson, P., Cardiac markers, *Br. J. Hosp. Med.* (London), 70, M84–M87, 2009.
225. Halliwell, B., Why and how should we measure oxidative DNA damage in nutritional studies? How far have we come?, *Am. J. Clin. Nutr.* 72, 1082–1087, 2000.
226. Petzold, A., Eikelenboom, M.J., Keir, G. et al., The new global multiple sclerosis score (MSSS) correlates with axonal but not glial biomarkers, *Mult. Scler.* 12, 325–328, 2006.
227. Pfuetzner, A., Weber, M.M., and Forst, T., A biomarker concept for assessment of insulin resistance, ss-cell function and chronic systemic inflammation in type 2 diabetes mellitus, *Clin. Lab.* 54, 485–490, 2008.
228. Lock, E.A. and Bonventre, J.V., Biomarkers in translation: Past, present, and future, *Toxicology* 245, 163–166, 2008.

Index